Resource Recovery to Approach Zero Municipal Waste

Edited by
Mohammad J. Taherzadeh
Tobias Richards

CRC Press
Taylor & Francis Group
Boca Raton London New York

CRC Press is an imprint of the
Taylor & Francis Group, an **informa** business

GREEN CHEMISTRY AND CHEMICAL ENGINEERING

Series Editor: Sunggyu Lee
Ohio University, Athens, Ohio, USA

Proton Exchange Membrane Fuel Cells: Contamination and Mitigation Strategies
Hui Li, Shanna Knights, Zheng Shi, John W. Van Zee, and Jiujun Zhang

Proton Exchange Membrane Fuel Cells: Materials Properties and Performance
David P. Wilkinson, Jiujun Zhang, Rob Hui, Jeffrey Fergus, and Xianguo Li

Solid Oxide Fuel Cells: Materials Properties and Performance
Jeffrey Fergus, Rob Hui, Xianguo Li, David P. Wilkinson, and Jiujun Zhang

Efficiency and Sustainability in the Energy and Chemical Industries: Scientific Principles and Case Studies, Second Edition
Krishnan Sankaranarayanan, Jakob de Swaan Arons, and Hedzer van der Kooi

Nuclear Hydrogen Production Handbook
Xing L. Yan and Ryutaro Hino

Magneto Luminous Chemical Vapor Deposition
Hirotsugu Yasuda

Carbon-Neutral Fuels and Energy Carriers
Nazim Z. Muradov and T. Nejat Veziroğlu

Oxide Semiconductors for Solar Energy Conversion: Titanium Dioxide
Janusz Nowotny

Lithium-Ion Batteries: Advanced Materials and Technologies
Xianxia Yuan, Hansan Liu, and Jiujun Zhang

Process Integration for Resource Conservation
Dominic C. Y. Foo

Chemicals from Biomass: Integrating Bioprocesses into Chemical Production Complexes for Sustainable Development
Debalina Sengupta and Ralph W. Pike

Hydrogen Safety
Fotis Rigas and Paul Amyotte

Biofuels and Bioenergy: Processes and Technologies
Sunggyu Lee and Y. T. Shah

Hydrogen Energy and Vehicle Systems
Scott E. Grasman

Integrated Biorefineries: Design, Analysis, and Optimization
Paul R. Stuart and Mahmoud M. El-Halwagi

Water for Energy and Fuel Production
Yatish T. Shah

Handbook of Alternative Fuel Technologies, Second Edition
Sunggyu Lee, James G. Speight, and Sudarshan K. Loyalka

Environmental Transport Phenomena
A. Eduardo Sáez and James C. Baygents

Resource Recovery to Approach Zero Municipal Waste
Mohammad J. Taherzadeh and Tobias Richards

Energy and Fuel Systems Integration
Yatish T. Shah

Cover: City image is copyrighted by Borås Energy and Environment and used with permission.

CRC Press
Taylor & Francis Group
6000 Broken Sound Parkway NW, Suite 300
Boca Raton, FL 33487-2742

First issued in paperback 2017

© 2016 by Taylor & Francis Group, LLC
CRC Press is an imprint of Taylor & Francis Group, an Informa business

No claim to original U.S. Government works

ISBN-13: 978-1-4822-4035-1 (hbk)
ISBN-13: 978-1-138-89275-0 (pbk)

Library of Congress Cataloging-in-Publication Data

Resource recovery to approach zero municipal waste / editors, Mohammad J. Tasherzadeh and Tobias Richards.
    pages cm -- (Green chemistry and chemical engineering)
  Includes bibliographical references and index.
  ISBN 978-1-4822-4035-1 (alk. paper)
  1. Waste minimization. 2. Refuse and refuse disposal. 3. Recycling. I. Tasherzadeh, Mohammad J. II. Richards, Tobias.

TD793.9.R47 2015
628.4--dc23   2015002852

Visit the Taylor & Francis Web site at
http://www.taylorandfrancis.com

and the CRC Press Web site at
http://www.crcpress.com

Contents

Series Preface

The subjects and disciplines of chemistry and chemical engineering have encountered a new landmark in the way of thinking about, developing, and designing chemical products and processes. This revolutionary philosophy, termed *green chemistry and chemical engineering*, focuses on the designs of products and processes that are conducive to reducing or eliminating the use and generation of hazardous substances. In dealing with hazardous or potentially hazardous substances, there may be some overlaps and interrelationships between environmental chemistry and green chemistry. While environmental chemistry is the chemistry of the natural environment and the pollutant chemicals in nature, green chemistry proactively aims to reduce and prevent pollution at its very source. In essence, the philosophies of green chemistry and chemical engineering tend to focus more on industrial application and practice rather than academic principles and phenomenological science. However, as both chemistry and chemical engineering philosophy, green chemistry and chemical engineering derive from and build upon organic chemistry, inorganic chemistry, polymer chemistry, fuel chemistry, biochemistry, analytical chemistry, physical chemistry, environmental chemistry, thermodynamics, chemical reaction engineering, transport phenomena, chemical process design, separation technology, automatic process control, and more. In short, green chemistry and chemical engineering are the rigorous use of chemistry and chemical engineering for pollution prevention and environmental protection.

The Pollution Prevention Act of 1990 in the United States established a national policy to prevent or reduce pollution at its source whenever feasible. And adhering to the spirit of this policy, the Environmental Protection Agency launched its Green Chemistry Program to promote innovative chemical technologies that reduce or eliminate the use or generation of hazardous substances in the design, manufacture, and use of chemical products. Global efforts in green chemistry and chemical engineering have recently gained a substantial amount of support from the international community of science, engineering, academia, industry, and governments in all phases and aspects. Some of the successful examples and key technological developments include the use of supercritical carbon dioxide as a green solvent in separation technologies; application of supercritical water oxidation for destruction of harmful substances; process integration with carbon dioxide sequestration steps; solvent-free synthesis of chemicals and polymeric materials; exploitation of biologically degradable materials; use of aqueous hydrogen peroxide for efficient oxidation; development of hydrogen proton exchange membrane fuel cells for a variety of power generation needs; advanced biofuel productions; devulcanization of spent tire rubber; avoidance of the use of chemicals and processes causing generation of volatile organic compounds; replacement of traditional petrochemical processes by microorganism-based bioengineering processes; replacement of chlorofluorocarbons with nonhazardous alternatives; advances in design of energy-efficient processes; use of clean, alternative, and renewable energy sources in manufacturing;

and much more. This list, even though it is only a partial compilation, is undoubtedly growing exponentially.

This book series (Green Chemistry and Chemical Engineering) by CRC Press/ Taylor & Francis is designed to meet the new challenges of the twenty-first century in the chemistry and chemical engineering disciplines by publishing books and monographs based on cutting-edge research and development to affect reducing adverse impacts on the environment by chemical enterprise. And in achieving this, the series will detail the development of alternative sustainable technologies that will minimize the hazard and maximize the efficiency of any chemical choice. The series aims at delivering readers in academia and industry with an authoritative information source in the field of green chemistry and chemical engineering. The publisher and its series editor are fully aware of the rapidly evolving nature of the subject and its long-lasting impact on the quality of human life in both the present and future. As such, the team is committed to making this series the most comprehensive and accurate literary source in the field of green chemistry and chemical engineering.

Sunggyu Lee
Ohio University

Preface

A population of 7 billion in the world means 7 billion waste producers. The widespread current practice of getting rid of municipal solid wastes (MSW) in the world is through landfill. These wastes represent a mixture of resources, but knowledge has not developed enough to enable their utilization in a proper and economical way. This results in an almost linear utilization of our resources, wherein the material passes through society only once before being dumped in a landfill; this practice is not sustainable in the long term. It means that we should aim for zero landfill and to completely recover our resources in order to realize a sustainable society. Although landfilling of all organic wastes is forbidden in Europe, there are only a few countries, such as Sweden, Germany, Belgium, and Switzerland, that have approached zero landfill, using a variety of technologies to recover resources from MSW.

This book provides a holistic approach to resource recovery from MSW toward zero waste. It is a complex subject with several technical, social, environmental, management, and sustainability aspects. However, there are cities and countries where zero waste is a reality, although continuous development is still ongoing. This book starts with an overview of solid waste management toward zero waste. It has several examples from Sweden and particularly from one city (Borås), where this topic has been on the agenda since 1986. After this, a discussion of sustainability aspects together with laws and regulations of waste management follows. One important choice, which is considered in Chapter 1, is whether people should separate their MSW at home or let machines and workers do it. When the waste is separated in different fractions, then we have several technologies to take care of them and convert them to different resources. Organic or biological wastes can be converted to compost or biogas and biofertilizers. We have combustion, pyrolysis, and gasification for the rejects. There are different recycling technologies, of which this book covers metals, electronic wastes, thermoset composites, papers, and fibers. Other technologies, such as glass recycling, are covered in the introduction (Chapter 1). However, in order to have good recycling, the recycling should be considered when the products are designed and produced. This is discussed in Chapters 9 through 12. Finally, if the materials are landfilled, then landfill mining should be considered. This is the topic of Chapter 14.

This book is designed to be suitable for teaching at the higher education level, as well as for researchers and companies and municipalities. We hope that it contributes to a better global environment and more sustainable societies.

Editors

Mohammad J. Taherzadeh earned a PhD in bioscience and an MSc in chemical engineering. He has been a professor in bioprocess technology since 2004 and research leader at the Swedish Centre for Resource Recovery (SCRR), University of Borås, Sweden. With about 50 researchers, the SCRR covers technical, environmental, and social aspects of sustainable resource recovery. Professor Taherzadeh is working on converting wastes and residuals to ethanol, biogas, animal feed, and biopolymers, focusing on fermentation development using bacteria, yeast, and filamentous fungi. He has to his credit more than 160 publications in peer-reviewed science journals, 12 book chapters, and 3 patents, and he is currently the main supervisor of more than 10 PhD  students and several postdoctoral fellows. Dr. Taherzadeh collaborates with several companies, and some of his research results have been industrialized. More information about him is available at www.adm.hb.se/~mjt/.

Tobias Richards has been a professor in energy recovery since 2010 at SCRR, University of Borås in Sweden. He is the leader of the group working on combustion and thermal treatment. Professor Richards focus area is treatment by thermal processes of different materials, especially mixed materials such as waste. His aim is to get valuable products such as electricity, heat, synthesis gas, and pyrolysis oil and, when necessary, destroy potential harmful substances. Professor Richards has to his credit 30 peer-reviewed and published articles and 2 book chapters and is currently supervising 5 PhD students.

Contributors

Dan Åkesson
Swedish Centre for Resource Recovery
University of Borås
Borås, Sweden

Adriana Artola
Department of Chemical Engineering
Escola d'Enginyeria
Universitat Autònoma de Barcelona
Barcelona, Spain

Raquel Barrena
Department of Chemical Engineering
Escola d'Enginyeria
Universitat Autònoma de Barcelona
Barcelona, Spain

Kim Bolton
Swedish Centre for Resource Recovery
University of Borås
Borås, Sweden

Maria José Zapata Campos
School of Business, Economics
and Law
Gothenburg Research Institute
University of Gothenburg
Gothenburg, Sweden

Joan Colón
Department of Chemical Engineering
Escola d'Enginyeria
Universitat Autònoma de Barcelona
Barcelona, Spain

Lisa Dahlén
Division of Waste Science and
Technology
Luleå University of Technology
Luleå, Sweden

Barbara De Mena
Das Technologie-Transfer-Zentrum
Bremerhaven
Water, Energy, and Landscape
Management Institute
Bremerhaven, Germany

Ulla Eriksson-Zetterquist
School of Business, Economics,
and Law
Gothenburg Research Institute
University of Gothenburg
Gothenburg, Sweden

Panagiotis Evangelopoulos
Materials Science and Engineering
Royal Institute of Technology
Stockholm, Sweden

Taina Flink
Stena Recycling AB
Gothenburg, Sweden

Xavier Font
Department of Chemical Engineering
Escola d'Enginyeria
Universitat Autònoma de Barcelona
Barcelona, Spain

Gergely Forgács
Chemical Engineering Department
University of Bath
Bath, United Kingdom

Christer Forsgren
Stena Metall AB

and

Chalmers Technical University
Gothenburg, Sweden

Per Frändegård
Department of Management and
   Engineering, Environmental
   Technology, and Management
Linköping University
Linköping, Sweden

Xavier Gabarrell
Department of Chemical Engineering
Escola d'Enginyeria
Universitat Autònoma de Barcelona
Barcelona, Spain

Ilona Sárvári Horváth
Swedish Centre for Resource Recovery
University of Borås
Borås, Sweden

Nils Johansson
Department of Management and
   Engineering, Environmental
   Technology, and Management
Linköping University
Linköping, Sweden

Maryam M. Kabir
Swedish Centre for Resource Recovery
University of Borås
Borås, Sweden

Efthymios Kantarelis
Materials Science and Engineering
Royal Institute of Technology
Stockholm, Sweden

Dimitrios Komilis
Department of Chemical Engineering
Escola d'Enginyeria
Universitat Autònoma de Barcelona
Barcelona, Spain

Joakim Krook
Department of Management and
   Engineering, Environmental
   Technology, and Management
Linköping University
Linköping, Sweden

Fredrik Niklasson
Energy Technology
SP Technical Research Institute
   of Sweden
Borås, Sweden

Anita Pettersson
Swedish Centre for Resource Recovery
University of Borås
Borås, Sweden

Hans-Joachim Putz
Paper Technology and Mechanical
   Process Engineering
Technische Universität Darmstadt
Darmstadt, Germany

Winfrid Rauch
Matthiessen Engineering
Besançon, France

Tobias Richards
Swedish Centre for Resource
   Recovery
University of Borås
Borås, Sweden

Kamran Rousta
Swedish Centre for Resource Recovery
University of Borås
Borås, Sweden

Antoni Sánchez
Department of Chemical Engineering
Escola d'Enginyeria
Universitat Autònoma de Barcelona
Barcelona, Spain

Samuel Schabel
Paper Technology and Mechanical
    Process Engineering
Technische Universität Darmstadt
Darmstadt, Germany

Gerhard Schories
Das Technologie-Transfer-Zentrum
    Bremerhaven
Water, Energy, and Landscape
    Management Institute
Bremerhaven, Germany

Mikael Skrifvars
Swedish Centre for Resource Recovery
University of Borås
Borås, Sweden

Mohammad J. Taherzadeh
Swedish Centre for Resource Recovery
University of Borås
Borås, Sweden

Mats Torring
Stena Recycling AB
Gothenburg, Sweden

Weihong Yang
Materials Science and Engineering
Royal Institute of Technology
Stockholm, Sweden

1 An Overview of Solid Waste Management toward Zero Landfill
A Swedish Model

Kamran Rousta, Tobias Richards,
and Mohammad J. Taherzadeh

CONTENTS

1.1   INTRODUCTION

Increasing waste generation in different societies is a challenge for both public health and the environment. The common practice for waste treatment is to dump the waste in landfill areas, which results in various environmental problems and high operation and maintenance costs as well as public protests in having the landfill in their vicinity (Tchobanoglous et al. 1993; Calvo et al. 2007). There have been numerous attempts to reduce the number of landfills in the world by introducing various waste treatment methods in different societies based on the infrastructure, culture, and the demands of the societies. For example, in Sweden, the strategy has been to develop incineration, recycling, and biological treatment to reduce the amount of waste sent to the landfill (Swedish Environmental Protection Agency 2005).

Municipal solid waste (MSW) is used in parallel with the biomass resources to produce heat and power as well as car fuel (Demirbaş 2001). Incineration or combustion is a common method used in industrial countries (Hall and Scrase 1998) where waste is used as a source of energy. However, from an environmental and technical point of view, there are some limitations in burning commingled waste, as the energy value of the waste is decreased by increasing the amount of wet fractions, i.e. food waste, in the waste stream. (Eriksson 2003; Finnveden et al. 2007). On the other hand, composting is commonly mentioned as a suitable biological treatment method for organic wastes. Moreover, the anaerobic biological treatment of the organic wastes results in biogas, which can be used as a car fuel or electricity source. In Sweden, 4.4 million tons of MSW were produced in 2012, of which about 2.3 million tons were combusted for heat and electricity production and around 0.67 million tons were treated biologically to produce biogas principally for car fuel and fertilizer for agricultural use (Swedish Waste Management Association 2013b).

The diversity of the materials, such as metal, paper, glass, plastic, and food, in the MSW on the one hand and the involvement of the different disciplines such as economy, environment, and social aspects on the other results in MSW handling being a complex system, which most of the developing countries have difficulty in affording. This chapter briefly describes this complexity, using a practical example from one municipality in Sweden. It also explains how this system can reduce the landfilling with the help of different treatments like energy recovery, biogas production, and recycling. This chapter gives an overview of an integrated solid waste management; while in Chapters 2 through 13, each topic is explained deeply.

1.2   INTEGRATED SOLID WASTE MANAGEMENT

The system of solid waste management is a complex process, since several disciplines such as environment, health, economy, society, legislation, and engineering are involved and interact in order to functionalize it (Nemerow et al. 2009). Figure 1.1 interprets the complexity of solid waste management by indicating with arrows the interactions and interrelations between different disciplines. It is obvious that without these interactions, it would be impossible to make this complex system successful. In addition, increasing the amount of solid waste and its composition by, for example, changing the consumption patterns could result in difficult challenges, both in industrial and developing countries, if there is no functioning and sustainable waste management. The need to manage waste in order to protect the environment and people's health is as important as the necessity of supplying energy and water to a society. In the meantime, the inhabitants, as waste producers, need a sufficient service for these issues, which should be provided by a related authority. To solve this problem, it is necessary to have an engineering system that can take into account all the aspects such as the needs of the society, good services, and environmental impacts as well as economic feasibility. On the other hand, a proper waste management system is usually considered as being too costly and too risky for investment and operation, and therefore has had difficulty in attracting investors. An effective

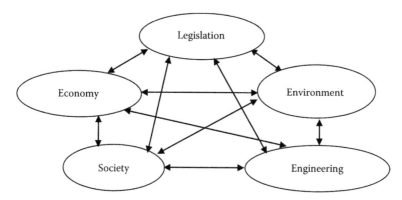

FIGURE 1.1 Interaction between different disciplines, which are involved in an integrated solid waste management.

waste management system is particularly important in developing countries, where the laws and regulations are not as clear or stable, and the public view toward waste is just "throw it away." Therefore, in order to create a proper system, some driving forces from political decisions and legislations are required.

Laws and legislations define the interactions between the different disciplines (Figure 1.1), which are imperative to support the technical system as an operative function. Although legislations are available in many countries, successful administration of such a complex system is a difficult task. How should the technical system be designed in order to follow the guidelines and to offer a sufficient service for the community? How can people be involved in the system as part of the operation? These questions support the importance of relations and interactions of the social and technical aspects in solid waste management.

Waste management in Sweden is an example of the administration of a complex system in a productive manner. In addition to many directives and legislations, which came into effect in the 1960s and paved the road to implement the present waste management system, technical and social interactions play a crucial role in the system. At a glance, the Swedish waste management example proves that the system can manage the relation between different disciplines properly. It has not only fulfilled society's expectations and reduced the environmental impacts of the waste, but also considered the economic aspects in the system. In addition to the legal and environmental guidelines, a good waste management system also has to have a strategy and stated goals to approach. The following waste management hierarchy is the strategy of the Swedish model that sets the goals and is the driving force for engineering and public attitudes.

1.3 WASTE MANAGEMENT HIERARCHY AS A STRATEGY

In order to fulfill all the expectations and needs of the MSW management system, having a strategy is essential. This strategy creates a framework for setting the goals and targets as well as lays out the plan for operating the system. It can be the guidelines for all the actors that are involved in MSW management. The popular strategy is the waste management hierarchy, which was established in order to guide the

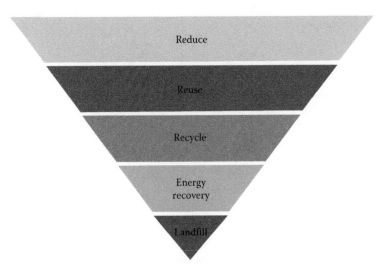

FIGURE 1.2   Different steps in the waste management hierarchy.

municipalities in how to manage their waste. Figure 1.2 illustrates the waste management hierarchy that was established on waste in the European Union (EU), namely, a
legal framework for the treatment of waste in the municipalities (2008/98/EC).

Reducing the amount of waste is the highest ranking goal of this strategy. This
can be achieved by changing the consumption patterns, designing for less material in
packaging, and so forth. The second one is reusing, which means using the products as
much as possible, for example, in secondhand markets or even repairing the products
in order to use them for a longer time. Thereafter comes recycling, which has a vital
role in saving energy and resources in material production. These first three steps are
commonly referred to as the "3Rs," and are usually stated as minimizing the material
flow by reducing the amount of the waste or circulating the materials in new products
by recycling in order to save both resources and energy. If it is not possible to follow
the 3Rs, then the waste should be converted into a valuable product such as energy.
This energy conversion can be from biological treatment of organic waste to produce
biogas as a source of energy or by preparing waste as refuse-derived fuel (RDF) for
the combustion plant to produce heat and electricity. The lowest rank of this hierarchy holds that dumping waste inside land should be avoided as much as possible.
Moving to the top of this hierarchy is the goal of each municipality in Sweden in its
waste management plan. However, there is some criticism about this hierarchy, which
asserts that "by organizing material circulation according to the European Waste
Hierarchy, incitements to decrease the rate of consumption diminish" (Hultman and
Corvellec 2012), and as a result achieving the first rank will be difficult. Nonetheless,
at the same time, it works perfectly to avoid landfilling.

This strategy makes the actors to rethink about the definition of the waste. What
is waste in fact? Is it garbage that we should get rid of as soon as possible or is it a
resource? Is it possible to make value-added products from the waste? Answering
these questions may help to understand such a complex system.

1.4 WASTE REFINERY VERSUS BIOREFINERY AND OIL REFINERY

Waste refinery is analogous to "biorefinery" or "oil refinery." This concept refers to the integrated treatment of different products, leaving in principle no waste and residuals. Waste refining demands rethinking about the wastes, that is, a mixture of the solid materials that is a "waste or resource." In other words, an "oil refinery" separates the mixture of oil fractions by targeting different markets, a "biorefinery" separates the biomass fractions to produce different materials for various markets, and a "waste refinery" separates the solid wastes—the mixture of different materials—into various products for different markets. An important aspect of all these three refineries is to leave no waste and residuals, but just products.

1.4.1 WASTE OR RESOURCE

Dictionaries define "waste" as materials that are garbage, useless or worthless. In *The Environment Dictionary*, "waste" is defined as "any material, solid, liquid or gas, that is no longer required by the organism or system that has been using it or producing it" (Kemp 1998). Jackson and Jackson (2000) mention that "waste is any movable material that is perceived to be of no further use and that is permanently discarded." Another definition is "waste is material perceived to have little or no value by society's producers or consumers" (Rhyner et al. 1995). Since this chapter focuses on solid waste, a definition of "solid waste" is provided here: "solid wastes are the wastes arising from human activities that are normally solid and that are discarded as useless or unwanted" (Tchobanoglous et al. 1993).

By looking at all these definitions, there is a common meaning that waste is something useless from a process. These unwanted materials can emerge from households after using the products or during the production process by manufacturers. If we want to omit the terms "useless" and "unwanted" from the definitions of "waste," we have to find a solution to use the wastes for various purposes. Admittedly, they cannot be waste if we can use them. So, what can they be? We will now look at the definition of "resource" in *The Environment Dictionary*: "Resources are any objects, material or commodities that are of use to society" (Kemp 1998). Accordingly, if there is a solution to use the waste in our society, we should call them "**resources**" instead.

There is another expression, which puts waste and resources together and explains that "there are no wastes, only residues that should be designed so that an economic use can be found for them" (Graedel and Allenby 1995). There are many examples that prove this last definition. For instance, molasses was once a waste product of the sugar factories. However, by looking at this by-product as a sugar resource for producing other products, molasses trade today has become big business. Molasses is used, for example, in the production of ethanol, citric acid, itaconic acid, butanol and acetone, 2,3-butanediol, dextran, baker's yeast, and food yeast (Olbrich 1963; Prakash, Henham, and Krishnan Bhat 1998). Another example is the solid residuals of the sugarcane plants called bagasse. This is used in the production of, for example, white writing paper, printing paper, kraft paper, cardboards, corrugated wrapping and newsprint papers, biogas, and animal feeds, as well as energy sources for the sugarcane factories (Yadav and Solomon 2006). Considering the annular production of bagasse is about 500 million tons in the world. If there was no application for this, managing of this amount of waste would result in an environmental disaster.

Considering bagasse and molasses as resources and not as wastes is a positive example of how to manage the industrial wastes. When the properties of the residuals are known and when they are separated as an individual material, it is usually easier to find applications for the residuals by developing the technology.

The question now is, how to consider the MSW as a resource? For more explanation about this waste, it is helpful to look closely at what are included in household wastes. MSW is usually a blend of food waste, paper, cardboard, plastics, textile, leather, yard waste, wood, glass, tin cans, aluminum, other metals, ashes, special waste (including bulky items such as consumer electronics, white goods, batteries, oil, and tires), and household hazardous waste (Tchobanoglous et al. 1993). These materials can be divided into the following five categories:

1. Recyclable material such as metal, paper, and plastic packaging, which directly go to the recycling industry after sorting
2. Organic waste such as food waste
3. Bulky waste such as white goods, furniture, and tires that should be collected separately and transferred to appropriate industries for treatment
4. Hazardous waste such as batteries, electronics, and chemicals that should be collected separately for appropriate treatment
5. Residual waste such as tissues and nappies that can be treated in a combustion plant.

If these materials were available separately, it would be possible to convert them into new products by recycling or by producing energy by combustion or anaerobic digestion. For example, food waste is one of the resources that can be converted into biogas. This can be done if the food waste is sorted from the other waste items. In addition, in order to make the recycling process more effective and economical, sorting the material is a crucial task. For example, it is possible to recycle glass many times if and only if we have the separated fractions of glass. In this case glass-recycling industries are willing to accept the materials for further treatment. This is very important because energy consumption for recycling the glass is 50% less than that required for the production of new glass materials. Accordingly, the separated materials from the waste stream have another definition, because they can be viewed as resources for other products and thus are not waste products.

1.4.2 Oil Refinery versus Waste Refinery

For a better understanding of how sorting materials in MSW can result in moving toward the top of the waste-management hierarchy, which is the goal of solid waste management, the oil refinery process is compared here with the waste refinery process.

Crude oil is processed in an industrial plant, usually through several distillation columns and cracking reactors in an integrated process, into valuable and useful products such as diesel fuel, gasoline, naphtha, asphalt base, heating oil, and other products that are used directly or as resource for another process (Fahim et al. 2009).

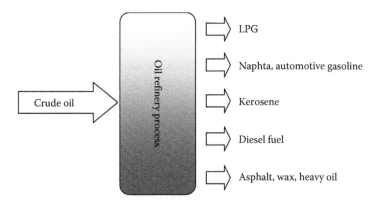

FIGURE 1.3 The scheme of separation in oil refinery process.

This separation occurs principally based on different boiling points of the oil molecules (Figure 1.3). All these products have particular applications and markets, for which the crude oil is not considered. If this separation did not take place, there would probably have been just one function for crude oil, that is, to burn. All the abovementioned materials from the oil refinery process are the final or intermediate products for a variety of materials and fuels.

The state of materials in MSW is comparable to the crude oil as both are a mixture of a variety of chemical molecules. If the five categories of MSW were separated, it would be possible to consider the definition of the waste from garbage to resource. Figure 1.4 demonstrates the waste refinery process and its products.

When the waste is commingled with food waste and other materials, it would technically be too difficult to treat it as different products, and the available possible solution is to just dump it into open landfills. The improved version of this dumping is known as controlled landfill, in which the leachate and other environmental aspects are controlled. On the other hand, burning this mixed waste is too complicated from a technical point of view, since its high water content results in a negative energy value of the wastes. That means when the water content of the waste is high, it is required fuel to burn the waste instead to get energy from the waste. Furthermore, burning plastics and hazardous wastes can result in serious environmental effects

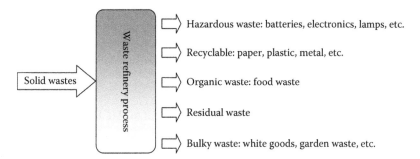

FIGURE 1.4 The scheme of separation in waste refinery process.

(Eriksson 2003). Both of these solutions are in the lowest ranking of the waste management hierarchy. However, if we separate the materials, as we do in an oil refinery process, we do have opportunities for movement toward the higher ranks in the hierarchy, and result in more effective solutions for waste management. Consequently, this separation is the crucial point in solid waste management, which prepares different materials in the waste for other value-added products.

Here, we are faced with another question. If the boiling temperature of different materials in, for example, distillation of oil refinery is the key factor for the separation, what would be the key factor to separate different materials from the MSW? To answer this question, we need to consider the discussion in Section 1.2. Solid waste management is a system and we should consider it totally as one package. In order to have the best sorting, all the functions and disciplines in waste management should interact properly. The best solution is waste separation at source, which is the most effective and least costly system (Tchobanoglous et al. 1993). It means that the citizens need to separate all the fractions in their homes. This requires a simple and convenient system as well as sufficient information provided in order to support the citizens' work in separation. More separation means more recycling, reusing, and energy recovery and less landfilling, which are the aims of the waste management hierarchy. In many developing countries, if the materials are not sorted at source, the material recovery facilities are usually established in the landfill site. This causes both social and environmental problems; also, the amount and quality of material that are picked from landfill are low.

It can be concluded that the main factor for determining productivity in waste refinery is how well the citizens have been educated and motivated to participate in the system. From a technical point of view, there are some solutions in order to make the system simple and understandable for the citizens. For example, in Sweden, hard- and soft plastics can be sorted together, which is a doable task for normal citizens. At the next stage, with the help of technology, different types of plastics could be sorted. The different methods of source separation of waste as well as its social aspects are discussed at length in Chapter 4.

The other successful separation process in Sweden is sorting PET bottles and aluminum cans with the help of deposit incentives. When customers buy a product canned in PET or aluminum, they pay a deposit of 1–2 SEK. The customers can return the empty bottles/cans to any supermarket in Sweden and get their deposit back. More than 88% of these materials are recycled in Sweden by this simple solution (Pantamera 2013). This separation results in huge savings in energy and resources to produce the aluminum cans and PET bottles. It also has a great economical potential for the recycling industry in this field. This example shows the importance of waste separation at source. It also shows how important it is to understand the necessity of the interaction between the different functions in waste management (Figure 1.1) in order to find a sustainable solution

1.5   SWEDISH WASTE MANAGEMENT IN PRACTICE

Sweden is one of the foremost countries that have considered the environmental problems in their political agenda (Engblom 1999). In 1969, the Environmental Protection Act (SFS 1969:387) came into effect to control the emission of the hazardous materials from different processes including waste treatment facilities. Thereafter, other laws merged together to establish the Environmental Codes (SFS 1998:808), which are the overall laws for environmental protection in Sweden (Swedish Environmental Protection Agency 2005). The laws and regulations about waste management will be further discussed in Chapter 3.

The combination of EU directives regarding waste (2008/98/EC) and existing Swedish legislations resulted in the Swedish government developing the legislation during the last decade. Producer responsibility, municipal waste treatment plan, local investment program, the landfill tax, ban on landfill disposal of combustible and organic waste, waste separation at source, and treatment of hazardous wastes are the milestones of the available legislations that came into effect during these years (Swedish Environmental Protection Agency 2005; Johansson et al. 2007). Another driving force for Swedish waste management was setting the goals in environmental policy for a sustainable Sweden. The government proposed the national quality objectives for this matter by merging 15 earlier legislations that concerned environmental protection (Government Bill 1997/1998). There is an association in Sweden called the Swedish Waste Management, which was formed to provide education and information, to exchange experiences, and to work with development and investigation as well as to monitor this field (Swedish Waste Management Association 2013a). The members of this association are all the municipalities and the related municipality companies as well as some private companies, which are active in the field of waste management (Swedish Waste Management Association 2013a). The association was founded in 1947, and it plays an effective role in developing and improving the waste management system in Sweden.

Municipalities in Sweden are responsible for establishing their own waste management plan in order to achieve the national goals in the environmental objectives. Municipalities can achieve this by establishing their own company or through joint collaboration with other actors in this field or even with other municipalities. The municipal waste management system is financed by the municipal waste tariff, which varies between municipalities. The citizens are obliged to follow the waste management system that is provided by the municipality, and pay the fee defined by the waste tariff. The waste tariff can be based on the volume or weight of the collected waste or a combination of both. The municipalities are responsible mostly for the household wastes, but collecting and treating the packaging material is the responsibility of the producers. In order to manage this, the producers established an organization in 1994, when the Ordinance on Producer Responsibility (SFS 1994a, 1994b) was introduced in Sweden.

Different categories of MSW in Sweden are collected in different ways. Food waste and residual waste are collected directly by the municipality from the households. Packaging materials and newspapers are collected by a "property-close collection" system or recycling stations. The bulky waste and hazardous waste are usually transferred to the recycling centers by citizens. All the municipalities in Sweden usually have several recycling stations and recycling centres. In addition, some municipalities have a dedicated food waste collection system (Swedish Waste Management Association 2013b). Therefore, the design of the material recovery/transfer facilities (MR/TF) varies in different municipalities, based on the

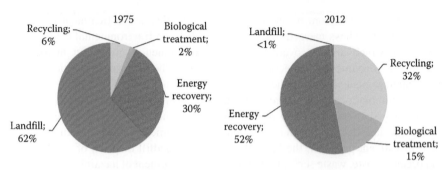

FIGURE 1.5 Comparison of the different methods in waste treatment in 1975 and 2012 in Sweden. (Data from Swedish Waste Management Association 2013b.)

separation methods used. If there is no food waste separation, the food waste and residual waste are transported to the incineration plant. In the case of separation of the food waste, it is transferred for biological treatment for biogas production, and the residual waste is transported to the incineration plant.

This waste management system in Sweden has reduced the landfilling dramatically from almost 1,500,000 tons in 1975 to less than 33,000 tons in 2012, which is less than 1% of the total MSW of this country (Swedish Waste Management Association 2013b). As Figure 1.5 illustrates, increasing the material recycling as well as biological treatment are the main reasons for reducing the landfilling. This can only occur if the waste is separated at the source. It is also important to note that the average waste generation per person was 317 kg in 1975, which increased to 460 kg in 2012. Although the increase of waste generation in Sweden means that the country has not been successful in reaching the first ranking of the waste management hierarchy, it has reduced the landfilling to less than 1% of the total MSW, which has been a tremendous achievement during the past 37 years.

In order to explain how a complex waste management system in a Swedish model can decrease the landfilling up to zero, the integrated system used in Borås, one of the Swedish municipalities, is discussed here. This system is known as the Borås model (Engblom 1999).

1.5.1 Waste Management Progress in Borås

Borås is a medium-sized Swedish city, located in the southwest of Sweden. It has about 50,000 households and 105,000 inhabitants (Boras.se 2013). More than 90% of the waste in Borås was transferred to landfill before the 1980s (Rousta 2008). In 1986, the first investigation for establishing an integrated waste management in the city was started. Then in 1987, the first waste management plan was formed and the second plan for the period 1991–2000 was established for the city in 1991. Reducing the landfilling was the main goal of this plan. In order to achieve this goal, the plan had some action points such as conducting waste separation at source, biological treatment, and energy recovery from the waste. The vital task was source separation, which started in 1988 with a pilot project involving 3000 households and took about 3 years. For this purpose, in-person communication with the citizens were used

TABLE 1.1
Important Events in Waste Management in Borås since 1986

Year	Event
1986	Distinguishing the problem
1988–1991	Conducting waste separation at source
1991	MR/TF
1995	Biogas and composting
2002	Upgrading biogas as fuel
2005	New waste combustion plant
2012	Planning for the new energy and environmental center

to describe the new model of waste sorting. The very first plan of the MR/TF called "Sobacken" was inaugurated in 1991, which was then developed into full-scale sorting with an optical sorting system within 4 years. A new and modern sanitary landfill was opened in 1992. In 1995, a biological treatment plant started to work to produce biogas and composting in MR/TF. The intermediate storage of hazardous wastes in MR/TF was started in 1998. The third waste management plan came into effect in 2001, covering the next 10 years. Following the same targets—that is, reducing the landfilling as well as reducing the total amount of waste generation—was also emphasized in this plan. In 2002, a new anaerobic digestion plant to produce biogas from organic wastes was kicked off, where the biogas was upgraded to vehicle fuel. The first public biogas station was opened in 2003 by developing the biological treatment plant. The new combustion plant for waste burning was opened in 2005. It provides district heat for the city, and also produces electricity in a combined heat and power (CHP) plant. The summary of the important events in this progression is shown in Table 1.1.

The fourth waste management plan applies for the period between 2012 and 2020. Nowadays, all the local transport buses in the city run only on biogas, produced from waste. The city has a motto: "a city free from fossil fuels;" waste management system is one of the important functions to achieve this target. The next step is to establish a combination system for material recovery, energy recovery, and wastewater treatment in one place called "Energy and Environmental Centre" in order to make the processes more effective. The result of this progression has been a reduction in the landfilling of MSW from 100,000 tons in 1990 to less than 200 tons in 2010 (Borås Energi och Miljö 2011). This reduction was achieved in spite of increasing waste generation during these years. The average proportion of different waste treatments in the city of Borås in 2009 is illustrated in Figure 1.6 (Borås Energi och Miljö 2011).

1.5.2 How the Borås Model Works

In the following section, the Borås model is explained in detail from a technical point of view. It includes the waste handling in the city and how the MR/TF (Sobacken) has been designed to manage the MSW by using different waste treatment methods.

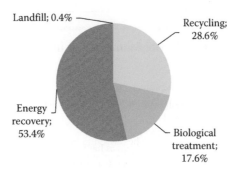

FIGURE 1.6   The proportion of different waste treatment methods in the municipality of Borås in 2009.

Then, the interaction of the social and technical aspects, which is vital to develop this system, is discussed.

1.5.2.1   Material Flow in Collection System

There are three main streams of MSW in Borås, which are collected separately (Figure 1.7). The first stream is the food waste and residual waste. This residuals fraction includes wastes such as nappies and tissues. The inhabitants sort these two fractions of waste into two types of plastic bags, black and white, which are provided by the municipality. All the food waste is sorted into the black bags, while the residual wastes are placed in the white ones. It is forbidden to put packaging materials, hazardous waste, lamps, batteries, electronics, and so on in these two bags. Then, the bags are collected together in one bin and transported to the MR/TF in order to be separated by optical sorting and prepared for further processes.

The second stream contains the packaging materials. Different containers have been provided for different waste fractions in this stream. The wastes that are collected by this stream are paper packaging and cardboards, plastic packaging, metal packaging, glass packaging in two separate containers for colored and transparent glasses, and newspapers. There are more than 80 sorting places in Borås, which are called recycling stations (Rousta 2008). The location of the stations and the type of container are designed based on the population and needs of different avenues. Some stations have all types of containers, while others have just for the major recycling wastes, that is, newspaper and colored and colorless glasses. The sorted materials are transferred to recycling industries, directly or through MR/TF. According to the producer responsibility law, managing the recycling stations means providing the containers, collecting the material, and recycling the material from the recycling station, which are all producer responsibilities. That is why a service organization with different actors from various recycling industries was formed for this purpose. In the majority of the municipalities in Sweden, the collection of the packaging materials is followed by this method, although there are some suggestions to improve it.

The third stream includes the bulky waste, hazardous waste, white goods, electronic waste, and so on. There are five recycling centers in Borås with suitable opening hours during the week (Rousta 2008), to receive the wastes transported by the citizens. There

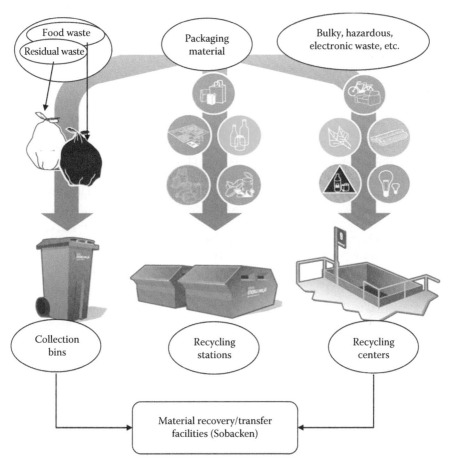

FIGURE 1.7 The collection system of households' waste in Borås. (Reprinted with permission from Borås Energi och Miljö AB, Borås, Sweden.)

is no charge for bringing the waste to these centers. There are different containers for different kinds of wastes at the recycling centers, such as WEEE (waste from electric and electronic equipment), garden waste, combustible waste, hard plastics, cardboards, metals, and ceramics. In addition, there are dedicated places for the white goods waste in these centers. The hazardous wastes such as lamps, batteries, sprays, paints, and solvents should also be sorted and collected in drums or small containers. The centers provide on-site instructors to help people in correct sorting. The sorted materials are then transported to the MR/TF for further distribution to specialized recycling industries or for appropriate treatment of, for example, hazardous wastes.

It is important to note that besides these three streams, medicine wastes are sorted separately by the inhabitants, and they are collected in pharmacies in order to transfer them to their dedicated companies for appropriate treatments. In addition, the previously mentioned deposition system for collecting PET, aluminum, and some glass bottles is also followed in the city.

1.5.2.2   Material Recovery/Transfer Facilities

Sobacken is the name of the MR/TF center in Borås, established in 1991. This center was designed to handle about 300,000 tons per year.

The main tasks at this center are (Rousta 2008)

1. To separate the white and black plastic bags using an optical sorting system
2. Biogas production from food waste and biological waste from industries such as restaurants, food and feed companies, and slaughterhouses
3. Intermediate storage for the hazardous wastes and other sorted waste from recycling centers and recycling stations in order to transfer them to the appropriate company
4. Shredding the combustible wastes from white bags as well as from recycling centers and industrial wastes and preparing the fuel for the combustion plants
5. Dumping the untreated waste from treatment activities, asbestos, and some fractions from construction and demolition waste in a sanitary landfill

Sobacken is the heart of waste management in the city of Borås. Figure 1.8 illustrates the material flow to this center in a simple way. The sorted materials that are transported to this place are converted through different treatments into value-added products, which go back to the society. For example, food waste is converted into biogas, which is used as fuel by the collective local transportation system in the city. The combustible part of the waste fuels the power plant that is producing district heat and electricity for the city. Collection of hazardous wastes from

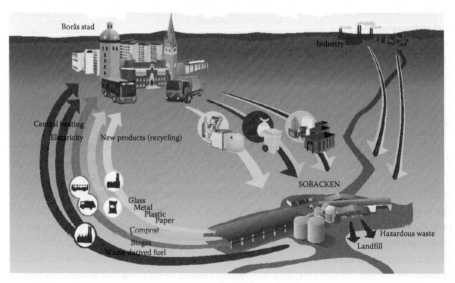

FIGURE 1.8   Schematic of material flow between Sobacken and society. (Reprinted with permission from Borås Energi och Miljö AB, Borås, Sweden.)

the waste stream reduces the environmental impacts from these materials to the society. Finally, just a little part of the waste (0.4%) is treated in a sanitary landfill (Borås Energi och Miljö 2011).

1.5.2.2.1 Biological Treatment

One of the main tasks in waste treatment in Sobacken is biogas production, treating about 18% of the total waste generation in the city (Borås Energi och Miljö 2011). The input to the digester is the raw material from two lines. One line from the black bags contains food waste, while the other line receives the biological wastes from industries, such as wastes from grocery stores, slaughterhouses, and food and feed productions and restaurants, as well as animal wastes, which are both solid and liquid. The food waste passes a pretreatment section where the plastic bags are opened, mixed with water, and separated from the plastic bags, and eventually other unsorted material using density variation. They are then transferred to a buffer tank. The waste from, for example, the slaughterhouse passes the hygienic tank with a retention time of 1 hour at 70°C or 10 hours at 55°C, and then pumped together with other industrial wastes to the buffer tanks. The total feed to the 3200 m³ digester tank is 150 m³/day. The time required for the biogas production is 21 days and the system works continuously at thermophilic conditions (55°C). The digestate is then pumped to digestate buffer tanks and passes through a decanting centrifuge to separate the particles from the water. A coagulating agent such as polyacrylamide is added to coagulate the colloid particles before the centrifuge. The solid fraction is then transferred to undergo a composting process, while the water is partly recycled and partly transferred to a dedicated wastewater treatment plant for denitrification.

The total biogas production from the digester and digestate buffer tanks is about 9600 m³/day, containing between 55% and 75% methane (Rousta 2008). The biogas is transferred to an upgrading unit in order to remove CO_2, H_2S, and humidity and thereafter to purify the methane to at least 97% in order to be used as a car fuel. Almost all the garbage trucks in the city, all the municipality cars, the local transport buses, and some private cars are fueled by this biogas, which is produced from the biological solid wastes (Rousta 2008). Since Chapters 5 and 6 of this book discuss the details of compost and biogas productions, a detailed discussion is avoided here.

1.5.2.2.2 Preparation of the RDF

The other activity in Sobacken is to treat the combustible waste in the waste boilers. There are two fluidized bed boilers for burning the waste in Borås, for which the feeding material needs to be less than 10 cm in the longest direction in order to operate efficiently. Otherwise, it can disturb the feeding and the mixing in the sand bed. The waste stream from the white bags as well as the combustible waste from the recycling centers and industrial waste are transported to this pretreatment area in Sobacken. There are two hammer mills to carry out the size reduction. After treatment, these materials are called RDF and they are ready to be fed into the boilers at the thermal treatment plant in Borås.

1.5.2.3 Thermal Treatment

The thermal treatment, that is, the combustion, of the waste in Borås is made in a CHP plant. It is called the Rya plant and has two 20 MW fluidized bed boilers. This corresponds to 7 tons of waste every hour. The waste is a mixture of household waste, roughly 30%, and industrial waste, about 70% (Johansson et al. 2006). In addition to this, special animal waste fractions are also used. This is the fraction where destruction must be certified, such as with animals having the BSE (bovine spongiform encephalopathy) disease. To produce electricity, steam is generated, which then passes through a turbine connected to a generator. Due to the difficult mixture of the waste material, the temperature and pressure are quite low for an electricity production unit; the temperature is around 400°C and the pressure is around 49 bar. Otherwise, there is a great risk of corrosion in the superheaters. The final pressure of the turbine is controlled with a condenser operating at around 100°C (temperature is dependent on the season), which warms the district heating water. With these conditions, the production is 40 MW of combined heat and electricity (Johansson et al. 2006). At the same location, there are also two 60 MW wood chip boilers that are connected to the same steam system and operate during the winters. Furthermore, the yearly production of the whole site is 160–170 GWh electricity and 600 GWh heat. The whole plant is located on a hill less than 2 km from the city center and close to residential housing as well as industrial and business sites. This location inside the city is possible due to the careful design of the boiler in combination with a suitable gas cleaning system (Johansson et al. 2007). The concept of CHP production increases the total efficiency of the boiler; specifically, it is possible to reach greater than 85% efficiency in energy recovery of the wastes or biomass.

1.5.2.3.1 Waste Boilers

The waste boilers in Borås are fluidized bed boilers. This means that there is a bed of sand particles through which air and recirculated flue gases pass. In addition, the particles are moved and rapidly mixed, which provides high heat and mass transfer rates and enables almost homogeneous conditions even though the incoming feed varies. A fluidized bed boiler is particularly beneficial when it comes to a wide range of load in the boiler (it gives low emissions at both high and low loads), and it can handle a variety of dry contents in the feed (Olsson et al. 2010). The waste is continuously fed into the boiler on top of the bed and reacts efficiently. However, there is a need for pretreatment of the waste before it enters a fluidized bed boiler. The reason for this is that it should be able to mix thoroughly with the particles and to enhance feeding through the chute.

When waste is combusted, there is a regulation in Sweden (and within the EU) that it has to be at least 850°C for a minimum of 2 seconds. This is to ensure complete destruction of hydrocarbons and especially chlorinated hydrocarbons such as dioxins and furans. The specific type of fluidized bed is the bubbling fluidized bed, which means that most of the particles stay in the bed. However, due to attrition and breakage of the sand particles, smaller particles are formed that leave the bed and move with the gas flow. If the bed temperature increases, some of the inorganic material in the ashes can sinter and form larger particles, which prevent fluidization; so, this situation must be avoided by a constant control of the temperature and can indirectly be measured by the pressure difference over the bed. To keep the bed temperature low enough when the material with the high-energy content is fed into the boiler, water is introduced. This lowers the overall energy efficiency, but decreases the need for fresh sand and reduces bed sintering (Niklasson et al. 2010).

1.5.2.3.2   Gas Cleaning

Even though the design of the modern waste boilers enables close to complete combustion of all incoming material, there is still a need for gas cleaning. The first step in the Borås plant includes SNCR (selective noncatalytic reduction) to decrease the amount of nitrogen oxides by introducing ammonia or urea, which converts the oxide to nitrogen gas.

Finally, lime and activated carbon are fed into the gas flow after passing the heat exchanger area. This will then be caught on bag filters together with smaller particles that passed the cyclone and the steric hinders in the boiler. Lime is added so that it can react with sulfuric compounds (typically sulfur dioxide) and form gypsum. The activated carbon traps different kinds of formed species; specifically, dioxins and furans are captured. Even though dioxides and furans are destroyed during the combustion, there is reformation at lower temperatures when other metals such as copper are present, and a final removal is therefore necessary. The flue gases (mainly CO_2 and H_2O) are then released into the atmosphere through a chimney.

1.5.2.3.3   Ash Handling

The remaining products after combustion are gaseous emissions and the solid residues are ashes. In this particular waste combustion boiler, there are four different ashes: bottom ash, boiler ash, cyclone ash, and fly ash. The bottom ash is removed by a constant removal of bed material. This is then treated (sieved) and the small particles are fed back to the bed, thus, decreasing the amount of fresh sand that is needed. The remaining bottom ash is sent for metal removal to a nearby recycling center (it is removed by magnets and eddy current). After metal removal, the remaining material has a rather low concentration of heavy metals and other species and can therefore be used as construction material on the landfill site at Sobacken. All other ash fractions are collected together and sent to Langøya in Norway, where they are neutralized and stabilized before they are used as reconstruction material in an old lime pit. In total, 5,000 tons of bottom ash and 11,000 tons of the other ashes are produced annually.

1.5.2.3.4   District Heating and District Cooling

The heat from the combustion is utilized by using water as a heat carrier. Subsequently, this heat is transported to the connected households and other buildings such as offices, hospitals, and shops, in what is called district heating. The district heating network extends 40 km out from the power plant and consists of insulated pipes under the ground. All users have a heat exchanger whereby the needed amount of energy from the system is extracted. The system works with a water temperature of around 90°C, but it can be adjusted depending on outer conditions. The usage of district heating lowers the electrical efficiency of the plant due to a rather high temperature (and thus high corresponding equilibrium pressure) at the end of the turbine but enables a much higher overall usage of the released energy.

In addition to the district heating network, there is also a much smaller district cooling network. This supplies cold water at 7°C and cools buildings such as supermarkets and hospitals during the warm summer months. Two different possibilities exist for the chill production. First is the compressor where electricity is fed into the equipment in the same manner as a refrigerator. The second possibility is by absorption cooler, where district heating is used to supply the needed heat to run the process. A drawback with the absorption cooler is the low thermal efficiency. However, this gives an opportunity to utilize heat during the warm periods when there are few other usages.

More details of the combustion of waste are discussed in Chapter 7.

1.5.2.4   Interaction of Different Disciplines in Integrated Solid Waste Management in Borås

The previous sections were about an overview of the model of waste management applied in Borås. Along with the legislations that have guided the system in an appropriate way, other aspects of the waste management—that is, social, economic, and technical—have developed well enough in order to fulfill the environmental goals of the society in this field.

"Borås Energi och Miljö AB" (BEM) is the name of the municipality company, which is responsible for the waste management in the city of Borås. The activities of this company are to collect the waste; produce biogas, electricity, and heat from the waste; manage the wastewater treatment; and provide drinking water for the inhabitants (Borås Energi och Miljö 2013). The annual sales of this company are around 1 billion SEK, and it has about 220 employees (Borås Energi och Miljö 2013). This proves that the waste in this definition is not just the useless material, but it is the resource for this application. It also shows that with an integrated management, it is possible to have an economical system for this matter, as present in Borås.

The overview of the Borås model shows that the progression of the technology to produce biogas, heat, and electricity from waste is crucial. The important point in this progression is how to consider the interaction with other aspects of the waste management. The system applied in Borås has been made as simple as possible. The complicated waste management demands consistently high operation and maintenance costs, which results in lower economical attraction for state or private investors. Besides the political guidelines, identifying the needs of the society and considering the extent to which the society can be involved in the system are the main points, which have been addressed in the Borås model in order to develop the technical system.

The interactions of the different disciplines that were discussed in Section 1.2 are applied here for the Borås model in terms of producing energy from the waste. Electricity and heat are the main forms of energy that are supplied to the households in Sweden. Burning the fossil fuels is condemned because of the greenhouse gas emissions. It is therefore necessary to replace the fossil fuel with another bioenergy source such as materials in the MSW. On the one hand, society needs energy, taking into consideration the environmental point of view; therefore, there is a need to find new resources. On the other hand, part of the wastes that are not recyclable and that are suitable for burning, such as white bags in the Borås model, should be managed in the proper way to avoid dumping in the landfills. This interaction requires the engineering actors to find a solution such as installing the fluidized bed incinerators for this fraction of the waste in the power plant.

The other example is the biogas from the waste in the city of Borås. Local transportation is crucial for a civilized society. Specifically in Borås, buses have made a great contribution. From an environmental perspective, fossil fuels such as gasoline and diesel are not suitable alternatives, while biofuels such as ethanol, biogas, and biodiesel could fulfill this requirement. On the other hand, a major fraction of MSW is suitable for use in biogas production. Again, this interaction requires the engineering actors to design and develop the facilities in order to fulfill the needs of the society as well as to manage the food waste in a technically and economically feasible process.

In these two examples from Borås, it is obvious that the system was developed based on the needs of the society to reduce the environmental impacts of the wastes and use waste material as a resource for the society. However, an important factor is the first step of the process, which is the separation of the waste at source. Otherwise, the process cannot be economically and environmentally feasible.

In Section 1.3, it was discussed that waste is a resource, if we can separate it into the right categories that fit the technology applied. The processes, which are applied in the city of the Borås, work successfully if the raw materials in each process are sorted correctly. This is the crucial task in this system, performed by the inhabitants in their homes. They are the ones who generate the waste, who pay the waste tariff to the waste management system, and who buy the products from this waste such as heat, electricity, and fuels, so they rely on the system's working properly. Quite simply, the system needs waste as raw material, this waste is generated by inhabitants in different activities, and so people are the waste producers. Accordingly, this waste should be managed, where people should separate the fractions at home based on the instructions. In addition, they should pay for this service as well as for the heat and electricity and the biogas that they consume. These three crucial roles require a huge consideration of the social aspects in such a system because the effectiveness of the system completely depends on how well people do their sorting in their homes.

In order to be successful with such a system, educating people about this field is necessary. Inhabitants should be aware of the environmental impacts of waste. They should also know how to sort the material as accurately as possible in order to have the system work properly. The biggest challenge here is reducing the amount of waste generation because the system needs waste as a resource for its activities. At the same time, the goal to move up in the waste management hierarchy should be taken into account. That is why different social activities are presented in this city in this field. Reducing the landfilling to 0.4% of the total MSW in Borås during the last 20 years required an engineering model, having the right interaction with society and the environment and at the same time being cost-effective and economical.

1.6   CONCLUDING NOTE

The Borås model proves that it is possible to utilize households' waste as a resource to recover material and energy; however, this is contingent on how the engineering system administrates different disciplines in the system. Separation of the hazardous waste and recyclable material from the waste stream as well as sorting the food waste and combustible parts allows for the opportunity to produce biogas from food waste and heat and electricity from residuals in an efficient way. This overview shows that the degree of success for the system depends on integrating the social, economic, political, environmental, and engineering aspects and the interaction between them.

REFERENCES

2008/98/EC. Council Directive (EC) 2008/98/EC of 19 November 2008 on waste and repealing certain directives. Official Journal of the European Union, L321/3, 22 November.

Boras.se. 2013. Borås statistic 2013 (cited January 10, 2013. Available from http://www.boras.se/forvaltningar/stadskansliet/stadskansliet/samhallsplanering/samhallsplanering/borasisiffror/statistik.4.82f0a312665003f0d800014558.html.)

Borås Energi och Miljö. 2011. Borås Waste Management Plan 2012–2020 Borås Energi och Miljö, Borås, Sweden.

Borås Energi och Miljö. 2013. About us. (cited January 10, 2013. Available from http://www.borasem.se/english/aboutus.4.63fbc1fa126f45b1ad78000138688.html.)

Calvo, F., B. Moreno, Á. Ramos, and M. Zamorano. 2007. Implementation of a new environmental impact assessment for municipal waste landfills as tool for planning and decision-making process. *Renewable and Sustainable Energy Reviews* 11(1):98–115.

Demirbaş, A. 2001. Biomass resource facilities and biomass conversion processing for fuels and chemicals. *Energy Conversion and Management* 42(11):1357–1378.

Englblom, G. M. 1999. *Waste Management, The Swedish Experience*. Stockholm, Sweden: Graphium Norstedts, Swedish Environmental Protection Agency.

Eriksson, O. 2003. Environmental and economic assessment of Swedish municipal solid waste management in a systems perspective, Stockholm, Sweden: KTH Royal Institute of Technology.

Fahim, M. A., T. A. Al-Sahhaf, H. M. S. Lababidi, and A. S. Elkilani. 2009. *Fundamentals of Petroleum Refining*. London: Elsevier Science.

Finnveden, G., A. Björklund, M. C. Reich, O. Eriksson, and A. Sörbom. 2007. Flexible and robust strategies for waste management in Sweden. *Waste management* 27(8):S1–S8.

Government Bill. 1997/1998. 1997/98:145. Swedish Environment Quality Objectives., edited by M. o. t. E. a. N. Resources. Government Office of Sweden.

Graedel, T. E. and B. R. Allenby. 1995. *Industrial Ecology*. Englewood Cliffs, NJ: Prentice Hall. pp. 83–187.

Hall, D. O. and J. I. Scrase. 1998. Will biomass be the environmentally friendly fuel of the future? *Biomass and Bioenergy* 15(4/5):357–367.

Hultman, J. and H. Corvellec. 2012. The European Waste Hierarchy: From the sociomateriality of waste to a politics of consumption. *Environment and Planning A* 44(10):2413.

Jackson, A. R. W. and J. M. Jackson. 2000. *Environmental Science: The Natural Environment and Human Impact*. Harlow: Addison-Wesley.

Johansson, A., E. Blomqvist, A. Ekvall, L. Gustavsson, C. Tullin, B. Andersson, M. Bisaillon, T. Jarlsvik, A. Assarsson, and G. Peters. 2007. Report: Waste refinery in the municipality of Borås. *Waste Management & Research* 25(3):296–300.

Johansson, A., M. Olofsson, E.-L. Wikström, A. Ekvall, L. Gustavsson, C. Tullin, T. Jarlsvik, A. Assarsson, B.-Å. Andersson, and G. Peters. 2006. The performance of a 20 MWth energy-from-waste boiler. *Proceedings of the 19th International Conference on Fluidised Bed Combustion*, Vienna, Austria, May 21–24, 2006.

Kemp, D. 1998. *The Environment Dictionary*. New York: Routledge.

Nemerow, N. L., F. J. Agardy, and J. A. Salvato. 2009. *Environmental Engineering: Environmental Health and Safety for Municipal Infrastructure, Land Use and Planning, and Industry*. Vol. 3. New York: Wiley.

Niklasson, F., A. Pettersson, F. Claesson, A. Johansson, A. Gunnarsson, M. Gyllenhammar, A. Victoren, and G. Gustafsson. 2010. Sänkt bäddtemperatur i FB-pannor för avfallsfoerbränning—etapp 2 [Reduced Bed Temperature in FB-Boilers Burning Waste—part II]. 2010, Waste Refinery, Sweden, Report WR-19.

Olbrich, H. 1963. *The Molasses*. Berlin, Germany: Fermentation Technologist, Institut für Zuckerindustrie.

Olsson, J., D. Pallares, H. Thunman, F. Johansson, B.-Å. Andersson, and A. Victorén. 2010. Förbättrad förbränningsprestanda vid avfallsförbränning i FB-pannor-Bäddynamikens inverkan på luft-/bränsleomblandningen, [Improved combustion performance of waste-fired FB-boilers-The influence of the dynamics of the bed on the air-/fuel interaction]. 2010, Waste Refinery, Sweden, Report WR-01.

Pantamera. 2013. Pantamera 2013 (cited January 10, 2013. Available from http://www.pantamera .nu/sv/v%C3%A4lkommen-till-returpack).

Prakash, R., A. Henham, and I. Krishnan Bhat. 1998. Net energy and gross pollution from bioethanol production in India. *Fuel* 77(14):1629–1633.

Rhyner, C. R., L. J. Schwartz, R. B. Wenger, and M. G. Kohrell. 1995. *Waste Management and Resource Recovery*. Boca Raton, FL: CRC Press.

Rousta, K. 2008. Municipality solid waste management: An evaluation on the Borås system. University of Borås, Sweden.

SFS. 1969. Miljöskyddslagen, *SFS. 1969:387*. Stockholm, Sweden: Miljö depardementet.

SFS. 1994a. Förordningen om producentansvar för returpapper, *SFS 1994:1205* [The ordinance of producers' responsibility for waste paper]. Swedish legislation, Stockholm, Sweden.

SFS. 1994b. Förordningen om producentansvar för förpackningar, *SFS 1994:1235* [The ordinance of producers' responsibility for packaging materials]. Swedish legislation, Stockholm, Sweden.

SFS. 1998. Miljöbalken, *SFS. 1998:808*. Stockholm, Sweden: Miljö depardementet.

Swedish Environmental Protection Agency. 2005. *A Strategy for Sustainable Waste Management: Sweden's Waste Plan*. Stockholm, Sweden: Swedish Environmental Protection Agency.

Swedish Waste Management Association. 2013a. *About Us* (cited January 16, 2013. Available from http://www.avfallsverige.se/in-english/).

Swedish Waste Management Association. 2013b. Statistics of households' waste. Swedish Waste Management, Malmö, Sweden (in Swedish).

Tchobanoglous, G., H. Theisen, and S. Vigil. 1993. *Integrated Solid Waste Management Engineering Principles and Management Issues*. New York: McGraw-Hill.

Yadav, R. L., and S. Solomon. 2006. Potential of developing sugarcane by-product based industries in India. *Sugar Tech* 8(2):104–111.

2 Sustainable Management of Solid Waste

Kim Bolton, Barbara De Mena,
and Gerhard Schories

CONTENTS

2.1 INTRODUCTION

Economic, environmental, and social aspects need to be taken into account in order for waste management to be sustainable (Goodland 1995). Economic aspects include, for example, financial motivation for inhabitants and companies to actively participate in waste management. When possible, waste should be seen as a resource that can generate economic wealth. Many types of waste, such as common plastics and electronic components, are environmentally hazardous and should be reused or recycled. A vision of this type of resource recovery is "cradle to the cradle" or "zero waste" and aims to minimize the environmental impact of the waste (Greyson 2007; Lehmann 2011). The social aspects of waste management are also critical for sustainable implementation. For example, in developed countries one's level of consumption is often equated with one's status or well-being; in contrast, in many developing countries some inhabitants survive by scavenging from waste disposal sites. These societal aspects need to be taken into account when developing strategies for sustainable waste management.

The three aspects of sustainable development mentioned above often provide different, contradictory perspectives of sustainable waste management. These perspectives need to be balanced when developing and implementing sustainable waste management strategies. This "balancing" is often based on cultural values and hence needs to be done by, or at least include, local stakeholders such as politicians, nongovernmental organizations (NGOs), companies, and citizens (Joseph 2006; Troschinetz and Mihelcic 2009). An example is the level of consumption in developed countries. A sustainable strategy may include lowering the level of consumption in order to reduce the environmental impact of manufacturing, selling, and using products. However, in addition to the sense of status and well-being that is often associated with consumption, many business models are based on the economics of scale (selling large amounts of cheap products). New business models and a change in consumer perspectives are required for sustainable reduction of consumption and the associated waste. A second example is the need for education and health care in developing countries, which are required for the social well-being of the citizens and require the availability of products such as electricity, medicines, and medical equipment. Environmental considerations would dictate that the electricity should preferably come from renewable sources and that the environmental impact of the medicines and medical equipment should be minimized. In contrast, economic constraints may require the use of fossil- or coal-based electricity and a lack of financial motivation may hinder sustainable management of medical waste.

As illustrated by these examples, there is no single or optimal method for sustainable waste management. Different conditions in various countries, such as cultural values, economic constraints, and levels of education and poverty, require different strategies and methods of implementation. Sustainable waste management strategies and implementation may even differ within a country. For example, the types and amount of waste streams in large cities are different from the streams in rural areas. Hence, different technical and social innovations may be required for these two geographical regions.

Sustainable waste management is therefore a large and complex challenge. It is also a topical and pressing challenge in both developed and developing countries. Many developed countries have suffered from economic crises and recession, and increased consumer expenditure is often thought to be a means to invigorate the economy. If this is not done in a sustainable way it will lead to increased consumerism and increased waste. Many developing countries are undergoing rapid urban and/or industrial growth and often have inadequate or nonexistent methods for waste management (Owusu and Afutu-Kotey 2010). This is not sustainable since it often has many deleterious consequences, such as the increased spreading of human diseases, and a large environmental impact, such as air, ground, and water pollution (Giusti 2009).

2.2 METHODS FOR SUSTAINABLE MANAGEMENT OF SOLID WASTE

As discussed in Section 2.1, there is no single, optimal method for sustainable waste management. This is also valid when one considers only the solid waste fraction. One needs to develop methods that are relevant to the local conditions, and analyze the strengths, weaknesses, opportunities, and threats of implementing waste management under these conditions.

In order to be sustainable, any strategy or method that is developed must include all of the sustainable development aspects that are relevant to the local conditions. These include the social, economic, institutional, legal, technical, and environmental constraints and possibilities that prevail in the geographical region. One needs to balance these aspects when developing a method for sustainable waste management. This is done best by involving all, or most, of the local and global stakeholders and leads to integrated sustainable waste management (ISWM; Joseph 2006; McDougall 2001; Morrissey and Browne 2004). Typical stakeholders are citizens and institutions that generate waste, municipal waste management organizations, policy makers, funding institutes, private enterprises, entrepreneurs, and members of the informal sectors such as waste scavengers. All of these stakeholders must understand their roles and responsibilities in the implementation of an ISWM method. Figure 2.1 illustrates the ISWM model.

The potential for success when developing and implementing ISWM models can also depend on the level of expertise that is required. In addition to the local stakeholders, it may therefore also be preferable to include nonlocal experts and researchers in ISWM (e.g., when they are not available locally). Examples are given in the case studies discussed in the next section. It is also important to learn from the successes and mistakes made by others when implementing waste management methods. This will be difficult if implementers of waste management methods do not provide an unbiased description of their experiences, including mistakes made and lessons learned.

Since there are several aspects included in ISWM, there are potentially several tools that can be used. For example, the environmental impact of the specific waste management method may be assessed using life cycle assessment (LCA). Economic and social aspects can be assessed using similar tools (Asiedu and Gu 1998; Hellweg et al. 2003). In addition, attempts have been made to combine these aspects and

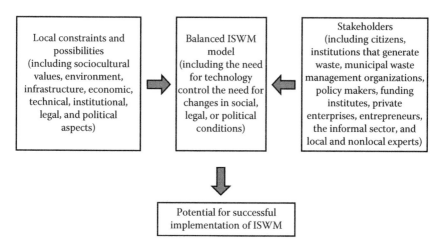

FIGURE 2.1   Illustration of the ISWM model.

develop sustainability criteria that cover all (or most) sustainable development aspects (Morrissey and Browne 2004; Zurbrügg et al. 2011). These criteria can be both qualitative and quantitative. One such model is the "Asian Guidelines of ISWM Assessment Methods" and is described below.

It should be reiterated that most of the tools that are used to assess the impact of waste management are often based on databases and social values that may not be relevant for the regions where the ISWM is being implemented. For example, the name life cycle *assessment* (as opposed to life cycle *analysis*) underscores the uncertainties that arise when assessing and comparing the environmental impacts of different waste management methods (Baumann and Tillman 2004). This is discussed in more detail in the next section.

2.2.1 LIFE CYCLE ASSESSMENT

2.2.1.1 Aim of LCA

LCA is a method to assess the environmental impacts of a process or a product (Baumann and Tillman 2004; Zaman 2010). An exhaustive study would include all impacts from the cradle to the grave. That is, one would include the effects of procuring the natural resources and using these resources during all stages of production and consumption (including reuse and recycling) as well as the environmental impacts incurred when disposing the product. However, in most studies one limits the scope of the LCA, and performs a cradle to the gate, gate to gate, or gate to grave study (Baumann and Tillman 2004).

Different types of LCA studies can be illustrated using solid waste management. One can imagine that a local municipality has identified a certain waste stream, such as cellulosic waste from the forestry sector, and wants to identify the disposal method that has the lowest environmental impact. Two methods that it would like to consider are the thermal conversion of the waste to energy (electricity) and conversion of the waste to viscose that can be used to produce new textile products. Hence, the municipality (who is purchasing the LCA and hence called the commissioner) has already limited the scope since it is only studying two methods (in a later LCA it may study other possibilities such as using this waste for fertilizer or for biological conversion to fuel). This is a comparative LCA since it compares the relative impacts of the two methods.

Since the same waste stream is used for the conversion to electricity or viscose, the municipality may wish to limit the scope to a gate to the grave study. That is, the environmental impacts that occur before the "gate" (such as the effect of fertilizers when planting and growing the trees and the use of fuel when felling the trees) are omitted from the study. The reasons why the municipality may choose to omit the impacts incurred when producing the waste include the facts that these impacts are the same for conversion to both electricity or viscose (and will hence not change the relative impacts), that this part of the study is too expensive and not expected to be significant, or that the data needed to assess the impact of this part of the process is not available (e.g., the quantity and type of fuel used for felling the trees).

The LCA may therefore begin with the impacts incurred when collecting the cellulosic waste and transporting it to the plants that produce electricity or viscose.

Subsequent stages include the impacts of electricity or viscose production (e.g., fuel for driving the production equipment and chemical emissions to air and water) and disposal of any by-products that cannot be used. However, in order to have a fair comparison of the two alternatives, the municipality may decide to include the positive impacts associated with using cellulosic waste for electricity or viscose production. For example, if this waste can generate 1 MW of electricity, then this electricity would not need to be generated from other sources such as fossil fuel. In this case one should calculate the impacts that would have been incurred if 1 MW had been produced from fossil fuel and subtract these impacts from the total impact of converting the waste to electricity. Similarly, if the viscose that is produced can be used to manufacture 1 ton of textile products, then these products would not have to be produced from other resources such as cotton. One should then calculate the impacts of manufacturing 1 ton of textile products using cotton and subtract these impacts from the total impact of converting the cellulosic waste to viscose.

It should be reiterated that LCA only assesses the *environmental* impact of a product or process. Economic and social aspects are not included, and other tools (such as life cycle costs) are needed to account for these impacts (Asiedu and Gu 1998; Hellweg et al. 2003). For example, LCA studies may yield results that support the waste management hierarchy, which can be phrased in several ways but often gives the following order of preference for waste management: reduce, reuse, materials recycling, composting, incineration with energy recovery, incineration without energy recovery, and landfilling. However, this hierarchy will not always be relevant due to economic or social aspects. In the example given above, the municipality may already have access to a plant for electricity generation from waste, but not for viscose production. Hence, even if viscose production has the lowest environmental impact, the municipality may choose electricity generation due to the economic advantages. Similarly, LCA does not take into account the working conditions of the cotton pickers, and when the municipality takes these social aspects into account, it may choose electricity generation from waste (even if viscose production has the lowest environmental impact).

2.2.1.2 LCA Method

Details of the LCA method are available in numerous publications (e.g., Baumann and Tillman 2004; Zaman 2010), and the method is briefly described here for the sake of completeness.

There are three types of LCA (Baumann and Tillman 2004). The example given above was a comparative study of two processes. In this type of LCA it is important to use the same method (e.g., same database) for both processes. One can also perform an LCA to create knowledge of the environmental impacts of a product or a process. This knowledge would allow the commissioner to identify stages in the process that have the largest environmental impact (called hot spots). The commissioner can then focus on these stages during process or product development. A third reason for performing an LCA is for marketing. These types of LCAs need to be audited.

There are four stages in an LCA study (Baumann and Tillman 2004). The first stage is called the goal and scope definition. The commissioner, usually with assistance from the LCA expert (called the analyst), decides on the context and purpose of the project. The commissioner also decides on system boundaries (e.g., gate to the grave in the example given above), the types of environmental impacts to include (global warming, eutrophication, etc.), and the level of detail (e.g., whether to use generic databases or to develop data that are more relevant for the specific process being studied). A simple flowchart is often constructed and the functional unit is also defined. For the example given above, a relevant functional unit may be 1 kg of cellulosic waste. The results of the LCA would then be the environmental impact of converting 1 kg of waste to electricity compared to converting the same mass to viscose.

The second stage, called inventory analysis, is usually the most time-consuming part of the study. The flowchart constructed in the goal and scope definition is drawn in detail. As shown in Figure 2.2, each part of the process can be shown as a box (or unit) in the flowchart, and data for all relevant inputs and outputs are collected for each box. The inputs that do not come from another unit—that is, those taken from outside the process—are resources used and the outputs that are not transferred to another unit are pollutants emitted. It can be noted that this is not a mass balance, since many outputs (such as water) are not considered environmentally harmful. All inputs and outputs are scaled to the functional unit. It is extremely important that the data used are relevant, valid, and reproducible.

In the third stage, called impact assessment, the environmental impact of the resources used and pollutants emitted is assessed. Although these assessment methods are often based on scientific research, they are complex and can depend on

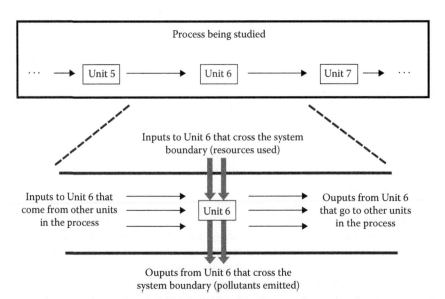

FIGURE 2.2   Illustration of three units that are part of an LCA flowchart. A unit can be, for example, collection of cellulosic waste or transport of the waste to a power plant. The inputs and outputs that are not contained within the process being studied are resources used and pollutants emitted. Examples of resources used are fuel (for transport) and of pollutants emitted are CO_2 and other harmful chemicals.

ethical values. For example, an emission (e.g., carbon dioxide) can lead to global warming and can also, via global warming, lead to extinction of plant or animal species. This is difficult to predict and hence to include in LCA models. This complexity is also underscored by the large number of assessment methods (Baumann and Tillman 2004), and the fact that some methods have a choice of weighting factors (which is a measure of how much a certain resource or emission impacts the environment relative to all other resources and emissions). The commissioner and analyst can select a weighting factor, depending on their perspective. For example, the Ecoindicator'99 method offers three different weighting factors based on cultural values: individualist (only proven environmental affects and a short-term view are taken into account), egalitarian (potential environmental effects and long-term view are taken into account), and hierarchical (in between individualist and egalitarian and based on some scientific facts).

The type of environmental impact, for example, global warming and eutrophication, can also be determined during the third stage. This is done by relating the resources used and pollutants emitted to a given environmental impact. This procedure is called classification and is illustrated in Figure 2.3.

The assessment results are interpreted in the fourth stage of the LCA. As mentioned above, this is an *assessment* and interpretation must be as objective as possible. It is important that all stages in the LCA are transparent and the strengths, weaknesses, and assumptions are clearly stated. A sensitivity analysis can also be performed, and if the uncertainty in any of the data may change the conclusions of the LCA, then one needs to return to an earlier stage (e.g., to increase the scope or to obtain more relevant data) and refine the assessment.

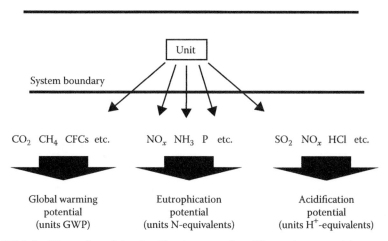

FIGURE 2.3 Illustration of the classification procedure. The environmental impact of the resources used and pollutants emitted, for example, from Unit 6 in Figure 2.2, are assessed. This involves a change in units, for example, from amount of CO_2 emitted in grams to its global warming potential (GWP).

2.2.2 ISSOWAMA Guideline on ISWM in Asian Developing Countries

2.2.2.1 Motivation

The challenges of solid waste management in developing and emerging countries are far from being solved. On the contrary, in those countries with limited resources and rapid urban and/or industrial growth, the problems have intensified at great speed. The general inadequate, when existing, methods of collection and disposal of solid waste in most Asian cities are causing important environmental and social harms, such as spreading of human diseases, environmental pollution in general and ground and water pollution in particular.

In order to raise awareness and to promote an adequate waste collection and treatment system, while simultaneously enabling economic growth of this sector in a technological efficient and sustainable way, new waste management systems must be established, which also take into account the informal sector. This integrated approach should comprise technical, environmental, legal, socioeconomical, and financial aspects, involving the key actors at different levels to ensure an effective implementation.

Therefore, addressing these issues, Asian and European solid waste experts joined forces to identify and overcome hurdles and practical difficulties. A group of 24 partners from Europe and Asia, under the leadership of ttz Bremerhaven (Germany) and interlinking with other local organizations, established the network ISSOWAMA—Integrated Sustainable Solid Waste Management in Asia. This project, which is described in Section 2.3.1, examined solid waste management issues in eight Asian countries to develop a performance assessment system for integrated management scenarios, which consists of a set of qualitative sustainability criteria along with quantitative impact indicators, enabling assessment of waste management strategies. The central focus was the integration of the sustainability aspects into the system, such as environmental impacts, economic considerations, and the social situation in the target region. It was important to integrate appropriate low-cost and efficient technologies with community-based management and their relevant governance, institutional frameworks, and socioeconomic constraints, linking waste treatment with poverty reduction and improvement of welfare of the population. Research institutions, municipalities, and other end users were targeted. These results were verified in different Asian cities, which was possible since the project consortium included countries in South Asia (Bangladesh and India), the Greater Mekong Subregion (Cambodia, Thailand, China, and Vietnam), and Southeast Asia (Indonesia and the Philippines).

An in-depth understanding of the strengths, weaknesses, opportunities, and threats in the specific local context was fundamental when developing a robust and sustainable solution. This not only greatly enhances mutual learning among practitioners, but also enhances replication and improvements in practice. Analysis of reasons for success or failure of waste management activities was the first step in this learning process.

In order to perform this analysis a solid waste management expert and an expertise network was established in order to coordinate, assess, and guide suitable research and strategic activities. The aim of these activities was to identify aspects like cost-effective treatment and sorting technologies, environmental impacts, gaps in technical knowledge, and socioeconomic and policy barriers to further execution. The network also proposed directions for future research and for local implementation.

2.2.2.2 The Guideline

The work described in the previous paragraph resulted in the ISSOWAMA guideline on integrated solid waste management in Asian developing countries, and assists decision makers and technical staff in different governmental levels as well as in municipalities, industry, NGOs, and other stakeholders dealing with waste management. It provides an overview of the impacts of, and influences on, integrated solid waste management.

The guideline also provides information about waste generation and composition in different Asian countries and introduces main elements of waste management systems, comprising collection and transportation, street sweeping, recycling, treatment, and final disposal. Aspects concerning hazardous waste management are also considered. The potential impacts on the environment (air, soil, and water), socioeconomic aspects, and health are included.

Furthermore, it provides a deeper insight in technologies for waste recycling, waste treatment such as composting or anaerobic digestion of the organic fraction of the municipal waste, waste incineration, or waste disposal on sanitary landfills. All technologies are briefly described and characterized regarding opportunities and limitations of applications, highlighting their strengths and weaknesses. The status of implementation in Asian developing countries is also described for each technology. Additionally, for each technology one case study in an Asian developing country is introduced to elucidate the potential of the technology and to report practical experiences.

Since many existing assessment methodologies for waste management projects and concepts are very complex and difficult to understand and apply for inexperienced practitioners, the ISSOWAMA guideline introduces a new—and simplified—case study assessment tool. It facilitates identification of challenges and improvement potentials in implementing waste management technologies from a case study to a new application. The tool is based on a questionnaire comprising the categories described below. The link to a practical case study helps to better understand the importance of the chosen categories and enables the user to transfer the experiences obtained in the case study to his or her own situation.

The following main categories of relevance to evaluate a case study are considered in the ISSOWAMA Guide of Solid Waste Management in Asian Developing Countries (Zurbrügg et al. 2011):

- Main purpose of the technology and waste streams of relevance
- Technical functionality, including in particular
  - Performance
  - Know-how requirements and availability
  - Availability of materials and spare parts
  - Adaptability on changing conditions

- Health impacts on
  - Workers and community
- Environmental impacts, considering among others
  - Emissions and nuisances
  - Compliance with standards
  - Resources efficiency
  - Materials recovery and valorization
  - Waste reduction/minimization
- Costs, finance, and economic aspects to make sure that the case is
  - As cost-effective as possible
  - Sustainable
- Social aspects and impacts, especially
  - Acceptance by the beneficiaries
- Organizational strength and institutional support to ensure
  - Acceptance of the system internally and at policy level
  - Adequate training of staff.
- Legislative and policy support to
  - Set the framework of applicable legislation
  - Avoid unnecessary bureaucracy

Finally, a general rating of the level of best practices helps when judging the practical relevance of the assessed system for a specific case.

For more detailed information, the reader should refer to the ISSOWAMA guideline. It is publicly available and can be downloaded from the project's website, www.issowama.net, free of charge.

2.2.2.3   Evaluation

The ISSOWAMA guide was discussed with relevant stakeholders in the frame of three thematic panel discussions organized in the Philippines, Thailand, and India. They were specially designed to discuss the ISSOWAMA approach with relevant stakeholders from different countries. In addition to these events, regional workshops on adapted waste treatment technologies and services were held in India, China, Bangladesh, Thailand, Vietnam, Cambodia, Indonesia, and the Philippines. They were specially organized for municipalities, SMEs, industries, and stakeholders from the informal sector, having the objective of dissemination to a broad audience. They also served as an instrument to introduce new relevant actors like local waste processors and regulators to the existent networks, as well as to raise awareness and promote good practices and environmentally sound technologies.

Both events, the thematic panel discussions and the workshops, have given deeper insight on the current status and started the dialogue between different stakeholder groups, which will contribute to the implementation of integrated solid waste management in the long term.

The main conclusions about the status of waste management in general and integrated solid waste management in Asian developing countries in particular are:

- The integrated solid waste management approach is suitable for the Asian developing countries. All feedback received from discussions in the workshops pointed out that integrating technologies and stakeholders is necessary.
- The main difficulties of this approach are engaging and coordinating all groups. Some informal groups (waste pickers, slum dwellers, etc.) are reluctant to participate in the dialogue that leads to a new waste management strategy. Some government groups also find it difficult to overcome the "business as usual" scenario and to engage in an open conversation with members of all groups. A strategy to achieve everyone's participation must be specific for each country.
- Public awareness is an issue in all countries. Practitioners and researchers in the sector agree that the public needs education in order to understand the problems that arise from inadequate solid waste management, the correct way to deal with solid waste management, and the opportunities it represents, such as recycling. The main hurdles in this regard are the lack of access to educational measures for some groups and lack of funds to develop comprehensive strategies.

In order to achieve success at the country level, deeper involvement of all levels of government is needed. Different government bodies have to make stronger efforts in applying the laws currently in force and to achieve transparency.

2.3   EXAMPLES OF ISWM

Two examples, or case studies, of ISWM are discussed below. The first example has been chosen since it illustrates a case study that addresses a large geographical region, namely Asian developing countries, and provides the guideline discussed in Section 2.2.2. In contrast, the second example concerns ISWM at the local level—the cities of Borås in Sweden and Yogyakarta in Indonesia—and a working partnership between these cities.

2.3.1   ISSOWAMA

Municipal solid waste management is a crucial issue for the fast growing urban settlements in Asia's developing countries. In Asia, about 4.4 billion tons of solid waste is generated per year, of which 790 million tons are of municipal origin. This situation is not satisfactory. When dumps, such as that illustrated in Figure 2.4, and landfills exist, they are neither secured nor properly organized. Deposition of hazardous waste is common. Lack of information at decision makers' level, lack of proper solid waste management concepts, and, last but not least, lack of finance are responsible for the current problems. The services for collection, transport, and processing are unique for Asia, but more or less similar in the different countries in Asia. The uniqueness is attributed to the typical waste composition, involvement of the informal sector, voluntary groups, private organizations, NGOs, and community-based organizations. But it can be assumed that up to 50% of residents of urban areas in low- and middle-income countries lack proper waste collection services (van de Klundert and Anschütz 2001).

FIGURE 2.4   Illustration of a dumping site with a waste picker in India.

Solving these problems by transfer of recent concepts and technologies for municipal solid waste management from Europe and implementation in Asia will not be successful and sustainable unless they are adapted to the situation in Asian developing countries. The main drawback is that the allocation of resources for the municipal solid waste management does not encompass the entire solid waste management situation and requires immediate attention of the governments and civic organizations to deal with the growing impacts on the environment. It is necessary to implement an integrated solid waste management strategy, which could pave the way for sustainable urban development in Asia with positive impacts on economic, environmental, and social aspects (Visvanathan and Trankler 2003). The solid waste management problem must be given utmost priority due to the negative impacts on the environment and threat to public health:

- Pollution of air due to uncontrolled burning of refuse
- Pollution of surface and groundwater by means of percolating leachate
- Pollution of soil due to release and introduction of toxic and hazardous compounds from waste materials
- Waste is a breeding ground for diseases threatening human health (injuries, infections, chronic diseases)
- Degradation of human dignity due to scavenging

An integrated and sustainable waste management system also has to consider social aspects properly. Sociocultural barriers, for example, can only be addressed by effecting a significant change in the mind-set and attitude of the different stakeholders: the decision makers, the business and informal sector, and the general public.

Potential technologies and services to be transferred to Asia must conform to local needs and be based on the current waste profiles. It should always be done through actual demonstrations, trainings, day-to-day hands-on management, and intensified monitoring.

Commonly, waste management and handling is conveniently assigned to the marginalized sectors in the community. Truth is, waste management is a basic service requirement and is a growing industry. It is therefore necessary that waste management personnel, from the waste handlers to the managers, are adequately trained and thoroughly equipped to adopt professional work standards.

The key factors for a successful and sustainable waste management system are

- Public awareness
- Capacity building at municipality level
- Availability of technologies
- Enforcement of regulations and legislation
- Entrepreneurship at private level (availability of micro credits, marketing, and trading activities to ensure profitability of services)

The development of a practical guideline to support implementation of integrated solid waste management strategies for Asian cities and rapidly growing regions was urgently needed. To achieve this aim, a performance-assessment system for alternative waste management scenarios, which consists of a set of qualitative sustainability criteria along with quantitative impact indicators, was developed. The sustainability aspects also had to be considered.

ISSOWAMA was a project that was funded within the seventh European Union framework programme (FP7) between January 2009, and June 2011. It is described at www.issowama.net and included solid waste management experts from Europe and eight Asian countries. These were Bangladesh, India, Cambodia, Thailand, China, Vietnam, Indonesia, and the Philippines, and hence represented regions from South Asia, the Greater Mekong Subregion, and Southeast Asia.

The main aim of the ISSOWAMA project was to develop a method to assess solid waste management strategies. The assessment method, which was described in Section 2.2.2, combines qualitative criteria and quantitative impact indicators. Of critical importance is that the method is both integrated and sustainable. It therefore considers the local conditions that will enable or hinder the solid waste management, includes all relevant stakeholders, and considers the environmental, social, and economic aspects of sustainability.

The project promoted international cooperation between research organizations, universities, and social and governmental stakeholders in a European and Asian context (local waste processors, local municipalities and policy makers, local NGO representatives, etc.). It consisted of six work packages (WPs). Four of the WPs focused on the development of the assessment method, the fifth WP was for disseminating the results of the project, and the sixth WP was project management.

In the beginning of the project the partners identified the current state of solid waste management research being done in the participating countries. This resulted in a database that includes the people involved in this research, technical data of the research, and the stakeholders that are considered and that participate in the research. This database, as well as the networks that can result from it, could be very important in developing future knowledge and strategies for solid waste management. It also helped to identify areas where nonlocal experts were required to complement the expertise of the local researchers.

The second stage of the project focused on identifying case studies within the participating countries. This included an assessment of the local conditions, and how these enabled and hindered solid waste management. To enable evaluation of the case studies, and to assist in comparing the evaluations, the criteria used in the evaluations needed to be established. These criteria were primarily based on the ISWM method, but also required that the consortium agreed on which waste streams were to be addressed. These were municipal solid waste (including construction and demolition wastes and wastes from incineration plants), healthcare waste, and e-waste as well as other hazardous domestic and urban waste.

The insight gained from these case studies was used together with an analysis of existing environmental impact assessment methods and impact categories to develop an ISWM method that was relevant for the Asia region. Hence, knowledge of the solid waste management technologies that are relevant to the Asian countries was developed, as well as their impact on the environment and human health. For example, it was revealed that a crucial aspect for the successful implementation of ISWM in Asian countries is the collaboration of stakeholders throughout the waste management chain. Collaboration therefore has to be organized at a large scale and government agencies must take responsibility for initiating the ISWM. In addition, the most important positive impacts of the solid waste management in Asian countries was safeguarding of public health and the environment, reducing the rate of resource depletion, and increasing the rate of reuse and recycling of valuable materials. All of these impacts also have beneficial economic impacts. These insights resulted in the Asian guidelines of ISWM assessment methods discussed above. The experiences of ISSOWAMA show the importance of international networks of experts sharing experiences and collaborating in developing solutions for one of the most serious problems in developing and emerging countries with growing economies and rapidly increasing population density in urban settlements, reaching dimensions hard to imagine for Europeans. The work has not been finished with the termination of ISSOWAMA. In the 30 months of its active duration, important initiatives have been started and hopefully useful information compiled for efficient and successful application by all experts dealing with solid waste management, particularly in, but not limited to, Asia. The activities need to be eagerly continued to reach a more subordinated goal: improvement of quality of life!

2.3.2 WASTE RECOVERY IN BORÅS, SWEDEN, AND A PARTNERSHIP IN YOGYAKARTA, INDONESIA

The city of Borås in Sweden was one of the pioneering cities for converting waste into useful products. A pilot program was implemented for 3,000 households in 1986, and was developed over a 10-year period to include all 100,000 inhabitants of the city (Taherzadeh and Engström 2013). Similar to the discussion given above for the

Asian region, all relevant stakeholders needed to be included in the development of the waste management system in Borås. This included law and policy makers, local companies (who needed new economic incentives and access to the preferred technology), the municipality, and the citizens. In fact, the citizens of Borås (as well as people visiting the city) have a key function in the waste management (cf. Chapter 1), since they need to sort their waste into organic waste and other waste (which can be further sorted into electronic waste, etc.). Education is crucial for motivating the citizens, since they need to understand the benefits to them and the environment.

Household waste is presently separated into 30 categories and recycled or converted to biofuel, electricity, or heat. Approximately 27% is recycled via private companies, 30% is converted to biofuel and compost, and 43% is combusted to produce electricity and heat (Taherzadeh and Engström 2013). The total production of biogas is 3 million m^3 per year, and this is upgraded to fuel, which is primarily used in the municipality transport system (buses, garbage trucks, and light vehicles). The vision of the city is to become totally fossil-free.

Research, education, and a close partnership between stakeholders is required for continued success of the waste management in Borås. The University of Borås offers education in resource recovery and research in this area is of strategic importance to the university. There is a strong partnership between the university, the city, SP Technical Research Institute of Sweden (SP—which is located in Borås), and private companies. One of the companies that play a central role is Borås Energi och Miljö (BEM—the company that is responsible for managing the city's energy and environment) since its strategies align with the city's strategies. In addition, pilot plants for waste treatment have resulted from research performed at the university, and these pilot plants are also used for ongoing research. It is important to reiterate that research and education is not only focused on the technological and environmental aspects, but also on social aspects such as consumer behavior (both when purchasing products and when managing the waste resulting from the purchase).

A partnership, called Waste Recovery—International Partnership (WR), was initiated in Borås in 2006. It consists of the University of Borås, the city of Borås, SP, BEM, and about 20 other companies and organizations with an interest in waste management. As illustrated in Figure 2.5, this cluster works with local clusters in other regions of the world to develop ISWM methods in those regions. The approach used by WR is very similar to the approach used to develop the Asian guidelines of ISWM assessment methods, and places focus on all of the local stakeholders. Education of relevant stakeholders is also emphasized.

One of the cities that have a partnership with WR is Yogyakarta in Indonesia. In particular, the Gemah Ripah fruit market in the Sleman region of Yogyakarta discarded about 4–10 tons of rotten fruit per day. This waste was transported and dumped 40 km from the market. Discussions were initiated between the University of Borås and the local university in Yogyakarta, called Gadjah Mada University. Through student and researcher exchange, initial plans were made for the conversion of this waste to biofuel. Determination of the initial dimensions of the bioreactor and its working conditions resulted from these types of exchanges. However, as mentioned above, successful implementation of the conversion of waste to biofuel required partnership between all relevant stakeholders (from Borås, Yogyakarta, and funding bodies).

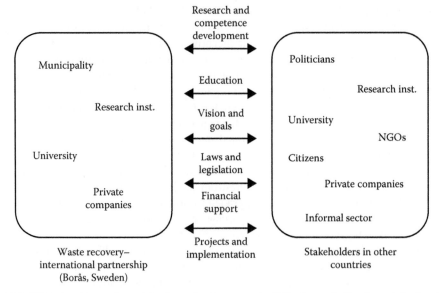

FIGURE 2.5 Illustration of the interplay between the WR cluster in Borås and the local clusters in foreign regions.

Funding from the Swedish Development Cooperation Agency, the Swedish Agency for Economic and Regional Growth, and Gadjah Mada University was used to initiate a network between WR and local governments, embassies, municipalities, companies, NGOs, and the workers at the fruit market. This was essential since Swedish technologies and values cannot be directly transferred to Yogyakarta. For example, the working conditions of a bioreactor in the Swedish winters is of little relevance to weather conditions in Yogyakarta, and the effect of implementing this technology on certain groups, such as scavengers, is not an issue in Borås.

Education throughout the implementation was also considered crucial to the success of the project. For example, the leader of the cooperative in charge of the market and people in the local municipality gained further education into waste management and the environment. The people that were responsible for the running of the bioreactor were educated in this, and the roles of each partner were clarified (e.g., who owned the bioreactor, who was responsible for its running, and who was responsible for paying for repairs when required). It was also ensured that the methane gas produced from the bioreactor could be used by the market.

Due to the successful collaboration between WR and the local stakeholders, as well as the successful partnership between the local stakeholders, the bioreactor was commissioned in 2011. Figure 2.6 is taken from the inauguration of the bioreactor, and shows representatives from three of the stakeholders in the project: Ewa Polano is Swedish ambassador in Indonesia; Mohammad Taherzadeh is professor from the Swedish Centre for Resource Recovery, University of Borås; and Mohammad Junaidi Chasani is VD of Bionat—the company that built the bioreactor. Approximately 4 tons of fruit waste per day is used as input to the bioreactor, which means that the 12 truckloads per week that needed to be transported to the dump have been reduced to one truck. In addition,

FIGURE 2.6 Inauguration of the bioreactor at the Gemah Ripah fruit market. Three of the stakeholders are shown in the figure: from left—Ewa Polano, Mohammad Taherzadeh, and Mohammad Junaidi Chasani. (Courtesy of Annie Andreasson.)

the 500 kWh of electricity per day that is generated is used to light the market place. Hence, in addition to the economic benefits of reduced transport and electricity costs, there are numerous environmental benefits that include reduced use of fossil fuel to transport the waste, reduced emission of methane to the air from dumping the waste, and reduced use of fossil fuel that was used to light the market place.

2.4 CONCLUDING NOTE

ISWM is done best by involving all, or most, of the local and global stakeholders. All of these stakeholders need to understand their roles and responsibilities. This chapter presented two of the tools that can be used when implementing ISWM: (1) LCA and (2) ISSOWAMA guideline on ISWM in Asian developing countries. This chapter also illustrates two case studies of ISWM, one of which addresses a large geographical region, namely Asian developing countries, and the other a partnership between two cities—Borås in Sweden and Yogyakarta in Indonesia. These two studies are examples of recent implementations of ISWM in developing countries, and this work must be intensified in order to achieve a satisfactory global impact.

REFERENCES

Asiedu, Y. and P. Gu. 1998. Product life cycle cost analysis: State of the art review. *International Journal of Production Research*, 36(4):883–908.
Baumann, H. and A.-M. Tillman. 2004. *The Hitch Hiker's Guide to LCA*. Studentlitteratur, Lund, Sweden.

Giusti, L. 2009. A review of waste management practices and their impact on human health. *Waste Management*, 29(8): 2227–2239.

Goodland, R. 1995. The concept of environmental sustainability. *Annual Review of Ecology and Systematics*, 26: 1–24.

Greyson, J. 2007. An economic instrument for zero waste, economic growth and sustainability. *Journal of Cleaner Production*, 15(13/14): 1382–1390.

Hellweg, S., G. Doka, G. Finnveden, and K. Hungerbühler. 2003. Waste and Ecology: Which technologies perform best? In *Municipal Solid Waste Management*, Eds., Ludwig, C., S. Hellweg, and S. Stucki. Springer, Berlin, Germany, pp. 350–398.

Joseph, K. 2006. Stakeholder participation for sustainable waste management. *Habitat International*, 30(4):863–871.

Lehmann, S. 2011. Optimizing urban material flows and waste streams in urban development through principles of zero waste and sustainable consumption. *Sustainability*, 3(1): 155–183.

McDougall, F. R. 2001. *Integrated Solid Waste Management.* Blackwell, Oxford.

Morrissey, A. J. and J. Browne, 2004. Waste management models and their application to sustainable waste management. *Waste Management*, 24(3): 297–308.

Owusu, G. and R. L. Afutu-Kotey. 2010. Poor urban communities and municipal interface in Ghana: A case study of Accra and Sekondi-Takoradi Metropolis. *African Studies Quarterly*, 12(1): 1–16.

Taherzadeh, M. J. and O. Engström. 2013. Converting waste to wealth. *SGI Quaterly*, 71: 4–5.

Troschinetz, A. M. and J. R. Mihelcic. 2009. Sustainable recycling of municipal solid waste in developing countries. *Waste Management*, 29(2): 915–923.

van de Klundert, A. and J. Anschütz. 2001. *Integrated Sustainable Waste Management—The Concept.* WASTE Assosiates, Gouda, the Netherlands.

Visvanathan, C. and J. Tränkler. 2003. *Municipal Solid Waste Management in Asia: A Comparative Analysis.* In *Proceedings of the Workshop on Sustainable Landfill Management*, Eds., Joseph, K., R. Nagendran, and K. Palanivelu. Allied Publishers Pvt Limited, Chennai, India, pp. 3–15.

Zaman, A. U. 2010. Comparative study of municipal solid waste treatment technologies using life cycle assessment method. *International Journal of Environmental Science and Technology*, 7(2): 225–234.

Zurbrügg, Ch., Y. Vögeli, J. van Buren, J. Potting, and V. Chettiyappan. 2011. *About Solid Waste Management, ISSOWAMA Project Deliverable 5.5—Simplified Guide on Solid Waste Management in Asian Developing Countries.* http://www.issowama.net/

3 Laws and Regulations Governing Waste Management

Institutional Arrangements Regarding Waste Management

Ulla Eriksson-Zetterquist and
Maria José Zapata Campos

CONTENTS

3.1 INTRODUCTION

While flying into São Paulo, parts of the structure, the infrastructure, and the distribution of this megacity revealed themselves to the traveler. In between the favelas climbing up the steep hills, the middle-class communities seem to prosper. Not visible from the air, but clearly so from the well-kept streets, some of the balconies and terraces of the apartments of the middle-class present themselves. For those who look very closely, piles of outdated television sets, computers, refrigerators, freezers, audio systems, and the like may be found on these very balconies and terraces. This is the waste of the middle class, who have exchanged old technologies for new, fashionable solutions. To those working with recycling—industry, informal recyclers, civil servants, entrepreneurs, legislators, and politicians, to name but a few—the outdated technologies on the balconies are not waste but products that can be processed into extremely valuable raw materials.

Similar to many other sources of raw material, however, the outdated electronic waste undergoes a number of transformations before being turned into such an asset. A few of the necessary transformations (Corvellec et al. 2013b) are listed here: unsustainable waste infrastructures (Corvellec et al. 2013c) have to be unlocked; the fragmentation of responsibilities across international, national, and regional levels, or public and private actors, has to be overcome (Davoudi 2009); new appropriate political instruments have to be developed and resources have to be allocated (UNEP 2010); new knowledge systems, capacities, and organizational structures (Nilsson et al. 2009) have to be put in place; technological innovations meant to recover various fractions from each other have to be developed; and legal reforms connecting waste streams with the markets while guaranteeing both the protection of the environment and public health should be established. Taken together, these different aspects of waste systems create what can be called an institutionalized setting, in which issues are closely intertwined with each other, taken for granted, and followed by a social and cognitive understanding of the system per se (Greenwood et al. 2008). In order to implement changes to the waste system, focusing on better recovery, reuse, or other processes, the legal issues form one part of many which have to be changed in order to rebuild the system into new sustainable solutions. As the title indicates, laws and regulations are part of the institutional arrangements of waste management; in order to understand these, the arrangement has to be explored.

A way to understand this interplay is introduced in the perspective organizing action nets, which will be used here as a frame of reference. Furthermore, we will also use the example of electronic waste to explore and illustrate the social processes supporting a society in which waste is handled in an efficient manner, as well as the role of waste laws and regulations that govern waste management in order to achieve this goal. One starting point is that laws and regulations do not operate as singular constructs, instead depending upon the institutional arrangement in order to work. Establishing a law prohibiting a certain product from being sent to landfill will not be helpful, if that law is not supported by an alternative system, including an efficient infrastructure for handling the waste and social incentives such as financial, habitual, and cognitive ones.

In order to provide a conceptual setting for understanding the context of laws and regulations, the Brazilian case briefly mentioned in the introduction will be set in a global perspective in order to more precisely discuss what waste is and how we conceptualize what can be seen as waste. In doing so, we present a framework discussing how infrastructure aimed at handling waste in society is organized in what have been described as lock-in situations. This framework emphasizes the need to include both materials and social systems when exploring the institutional arrangements of waste management. We then present a general overview of how societies are expected to handle waste, as well as how this changes over time and place. With this as a background, the role of laws and regulations governing waste management will be elaborated on. As a specific example, we introduce the case of electronic waste as a rapidly growing waste fragment in today's societies. This is followed by the introduction of different theoretical approaches to the study of waste regulations and laws as institutional arrangements of the waste management system. Next, the practices of laws and regulations in waste from electrical and electronic equipment (WEEE) are explored, followed by a discussion of occupational risks and hazardous environments relating to WEEE.

3.2 CONCEPTUALIZING WASTE

Given the introduction, waste can easily be understood as a problem that is out there for someone else to handle. To nuance the problem of turning electronic waste into raw materials, the image of the Brazilian balconies can be compared with waste management in other local contexts. A Western household typically includes some space set aside for old cell phones, remote controls, and toys from McDonalds, and yet another space in which old laptops are piled up. In a similar vein, an office or workplace that has been in operation for a number of years and has a number of employees will most likely include corresponding spaces for old computers, laptops, TV sets, old tape recorders, and microphones, to name but a few. The major difference vis-à-vis the middle-class households in Brazil is that the Western storage spaces are often even more out of sight of the street, and thus even more easily forgotten about. Yet, both spaces fulfill similar functions, that is, that of a transitional space wherein the social life of things (Appadurai 1986) is about to be decided sooner, or more probably, later. Things are stored in these spaces "in between," being dispossessed of their previous social roles and staying silent ready to gain a new social life, a return to life, as a reused thing with new ownership, meanings and, eventually, new affordances attached to them; or eventually as new raw materials, disassembled from former objects. Gregson et al. (2007) have shown how the practice of saving and wasting is critical to materializing identities and to the key social relationships of family and home. They argue that understanding the increasing amount of waste production requires a focus on the social and emotional relationships of households, and not on the trajectories of the things themselves. Gregson et al. (2007) claimed that, while we as a society are producing more and more waste, more and more things, for example, electronic waste, are being kept. This has obvious consequences for the accommodation and storage of things (Cwerner and Metcalfe 2003), for example, the creation of new spaces such as those balconies we noted with surprise during our visit to São Paulo. Discarding things cannot, however, automatically be equated with waste disposal or generation, in the same way that discarded things do not necessarily become waste. It is at the very moment when discarded things are disposed of at garbage sorting rooms, in bins, or in other authorized, or illegal, places that they become waste. Thus, one question is how do we convince householders to get rid of their possessions and reintroduce them via upcycling or recycling processes? Since disposing and wasting are not just physical acts involving material things, but cultural and social too, this is not an easy task (Douglas 1966).

Later on, when things are turned into waste, new challenges emerge. Or as expressed by Mary Lawhon (2013, p. 700): "Like so many other 'disposable' parts of our lives, most people give little or no thought to what happens after our various and diverse commodities turn into waste." This neglect may, however, be about to change, at least when listening to those called the "urban miners." Disposed of items may be about to see new light.

There is a great potential in waste. For instance, WEEE is partly made from valuable metals such as gold and platinum, rare earth metals, and other valuable parts made of copper. One consequence of this may be that all these spaces can be seen as a first step in the process of turning the forgotten stuff into an asset. To this economic dimension, the social and environmental dimensions can be added, calling for further aspects of an effective waste management process.

But how can societal actors collaborate to arrange such a first step and such a process? Can new laws be the solution? Or is it better regulations, which state that all old objects should be returned to undergo the recycling process? The brief answer is no—most societies have a number of formal and informal rules and regulations governing how to handle the various dimensions of waste; yet, the recycling processes are hampered. Additionally, one common pattern worldwide is that electronic waste, in particular, has a great potential to be handled more efficiently and in a more sustainable manner. When scrutinizing the various contexts of laws and regulations, a more complicated picture emerges in which various actors, the connections between these actors, and institutions sometimes enable, but sometimes hinder, an effective social process for handling waste.

3.2.1 UNDERSTANDING SOCIOTECHNOLOGICAL ARRANGEMENTS FOR WASTE MANAGEMENT: THEORETICAL APPROACHES

The institutional arrangements hindering sustainable waste recovery can be understood in various ways. In a research overview of infrastructure lock-in situations, Corvellec et al. (2013c) explored four different rationales concerning waste incineration. The four rationales were institutional, technical, cultural, and material. Institutional rational includes legislation and political dimensions regarding how to govern waste. Technical rationales include, for instance, incineration plants, infrastructure for district heating, and other initial economic investments of an enormous nature. Cultural rationales are based on the stories and narratives describing the benefits of a certain waste solution, for example, being an "efficient, profitable, and environmentally sustainable way to process waste" (Corvellec et al. 2013c, p. 36), preferably followed by a story about a preceding crisis (e.g., landfills). Material rationales are based on the routes for collecting waste, transforming energy into warm water, and the physical space of a waste management plant. As each of these is mutually independent, they are difficult to change. Being interrelated, they may create a lock-in situation.

The infrastructure lock-in situation touches upon another dimension of waste systems, that is, how they are made up of social and technical systems and the relationship between these systems. These aspects and the relationships between them can be understood by applying an action net perspective (Czarniawska 1997, 2004, 2007; Corvellec and Czarniawska 2014a). When exploring processes in which novel ideas are adopted by institutionalized settings, and potentially turned into sustainable innovations, the perspective of organizing in action nets, as developed by

Czarniawska (1997, 2004), has benefits. In brief, organizing in action nets pays attention to how actions connected to each other jointly contribute to the establishment of new ideas in institutionalized contexts. This perspective is partly based upon the sociology of translation in which it is suggested that technology and humans have to be studied using a symmetrical perspective, partly upon the assumption that innovations (e.g., new technology to recover waste fractions or new infrastructure to collect waste from the hidden drawers, boxes, and balconies mentioned earlier) occur in institutionalized environments. In the following sections, the consequences of these perspectives will be further developed.

In the symmetrical perspective, social and technological aspects are both seen as actors in the construction of networks. An innovation, that is, regarding how to prevent waste, can become a sustainable solution, if these networks have many allies that are, preferably, influential (Latour 1987). Both human (in this case, for instance, producers, consumers, politicians, managers) and nonhuman actors (in this case, for instance, WEEE, valuable parts, infrastructure, systems regulating how to pay for waste disposal) are ascribed the ability to both carry and contribute to new connections and new associations, which together constructs the network (Latour 2005). A key part concerns translation. Here, translation focuses on the process in which an object is transferred from one context to another, and thus transformed in order to fit into the new context (Serres 1974; Knorr Cetina 1981). Through these translations, the social and the technical are changed in a state of constantly ongoing interplay; they are consecutively constructed in order to support each other in the establishment of the network, which is a prerequisite of enabling the idea to result in a sustainable solution (Latour and Woolgar 1979/1986). By following the associations of the actors (waste systems, production systems, or consumption patterns, to name but a few), by following the actors themselves (i.e., technology for recovery, decisions about infrastructure, practices regarding how to sort out the fractions of waste), and by trying to go beyond what seems to be an obvious causal relationship, it is possible to study how innovations in the field of waste regulation manage to become established in institutionalized contexts.

The ethnographic study of infrastructure initiated by the work of Star (1999) can also contribute toward shedding light on infrastructure lock-in. Star and Ruhleder (1996) argue that the absence of infrastructural flow creates visibility, just as their continuous normalized use creates taken-for-grantedness and invisibility (Star 1999; Graham and Thrift 2007). In smoothly functioning cities, electricity, water, and waste collection infrastructure networks are only noticed when they break down. In addition, most urban infrastructures are located beneath the surface of cities, or at their fringes, which also makes them literally invisible (Kaika and Swyngedouw 2000). Even when some nodes of the infrastructure are materially visible, such as the balconies of middle-class householders in São Paulo, they tend to be invisible to the gaze of the ordinary citizen—but not to the eye of the tourist, of course. The material and social invisibility created by infrastructures can contribute to both environmental problems and socioecological changes (Monstadt 2009). For example, the display of containers for collecting batteries in supermarkets can make batteries forgotten about at home in a drawer visible again. Similarly, recycling containers succeed in making the different materials and assets still contained in waste visible again by forcing users to look at their waste. These recycling containers are designed to transform waste "from ambiguously, undefined things into named, understood, transparent and performative materialities" (Corvellec and Hultman 2013, p. 145). Making infrastructures and waste visible can thus contribute toward creating new disposal trajectories, which recover the forgotten e-waste stored in drawers and basements, and on balconies; but which also recover e-waste that is wrongly disposed of.

From cultural studies, infrastructures have also been approached under the prism of crossing points (Hawkins 2006, 2009). Hawkins (2006) describes how the sound of breaking glass in a recycling container signals the moment when empty bottles are reborn in the recycling economy. At the exact moment when the owner of the bottle throws it into the container, it passes a crossing point in the waste chain. The bottle is not a bottle anymore, but glass that is animated and transformed within the recycling economy. The transformation of empty bottles into recyclable glass does not mark the moment when bottles die, but the moment when new use and value are added—"a beginning, not an ending" (Hawkins 2006, p. 93). The transformations enacted in waste infrastructures, such as a recycling container, activate the materials and energy existing in the waste. However, they can also enact less-benign animations, such as pollutants and emissions that are hazardous to the environment and to human health (Gregson et al. 2010). The reused goods, and the recovered materials and energy, can also gain new economic value when they are connected to corresponding new and emerging markets. In this material metamorphosis, these goods and materials gain new meanings and uses associated with their new owners.

To sum up, from this perspective, infrastructures associated with waste management emerge as crossing points, places for physical, material, semiotic, etymological, symbolic, social, legal, and organizational transformations. Waste infrastructures are also crossing points between different organizational worlds, for example, the collective actions of citizens and householders and of the local authorities responsible for waste collection meet at recycling stations (Zapata and Zapata Campos 2013).

The transformations enacted at crossing points can have various implications for the environment, the economy, and human health, implications that are not always positive. For example, *catadores* (waste picker) cooperatives in São Paulo collect WEEE from householders and disassemble it at their waste recycling centers. However, insufficient work legislation; the difficult, risk-prone, and unhealthy conditions in which they work; and the lack of appropriate technology threaten both the health of the operators (Gutberlet 2010, 2012) and the quality of the disassembled materials.

Taken together, these assumptions create the foundation for studying processes in which new ideas and trends concerning waste management become established in organizational contexts. The processes studied cover development from a new idea to a new product or a new service being adopted by new markets and/or governmental organizations. Organizing is seen as collective actions connected to each other. Usually, these actions are seen from the premise of being legitimate at a given point in time, in a given context—that is, they are institutionalized. Following the actions that are connected to each other constructs an action net; it is possible to follow how a new idea, as an innovation in the field of waste management, enrolls allies and is translated in order to adjust to new contexts (Czarniawska 1997, 2004). These contexts can be within one particular organization, but also in between organizations or on a market.

It can be noted that actions, as used in the action-net-perspective, resemble the contemporary concept of practice in fashion. One difference, however, is that practices generally refer to what people do, and less often to how artifacts contribute to these practices. As noted by Joerges and Czarniawska (1998), as well as by Pinch (2008), social norms and values are inscribed in machines. This means that, when an actor uses a machine, he or she will more or less consciously repeat the norms inscribed by the producer of the machine. In the waste context, this means that norms and values, for instance, are inscribed into the infrastructure, upon which people will act. Regulation cannot, for instance, be separated from the practices provided by the infrastructure. If the law prohibits storing e-waste at home, and the infrastructure fails to enable ways to collect this particular waste, the law will be less efficient. By paying symmetrical attention to the actions of humans and nonhumans in an action net, organizing in action nets will include a multitude of the dimensions of social practices in organizing.

3.3   FROM LANDFILL TO RECYCLING AND REDUCTION

Different practices for handling waste have evolved and been in practice over time. Traditionally, handling the waste of societies has been seen as a public responsibility. This governance is most visible on the occasions when a society fails to take this responsibility (Corvellec and Hultman 2012; Zapata Campos 2013). The management of waste changes over time. The main model of waste management dominating the twentieth century was landfill. However, as the consumption of disposables such as paper, plastics, and similar products grew immensely after WWII, and as environmental concerns started growing in importance from the 1960s and on, this model was questioned. As landfill does not solve the waste problem, rather it preserves waste, including potential environmental hazards due to leakage issues, this model gave rise to doubt. Alternative models were developed, for example, incineration. In Sweden, for instance, the introduction of incineration as a waste management model led to a reduction in the household waste being sent to landfill from 100% in the early 1970s to 3% in 2010 (Corvellec and Hultman 2012). This model has also been questioned as it only leads to the disposal of waste and not to a solution of the waste problem itself. This has led to yet another model, namely, wasting less. According to this model, the product cycle per se is in focus. Corvellec and Hultman (2012, p. 299) described this as follows:

> For wasting less, one should either not use resources that produce such material in the first place, or recover all possible value from it by redefining of the meaning of waste and changing the trajectories of material flows through society. Wasting less is a multi-entry narrative that connects to the social critique of an overflow society, the economic project of using resources diligently, the technological promise of bringing solutions to problems, and ecological concerns about climate change and peak metals. By so doing, it emphasizes the multidimensionality of our relationships to waste.

However, each of these models brings advantages as well as disadvantages. The incineration of household waste not only reduces the volume of waste sent to landfill, but has also the added advantage that the energy recovered can be used in district heating systems. About 50% of Sweden's household waste is transformed into heating system energy (Waste Management Sweden 2013). One disadvantage of this system is that this sort of waste reduction inevitably leads to potential assets being destroyed. Additionally, the incineration process also results in ash, which may be environmentally harmful and which will be deposited in landfill. Wasting less, on the other hand, has the benefit of reducing waste per se, and thus also any negative environmental effects (Corvellec and Hultman 2013). Another dimension of wasting less is recovering, which is founded on the principle of reusing and transforming former waste—such as TV screens—into valuable parts that may be useful in production processes. Thus, the TV screens on São Paulo's balconies, and the cell phones in Western drawers, which have been hidden and forgotten about for a long time, would be better off being treated as potential treasure. This complies with the idea in the legislation focusing how to manage WEEE.

According to a mandate from 2003, the European Union (EU) aims to "change waste into a resource that can be reused in production processes" (Stowell 2013, p. 108). More specifically, the EU promotes waste handling in terms of the waste hierarchy, namely, reduction, reuse, recycling, and the recovery of e-waste. To this, refurbishing or repairing can be added; however, the parts of the waste hierarchy are not to be found in the EU's WEEE directive. The goal is to both protect and preserve the natural environment and resources (see Stowell 2013). Another dimension of this is that, instead of extracting "virgin" resources, "urban mining" includes the potential to exploit precious minerals.

At present, the governance policy for managing waste is expressed in terms of a "waste hierarchy," according to which landfill is the first step, the least preferable, followed by energy recovery, material recycling, product reuse, and the most sought after highest step in the hierarchy; waste prevention (Corvellec and Hultman 2013). As mentioned above, the waste hierarchy comes with various perspectives on waste as a material. In landfill, the materialized waste is generally a problem, taking up too much space, combined with the production of more greenhouse carbon emissions than other methods, as well as potential environmental hazards. During the next step, the same waste material becomes valuable as a possible energy source. When we incinerate this particular waste, however, the reuse and recycling potential will be lost forever. Furthermore, waste prevention has a bearing on the economy per se, with implications for the production of goods and services—the usage of less material—as well as consumption. If consuming less, the consequences for the economy per se will be questioned. This development of various views on waste was summarized by Corvellec and Hultman (2012, p. 303) thus:

> From having been something to get rid of and made to cognitively and geographically disappear, waste is increasingly framed as both a value reservoir that is worth competing for and something unwanted to be minimized.

However, a consequence of each system relates to the locking-in effects. If household waste is seen as a resource for energy recovery through incineration, efforts to change over to a system in which household waste is turned into biogas will be reduced. A material solution for handling waste in terms of, for instance, incineration comes with social organizing, for example, how to sort and handle household waste. The social and material dimensions of waste "are reciprocally and temporally constituted" (Corvellec and Hultman 2012, p. 306). In other words, the urban infrastructure and the sociotechnological system are interconnected. Systems for energy, water, materials, and services, for instance related to transportation, will thus be stabilized and made sustainable. One shortcoming, however, is that the actual stabilization and maintenance inevitably prevents innovation, as well as the emergence of alternative solutions (Corvellec et al. 2013a).

On the other hand, wasting less will be followed by other commitments. The prevention waste model will change former waste into new products, and these will then be transferred back into economic flows, "[W]aste organizations need to develop new technical and social competencies, invent new business models, and offer waste management services that correspond to the narrative that waste is no longer a problem but a resource" (Corvellec and Hultman 2012, p. 306).

Even if wasting less may seem to be an alternative to reducing waste, it will entail other issues. As long as business models, or waste management services, are not developed, the solution of wasting less will be hard to realize. Additionally, there is no specification of the content in waste prevention (Corvellec and Czarniawska 2014a). It is, to follow the previous argument, locked in with other societal solutions and systems, which unfold according to certain rationales.

3.4 LAWS AND REGULATIONS GOVERNING WASTE MANAGEMENT

As the waste management system is under continuous development, a number of laws and regulations are applied in this area. Each of them will, in their own way, both strengthen and hinder shifts in the management system per se. Here, we will present examples of the role of laws and regulations in national waste management systems and their consequences for the processes of material and energy recovery.

There are a number of studies of the efficiency, shortcomings, and advantages to be found in already-established regulations and laws governing waste recycling systems. One example is the role of taxes in combination with bylaws and their reinforcement in order to promote the diversion of solid waste management away from landfill to recycling and incineration. In the case of Sweden, landfill was predominant until the mid-1980s. Its disappearance (1% in 2012) accelerated with the introduction of landfill taxes in the 2000s (European Environment Agency 2013). The landfill tax had a significant impact on the amounts of municipal solid waste sent to landfill and has also led to a solid decrease in landfill down to 1% in 2012. Despite taxation having radically affected this landfill reduction, landfill diversion is also attributed to additional measures within the regulation system. For example, the landfill tax was coupled, in 2001, with a landfill ban on combustible waste. Furthermore, the landfill tax was increased during the years that followed. Similarly, in 2005, a new landfill ban on organic waste was introduced. The combination of regulatory and taxation measures succeeded in halving the amount of landfill up

until 2004. In 2009, all landfills not complying with the requirements in the regulation on landfill were closed down, and as a result, the number of landfills in operation
was almost halved (Swedish Environmental Protection Agency 2010). The decline
in landfill has affected the growth of both waste-to-energy and material recycling
(Zapata et al. 2014).

Another example of studies of the challenges and advantages of the different laws
and regulations is how to finance waste recycling systems. The two general models
for financing recycling are advanced recovery fees (ARFs) and extended producer
responsibilities (EPRs, individual or collective; Plambeck and Wang 2009). Lawhon
(2012) elaborated upon what she defined as two different sociotechnical systems
for handling waste. Individual systems hold the manufacturer responsible for the
produced waste, and respond to the ARF system. The idea behind the system is that
producers will redesign their potential waste in order to decrease recycling costs. As
recyclers compete with each other over costs in the individual systems, in order to
receive waste, this system is said to be the most cost-effective. In this system, collection arrangements are established separately from each other, with the system being
funded by fees that are paid when the material is recycled. Recyclers compete over
the cost of taking care of waste. The next system is the collective—corresponding
to the EPR system—in which manufacturers collaborate by joining single collection
systems. These will be funded by fees that the consumers pay in advance for the collection process. The collective system can lead to higher costs as it is noncompetitive.
Such costs may be the result of inefficiencies, or a system that is better at recovering
material and is more environment-friendly. The policy makers translate these different issues into two questions. The first question is, "Which system architecture will
drive the most recovery?," and the other is, "What will be [the] most economically
efficient?" (Lawhon 2012, p. 75). Each of these questions relates to further issues, for
example, how to use the waste governance system to increase labor-market opportunities, entrepreneurship, the extraction of valuable fractions, and environmental and
social development aspects. In relation to the governance perspective, it is of interest
to ask how producer responsibility can be used in such a situation. As pointed out
by Lawhon, existing manufacturers of valuable fractions may not want to give up
such resources to producers. EPR may then lead to decreased local control and, in
doing so, decreased interest in entrepreneurial activities. If connected actions contribute to the potential to get hold of valuable resources, local control, or potential
entrepreneurial activities, it will take a considerable amount of other actions, namely,
discussions, negotiations, and conflicts, in order to enroll interest and support for
another system.

The introduction of collective systems in order to handle beverage containers in
the United States displays dimensions of actors connected into such systems. The
collective system, such as the "bottle bill" recycling system that is applied to 11 states
of the United States, provides one example with many dimensions. In the "bottle
bill" recycling system, consumers are encouraged to recycle beverage containers;
in return, they receive a monetary incentive. As a result of this system, California
reported an increased recycling rate during 2006, when 60% of all beverage containers were recycled (Kahhat et al. 2008). An example of a version of the individual system is end-of-life vehicles, due to which 95% of U.S. cars were sent for dismantling,

shredding, and recycling in the early 2000s. The processes were handled by, among others, an estimated 6,000–7,000 dismantlers (Kahhat et al. 2008). Another example is the pay-as-you-throw (PAYT) solid waste collection programs. In traditional flat-rate waste collection programs, residents pay a fixed fee for waste collection, regardless of how much waste they generate. PAYT programs break with this tradition and charge households for solid waste collection services in proportion to the amount of waste that they throw away, following the principle "polluter pays." PAYT programs introduce a financial incentive for residents to increase their waste separation and recycling, and thus to lessen the volume of material that they send to landfill and incineration (Bilitewski 2008). For example, in the city of Gothenburg, the introduction of the PAYT program has succeeded in reducing by 20% the weight of the mixed waste produced by householders living in detached houses (Corvellec et al. 2013c). Corvellec et al. (2003c) explained how the new waste invoice was designed by politicians and waste managers in Gothenburg as a soft policy technology for controlling householders' behavior at a distance, by specifying side by side the exact amount of waste each household produces with the cost of the collection service and their environmental consciousness.

The EPR and ARF systems have also been related to the rate of new products being introduced. If prices are higher, for example, by a "fee-upon-sales" of disposal systems, consumers will expect to use the product for a longer time. Following this, new products will be introduced less frequently. However, in order to increase product development designed for recyclability, "fee-upon-disposal" and an EPR system have been found to increase recycling but also to increase the rate at which new products are introduced (Plambeck and Wang 2009). Following the systems hence provides some insights into models and rationales regarding how to handle waste, but rather limited ones.

Another way to explore how to handle waste is to follow a specific waste group. Electronic waste is of specific interest as its new technologies encompass a fast-growing market, an often-underdeveloped infrastructure for collecting the specific waste, and a material that can easily be transported from one continent to another in order to be recycled.

3.5 RECYCLING WEEE—A GLOBAL ISSUE

Even though waste may appear to be a local problem needing to be solved, it is indeed a product that illustrates well our contemporary globalized patterns. As a consequence of the widespread use of information and communication technologies (ICTs), electronic waste is increasing rapidly. Hence, a waste stream called e-waste is claimed to be among the fastest growing waste streams worldwide at 3%–5% per year (Stowell 2013). Another measurement is a figure ranging "from 20 to 50 million metric tons" (Lawhon 2013, p. 704). Furthermore, e-waste is expected to make up about 5% of all consumer-generated waste. As ICTs develop and advance, consumers, for instance, are increasingly changing over to LCD television screens or buying new products even though these are not essential (Kahhat et al. 2008). Moreover, as new products are expected to be developed, both consumers' willingness to pay and revenues shared by manufacturers will be depressed. Products are thus expected to be short-lived, and will be introduced while their functionality is still showing flaws (Plambeck and Wang 2009). Simultaneously, the amount of e-waste from obsolete technological solutions is increasing. Yet, the categorization and subsequent measurement of "what e-waste is" remains unclear (Lawhon 2013).

In Brazil, electrical and electronic equipment is estimated to make up 4.1% of that country's GDP. While the market for TVs, washing machines, and the like is expanding in the emerging markets, the market pattern for similar products in the developing countries "is said to be mature" (Araújo et al. 2012, p. 336). As an example of this, sales of TVs are decreasing in countries such as Japan (by 4% between 2000 and 2008), and are estimated to increase slowly in the developed countries (by 0.3% between 2011 and 2015), and are increasing more clearly in the developing countries (by 6% between 2011 and 2015). In the year of 2006, an U.S. estimation expected 400 million products to be sent to e-waste yearly. Estimated in tonnage, those products would comprise more than one million metric tons per year (Plambeck and Wang 2009).

Computers and cell phones, on the other hand, are still expanding internationally; this is seen as a fast-growing nonmature market. In Brazil, 13% of households had a computer in 2001, with this figure increasing to 35% in 2008. During the same period, the number of households with cell phones rose from 8% in 2001 to 42% in 2008. As income levels rise, along with technological developments that render computers and cell phones cheaper, Araújo et al. (2012) anticipated a large potential for these two products. In a discussion about product life cycles, Araújo et al. (2012) suggested that, even if the market becomes saturated, further sales can be expected as new products featuring additional functionality are developed.

Yet another problem occurs, however, when the increasing level of e-waste has to be dealt with. The increase in e-waste has not, for instance, been matched by processes of waste disposal such as the collection, recycling, or reuse of obsolete products (Kahhat et al. 2008). Estimating the potential WEEE recycling is difficult, as the system is complex and calculative data is hard to obtain. In order to estimate the number of WEEE products being left in hidden and forgotten spaces, for example, balconies, scenarios for life cycle dynamics have been used. These scenarios turned out to be problematic for a number of reasons, relating to a lack of information and product development. There was, for instance, a lack of information in regard to generation volume, discard rates, and distribution. The lifetime mean decreased over the time periods set out to measure life cycles, making it hard to measure, for instance, the lifetime of a computer. In addition, products such as computers changed in both weight and format (as laptops came to replace desktops; Araújo et al. 2012). In short, this creates problems for actors striving to establish e-waste management strategies.

One estimate of the volume of WEEE in Brazil's waste management is reported to represent 0.64 kg/year/capita, while another estimate anticipates 3.4 kg/capita (Araújo et al. 2012). Even though the market for WEEE products is expanding, about 2% of these products are estimated to be included in the recycling process, the rest being left in the forgotten, hidden spaces. As a response to this, mobile operators have introduced proactive programs such as "Recycle Your Cell Phone." Between 2007 and 2009, 0.2% of the cell phones sold in Brazil (totaling 301 million since the start of global system for mobile communications [GSM]) were returned. Of these, 10%

could be reconditioned and reused. One question is how to organize these streams of e-waste for reuse, and how to separate them from other types of e-waste. In order to solve these issues, a more general idea of how to organize the management of waste, from reducing to recovering, is necessary.

3.6   LAWS AND REGULATIONS IN GLOBAL PATTERNS

In parallel with the increase in e-waste, issues and practices regarding how to handle this actual e-waste are being emphasized across the globe. However, markets differ in their maturity, figures for WEEE estimation are difficult to make accountable, and regulations for e-waste management vary. This leads to a situation in which the development in relation to different regional market conditions has to be taken into account (Kahhat 2008; Araújo et al. 2012). That is, there are differences in regulations whether you are in Western Europe, China, the United States, South Africa, or Brazil. Regional solutions and how these interact with local practices will be presented in this section.

WEEE contains hazardous substances, for example, lead, mercury, polybrominated biphenyl, polybrominated diphenyl ether, and polychlorinated biphenyl, to name a few. These are known to have severe effects on both health and the environment. For instance, if lead leaches out into the soil, it will be absorbed by crops during agricultural production, resulting in potential health problems for many.

In Europe, the EU's WEEE regulation from 2002 was developed as a specific piece of legislation aimed at diverting "WEEE from landfills and increasing the recovery of materials by imposing extended producer responsibility for electrical and electronic equipment" (Araújo et al. 2012, p. 337). More specifically, this regulation was elaborated into a new directive in 2012, intended to "preserve, protect and improve the quality of the environment, protect human health and utilize natural resources prudently and rationally" (European Commission—WEEE Directive 2012). As a part of this, ambitious goals for collection rates are established. This states that:

> From 2019, the minimum collection rate to be achieved annually shall be 65% of the average weight of WEEE placed on the market in the three preceding years in the Member state concerned, or alternatively 85% of WEEE generated on the territory of that member state.

A number of the more recently joined member states have reduced and postponed collection rate goals due to their underdeveloped infrastructures (European Commission—WEEE Directive 2012).

In Japan, the focus as regards handling e-waste was put on the manufacturers and importers in 1998, who were then required to recycle household e-waste such as TVs, refrigerators, washing machines, and air-conditioners. Some of the cost of these processes is paid by the consumers, who are obligated to pay an end-of-life fee on these products. Computers, for both business and personal use, were added to the list during the early 2000s. An EPR law was enacted in South Korea in 2003, targeting air-conditioners, TVs, and PCs. The system is financed via deposits paid into recycling funds, managed via accounts with the government. Recycling rates are reported to have increased as a result of this law being enacted. Taiwan has enforced a system by which manufacturers and importers have to pay a recycling fee, which is transferred to recycling facilities (Kahhat et al. 2008).

While these are examples of countries with regulations governing how to handle e-waste, other examples are also to be found. In the United States, 80%–85% of the "e-waste ready for end-of-life management ended up in U.S. landfills" between 2003 and 2005 (Kahhat et al. 2008, p. 958). As a consequence of the environmental hazards relating to landfill, and the potential assets in e-waste, exporting has become one way to handle U.S. e-waste. In doing so, China has become one of the largest recipients of e-waste.

These imports have, however, led to the implementation of new laws and regulations regarding WEEE in China. Southeast China has suffered immensely negative environmental effects from WEEE recycling (Kahhat et al. 2008; Araújo et al. 2012). As a consequence, Chinese laws have been enacted forbidding WEEE imports into China. Additionally, policies intended to strengthen EPR have been suggested as a way to handle recycling processes in a more environment-friendly way (Araújo et al. 2012). In spite of these laws, e-waste is still being imported into China. Further, the growing export trend has also come to include developing countries such as India (Kahhat et al. 2008). Meanwhile, various systems of recycling e-waste have been introduced in the United States. These are financed either via fees paid by the consumers or by means of the costs being shared between the municipalities and the manufacturers.

In the United Kingdom, the 1975 Waste Directive and the 1991 Hazardous Waste Directive have "been translated into a regulation as a mechanism to control the conduct of those engaged in EEE. The Regulation is broken into different parts each section outlining the responsibilities of the different actor groups e.g. producer obligations, approved facilities obligation, exporters and UK Government" (Stowell 2013, p. 110). As a result, operations handling the recovery of computers, for instance, are managed via a licensing system that is supposed to ensure that organizations handling such waste neither endanger the employees operating the processes nor cause environmental degradation. Organizations working in this area can be separated into three different groups: those operating computer recycling, those involved in asset recovery, and those offering environmental services and waste management. As a consequence of the new waste directives, practices concerning how to handle and store parts originating from computer recycling have been revisited. In order to include impermeable surfaces and to prevent spillage and the risk of polluting water sources, new compulsory storage facilities have been developed. More specifically, in order to handle recycling operations, organizations need specific scales for measuring waste weight, specific spaces for storage and disassembling, and "closed units to house hazardous waste, containers, cables, plastics and metals and racking to store computers awaiting the next phase of recovery" (Stowell 2013, p. 113).

As it is difficult to measure e-waste, as noted above, in a similar vein, it is difficult to follow e-waste flows. In contradiction to media reports about waste flows going a north–south direction, researchers have suggested, that "most e-waste flows are not North-South" (Lawhon 2013, p. 704). Due to both economic reasons and issues of environmental justice, it is of great interest to eliminate the processes which are sometimes labeled "e-waste dumping" or "toxic colonialism" (Lawhon 2013). As a result, South Africa handles e-waste that results from electronic equipment used within that country, e-waste purchased from or donated by "the global North," and e-waste imported from other African countries, in addition to multinational companies such as Hewlett-Packard importing e-waste into South Africa from other African countries due to South Africa having higher processing standards. In the global circulation of goods, South Africa exports some e-waste to Asia (for what is expected to be a less environment-friendly recycling process), and to Europe where valuable components are smelted.

In another study, Lawhon interviewed stakeholders in governance organizations involved in the South African e-waste system. The joint agreement among the interviewees tells of some of the various dimensions of e-waste. Lawhon (2012, p. 77) summarized them thus: (1) e-waste is a problem and opportunity, (2) producers should be responsible for waste, (3) there is a need to create jobs and make cheap electronics available, (4) there is a need to protect the environment and reduce waste, and that consequently, and (5) it is likely that an external subsidy is needed for responsible e-waste recycling/refurbishing to be economically viable.

Policies developed in South Africa for managing e-waste have led to a weakening of state involvement in waste governance. On the other hand, this development has provided industry with opportunities to choose how it would like to manage e-waste.

One problem with the laws and systems governing e-waste is that they have to satisfy a number of aspects, or as framed in this chapter, the institutional arrangement per se. Recycling has to be completed in the proper manner; the consumers need an incentive—financial or social—to move their things from the hidden spaces (balconies, boxes, storage) to the collection system. According to Kahhut et al. (2008, p. 960), a market is needed where "firms compete to offer more efficient reuse and waste management services." However, laws and regulations do not constitute a final solution to the problem of how to handle e-waste. Practices resulting from these laws and regulations lead to yet unanticipated aspects, which will affect the waste system as such.

3.7 LAWS IN PRACTICE—OCCUPATIONAL RISKS AND HAZARDOUS ENVIRONMENTS

A further dimension of how to handle e-waste comes from day-to-day operations on the waste itself. In a paper discussing occupational practices and environmental risks when recycling computers, Stowell (2013) listed various suggestions for dealing with the risks of handling waste that occur in the developing countries. For instance, there have been suggestions for governments to enforce best practices, to facilitate legislation and make it less complex, to create "safer" recycling infrastructures, and/or to hold corporations accountable for the manufacturing and generation of waste (Stowell 2013, p. 109). What can be noted is the fact that many of these suggestions involve the use of improved laws and regulations when handling e-waste. In practice, however, laws and regulations are only one actor among many establishing sustainable and efficient e-waste recovery.

One major problem when disassembling electrical waste is the working practices of personnel, as well as inappropriate techniques applied by the informal businesses that often handle recycling in the developing countries (Araújo et al. 2012). This occurs when handling both local and global electrical waste (which may be shipped to the developing countries due to cheaper labor costs there and due to other expenses relating to the production process of extracting reusable materials). The recovery of computers, for instance, includes operations for handling both toxic and renewable resources such as metals, solvents, acids, and other materials. Chemical components that are transported, stored, and extracted expose, during all these parts of the process, the people operating the processes to hazards. However, even though regulations and laws strive to ensure the protection of both human health and the environment, it has turned out to be difficult to ensure that people, for instance, in the United Kingdom who are dealing with phosphorus removal, actually comply with these regulations in their day-to-day practices. Workers, in spite of knowledge and regulations, declining to use protective gear and thus putting their own health at risk is one part of the process, another part being pollution that is more hidden. For instance, glass with minor shares of lead is used in CRT monitors. Exposure to lead is known to cause severe problems to the central nervous system, blood disorders, and so forth, both in humans and in other mammals. After extracting and vitrifying this glass, lead is still to be found in the end product, even though this is at minimal levels. This product is then used in "tarmac for roads, aggregate blocks, golf courses, anti-slip coatings, tiles and recycled plastics" (Stowell 2013, p. 115).

Nonetheless, other countries have taken action differently. In South Africa, for instance, there was no explicit policy on e-waste as of early 2012. This does not mean that there is no legislation covering the processes of managing hazardous substances and waste. Rather, these issues are covered in the "2008 National Environmental Management Waste Act" (Lawhon 2013, p. 707). According to this, the waste hierarchy ought to be applied in combination with other acts. These include, for instance, the control of hazardous waste, pollution, and precious metals (see Lawhon 2013).

This is one example of how the implementation of laws and regulations does not necessarily lead to compliance in practice. Actions are connected to each other in ways that handle the waste, but not as yet in compliant and sustainable ways. However, if practices are implemented, risks may arise when new actions collide with former ones, for example, the reuse of glass at golf courses. Additionally, the complexity of computers can include unknown chemical, and hence environmental, effects during processes of recovery.

3.8   CONCLUDING DISCUSSION

In this chapter, the institutional arrangement regarding laws and regulations in the waste sector has been outlined. Practices following a specific law or regulation, yet if these practices are obedient, compliant, and progressive, will still have the potential of including further actions and actors. If all but one of these fails to follow the new regulations, both environmental and human risks will still arise. The waste system per se is made up of social and technical aspects, as well as the relationships between them; what we have defined here is the organizing of an action net.

The laws supporting more sustainable e-waste recycling are founded on the best of intentions. Nevertheless, in order to function in a sustainable way, the other actors affected, for example, corporations manufacturing products that will end up in the waste system and the people working with recycling, will have to comply with the regulations. If not, there is a considerable risk of the hazardous content in e-waste being distributed further. Even more, yet if acting in obedience with all established regulations, the new product made from the reused material will run the risk of being contagious in its new environment. It could also be the case that contemporary technology has not yet found the risk built into the present models of recycling. The advantage of the action net perspective is that there is no given beginning and no given end to the analysis; thus, actions connected to the waste system can be traced in several dimensions.

A further opportunity provided by this perspective relates to the fact that the content of waste prevention is unclear or, more specifically, open to interpretation. Added to this is the present understanding that exists in the market. As formulated by Araújo et al. (2012, p. 337); "it appears as neither the informal nor the formal market has yet realized the economic and social potential of WEEE recycling." Given this, it is not only the practices appearing as a result of the regulations but also the possible various market activities have to be taken into account.

REFERENCES

Appadurai, A. (ed.) (1986) *The Social Life of Things*. Cambridge: Cambridge University Press.
Araújo, M. G., Magrini, A., Mahler, C. F., and Bilitewski, B. (2012) A model for estimation of potential generation of waste electrical and electronic equipment in Brazil. *Waste Management* 32:335–342.
Bilitewski, B. (2008b). Pay-as-you-throw—A tool for urban waste management. *Waste Management*, 28(12), 2759.
Corvellec, H., Ek, R., Zapata, P., and Zapata Campos, M. J. (2013a) From waste management to waste prevention. Closing implementation gaps through sustainable action nets. Application to the Swedish Research Council FORMAS. http://ism.lu.se/fran-avfallshantering-till-avfallsforebyggande/from-waste-management-to-waste-prevention.
Corvellec, H. and Hultman, J. (2012) From "less landfilling" to "wasting less." Societal narratives, socio-materiality, and organizations. *Journal of Organizational Change Management* 25(2):297–314.
Corvellec, H. and Hultman, J. (2013) Waste management companies: Critical urban infrastructural services that design the sociomateriality of waste. In Zapata Campos, M.J. and Hall, C.M. (eds.), *Organising Waste in the City*. Bristol: The Policy Press, pp. 139–157.
Corvellec, H., Zapata Campos, M.J., and Zapata, P. (2013b) Extending the realm of accounting inscriptions: Pay-As-You-Throw (PAYT) solid waste collection invoicing and the moral enrolment of residents in environmental governance. *Accounting, Organisation and Society conference on "Performing business and social innovation through accounting inscriptions,"* Galway, Ireland, September 22–24, 2013.
Corvellec, H., Zapata Campos, M. J., and Zapata, P. (2013c) Infrastructures, lock-in, and sustainable urban development: The case of waste incineration in the Göteborg Metropolitan Area. *Journal of Cleaner Production* 50:32–39.
Corvellec, H. and Czarniawska, B. (2014a) *Action Nets for Waste Prevention*. GRI-rapport 2014:1. Sweden: University of Gothenburg. http://gup.ub.gu.se/publication/194453-action-nets-for-waste-prevention "\t" _blank.
Corvellec, H. and Czarniawska, B. (2014) *Action Nets for Waste Prevention*. In K. M. Ekström (Ed.), Waste management and sustainable consumption: Reflection on consumer waste: 88–101. Oxford: Earthscan-Routledge.

Cwerner, S. and Metcalfe, A. (2003) Storage and clutter: Discourses and practices of order in the domestic world. *Journal of Design History* 16:229–239.

Czarniawska, B. (1997) *Narrating the Organization. Dramas of Institutional Identity*. Chicago, IL: The University of Chicago Press.

Czarniawska, B. (2004) On time, space and action nets. *Organization* 11(6): 777–795.

Czarniawska, B. (2007) *Shadowing and Other Techniques for Doing Fieldwork in Modern Societies*. Malmö, Sweden/Copenhagen, the Netherlands/Oslo, Norway: Liber/CBS Press/Universitetsforlaget.

Davoudi, S. (2009) Scalar tensions in the governance of waste: The resilience of state spatial Keynesianism. *Journal of Environmental Planning and Management* 52(2):137–156.

Douglas, M. (1966) *Purity and Danger: An Analysis of Concepts of Pollution and Taboo*. London: Routledge.

European Environment Agency (2013) *Municipal Waste Management in Sweden*. ETC/SCP working paper. Denmark: European Environment Agency.

European Commission-WEEE Directive (2012) *Article 7. Collection Rate*. http://eur-lex. europa.eu/legal-content/EN/TXT/?uri=CELEX:32012L0019 (accessed on June 5, 2014).

Graham, S. and Thrift, N. (2007) Out of order: Understanding repair and maintenance. *Theory, Culture & Society* 24(1):1–23.

Gregson, N., Metcalfe, A., and Crewe, L. (2007) Identity, mobility and the throwaway society. *Environment and Planning D: Society and Space* 25:682–700

Gregson, N., Watkins, H., and Calestani, M. (2010) Inextinguishable fibres: Demolition and the vital materialisms of asbestos. *Environment and Planning A* 42:1065–1083.

Greenwood, R., Oliver, C., Sahlin, K., and Suddaby, R. (eds.) (2008) Introduction. *The SAGE Handbook of Organizational Institutionalism*. London: Sage, pp. 1–46.

Gutberlet, J. (2010) Waste, poverty and recycling. *Waste Management* 30(2):171–173.

Gutberlet, J. (2012) Informal and cooperative recycling as a poverty eradication strategy. *Geography Compass* 6(1):19–34.

Hawkins, G. (2006) *The Ethics of Waste*. Lanham, MD: Rowman and Littlefield.

Hawkins, G. (2009) The politics of bottled water: Assembling bottled water as brand, waste and oil. *Journal of Cultural Economy* 2(1/2):183–196.

Joerges, B. and Czarniawska, B. (1998). The question of technology, or how organizations inscribe the world. *Organization Studies* 19(3): 363–385.

Kahhat, R., Kim, J., Xu, M., Allenby, B., Williams, E., and Zhang, P. (2008) Exploring e-waste management systems in the United States. *Resources, Conservation and Recycling* 52:955–964.

Kaika, M. and Swyngedouw, E. (2000) Fetishizing the modern city: The phantasmagoria of urban technological networks. *International Journal of Urban and Regional Research* 24(1):120–138.

Knorr-Cetina, K. (1981). *The Manufacture of Knowledge: An Essay on the Constructivist and Contextual Nature of Science*. Oxford New York: Pergamon Press.

Latour, B. (1987) *Science in Action: How to Follow Scientists and Engineers Through Society*. Harvard University Press.

Latour, B. (2005) *Reassembling the Social: An Introduction to Actor-Network-Theory*. Oxford: Oxford University Press.

Latour, B. and Woolgar, S. (1979/1986). *Laboratory Life: The Construction of Scientific Facts*. Princeton, NJ: Princeton University Press.

Lawhon, M. (2012) Contesting power, trust and legitimacy in the South African e-waste transition. *Policy Science* 45:69–86.

Lawhon, M. (2013) Dumping ground or country-in-transition? Discourses of e-waste in South Africa. *Environment and Planning C: Government and Policy* 31:700–715.

Monstadt, J. (2009) Conceptualizing the political ecology of urban infrastructures: Insights from technology and urban studies. *Environment and Planning A* 41:1924–1942.

Nilsson, M., Eklund, M., and Tyskeng, S. (2009) Environmental integration and policy implementation: Competing governance modes in waste management decision making.
Environment and Planning C: Government and Policy 27(1), 1–18.

Pinch, T. (2008) Technology and institutions: Living in a material world. *Theory and Society*
37: 461–483.

Plambeck, E. and Wang, Q. (2009) Effects of e-waste regulation on new product introduction.
Management Science 55(3):333–347.

Serres, M. (1974) *La Traduction: Hermes III. Paris.* Les Éditions de Minuit.

Star, S. L. and Ruhleder, K. (1996) Steps towards an ecology of infrastructure: Design and
access for large information spaces. *Information Systems Research* 7(1):111–134.

Star, S. L. (1999) The Ethnography of Infrastructure. *American Behavioral Scientist* 43:377–391.

Stowell, A. (2013) Environmental risk: The impact of the UK WEEE regulation—An incentive to change occupational practice? *Business Review* 60(4):107–121.

Swedish Environmental Protection Agency (2010) Effekter av deponiförordning ens införande.
Swedish Environmental Protection Agency: Stockholm, Sweden. http://www.
naturvardsverket.se/Nerladdningssida/?fileType=pdf&downloadUrl=/Documents/
publikationer/978-91-620-6381-8.pdf (accessed on March 19, 2015).

UNEP (2010) *Waste and Climate Change. Global Trends and Strategy Framework.* Osaka,
Japan: UNEP.

Waste Management Sweden (2013) *Towards a Greener Future with Swedish Waste to Energy.
The World's Best Example.* Malmö, Sweden: Avfall Sverige. http://www.avfallsverige.
se/fileadmin/uploads/forbranning_eng.pdf

Zapata Campos, M. J. (2013) The function of waste urban infrastructures as heterotopias
of the city: narratives from Gothenburg and Nicaragua. In Zapata Campos, M.J. and
Hall, C.M. (eds.), *Organising Waste in the City.* Bristol: The Policy Press, pp. 41–59.

Zapata Campos, M. J., Eriksson-Zetterquist, U., and Zapata, P. (2014) Waste socio-technological
transitions: From landfilling to waste prevention. In Loschiavo dos Santos, M.C.,
Walker, S., and Lopes Francelino S. (eds.), Design, Waste and Dignity. Sao Paulo: Olhares.
pp. 309–324.

Zapata, P. and Zapata Campos, M. J. (2013) Urban waste. Closing the loop. In Pearson, L. J., Newton,
P., and Roberts, P. (eds.), *Resilient Sustainable Cities.* London: Taylor & Francis/Routledge.

4 Source Separation of Household Waste
Technology and Social Aspects

Kamran Rousta and Lisa Dahlén

CONTENTS

4.1 HOUSEHOLD WASTE SEPARATION

In Chapter 1, it was discussed that separation of different materials in the waste stream is necessary for sustainable solid waste management and efficient resource recovery. When waste materials are kept clean and separated properly, they can be viewed as favorable raw materials for further processing, namely, material recycling or efficient energy recovery. This often happens in the industrial waste management; moreover, it has been proved that minimized waste generation saves money as well as increases the production efficiency. However, when it comes to municipal waste, it is more complicated because of the diversity of the materials. For example, in the local informational brochure, found in any city in Sweden (the so-called sorting guide), there are more than 200 items that should be sorted in more than 10 waste categories, which is not a simple task for the citizens.

In principle, waste separation can be accomplished in two ways: (1) commingled collection of waste with subsequent mechanical and/or manual sorting at material recovery facility (MRF) and (2) separate collection of different waste fractions, readily sorted where the waste was generated, that is, at source (source separation). The different waste separation methods can be combined and adjusted from no source separation at all to separate collection of multiple fractions at source. Separation of commingled waste at MRFs requires rather heavy investment in machinery such as mills, cutters, screens, magnetic separators, float–sink separators, cyclones, drum separators, and so on as well as consideration of the work environment and risk of contaminants for the workers during manual sorting (Figure 4.1), but then there is no need to involve people where the waste was generated.

By contrast, source separation requires management of separate waste bins and moreover investment in involvement of the inhabitants, who are supposed to carry out the task. In other words, sorting at MRF relies mainly on technical aspects and/or manual work under troublesome conditions, while source separation relies mainly on behavioral aspects at the household level.

Furthermore, a crucial issue in waste separation is the quality of separated materials. Purity is important, often decisive, for further processing of the material. The quality of separately collected material is generally higher than materials separated at MRF, because it has not been mashed together with other wastes. Thus, the source-sorted recyclables will be more valuable and more likely to replace pure raw materials in production processes. When recycling replaces extraction of raw materials, it is proven to significantly save resources and reduce environmental impact, even compared to incineration with energy recovery and also when long distance transport of recyclables is taken into consideration (Sundqvist 2005).

FIGURE 4.1 Manual sorting of commingled waste. (Courtesy of Mohammad Taherzadeh)

In summary, more fractions separated at source need less investment at MRF but more investment on how to involve the inhabitants. Successful involvement will generate high quality of collected materials. If the goal is to implement source separation schemes in the waste management system, the involvement and active contribution of the citizens must be thoroughly understood. Therefore, it is necessary to consider the social and behavioral aspects. In the following sections, the role of the inhabitants and factors that can influence participation in a source separation scheme will be discussed. Some typical technical systems for separate collection of sorted wastes will also be described and recommendations will be given on how to design and evaluate a source separation scheme related to the entire waste management system.

4.2   ROLE OF HOUSEHOLDERS IN WASTE SEPARATION

Consumption is increasing in the world and waste generation is directly correlated to the consumption level. For instance, in Sweden, the household waste increased from 317 kg per person in 1975 to 461 kg per person in 2013. (Swedish Waste Management Association 2014). The waste amount varies among different countries (Hoornweg and Bhada-Tata 2012). Developed countries generate more waste per person per year compared to developing countries, but because of the difference in population, the total amount of waste for both types of countries is large and increasing. Increased waste generation results in issues for society in both social, environmental, and economic aspects. Therefore, a municipality should have a similar approach to waste management as in supplying energy, water, and other services, that is, a part of necessary infrastructure. Usually in developed countries, all these services are financed through tariffs paid by the inhabitants. For example, in Sweden, all property owners are obliged to pay the municipality for waste collection and treatment. The tariffs are decided locally and vary from town to town, more or less designed with incentives for minimized waste generation and maximized sorting of recyclables for separate collection. The inhabitants are in fact obliged by law to sort the waste at source; however, there is no penalty for neglecting this task, and even if there was, it would not be easy to implement. Plastic, paper, glass and metal packaging, newsprints, hazardous waste, and food waste, are the raw materials that can undergo further processing in order to recover resources from household waste. The inhabitants consume the goods and generate waste, which they are expected to separate for facilitating further process. The householder is then not only a consumer but also a performer character in the waste management system. Simultaneously, the householders pay the waste tariff, which implicates another role for them as a customer in the waste management system. The question is how to find a balance between these three roles (consumer, performer, and customer) in order to achieve efficient source separation of household waste (Figure 4.2). Inhabitants need clarifying answers in order to overcome this confusion. The answers should be explicit and rational as well as increase their awareness and willingness to contribute in the waste management system. If this occurs, there will be an obvious participation in source separation schemes by inhabitants.

• They need my waste for their process; why should I pay for handling it?

• I pay for the service; why should I separate my waste?

• What is my responsibility?

• How is my job accounted for in the system?

• ...

FIGURE 4.2   Citizens' confusion in their roles in source separation.

4.2.1   FACTORS THAT INFLUENCE PARTICIPATION IN SOURCE SEPARATION

Before exploring the different factors that may affect householders' participation in source separation schemes, we explain briefly how behavior forms from a theoretical perspective. For understanding the relation between attitude and behavior, Ajzen and Fishbein (1980) developed the theory of reasoned action (TRA). In this theory, the most important elements prior to behavior are intentions that are influenced by attitudes, referring to beliefs about the particular behavior, and by subjective norms, referring to an individual's awareness about particular behavior influences by others in one's environment (Ajzen and Fishbein 1980). After numerous studies, Ajzen 1985 found that TRA could not describe the circumstances, which are not consciously under a person's control. He added a third element to TRA, perceived as behavioral control; thus, the theory of planned behavior was framed. This third determinant describes that behavior can also be a function of how a person perceives the difficulty of the behavior as well as how a person believes/feels his or her act can be successful (Ajzen 1985). In terms of environmental behavior, Ölander and Thøgersen (1995) have developed a model based on these two and also other theories in this field. In this model, which is called motivation–ability–opportunity–behavior model (Figure 4.3), they explained more clearly how behavior such as separation of waste at source could be formed. According to Ölander and Thøgersen (1995), the *intention* is the immediate antecedent for the behavior, but nothing will happen without *ability* and *opportunity*. *Motivation* of the actor is the first element of the model. Motivation alone can promote intention, but the motivated person must be able to perform the behavior. Therefore, the *ability* concept, which includes habit and task knowledge, was added to the model as both a moderator for intention and a determining factor for behavior. The second parameter that leads intention to actual behavior is *opportunity*. Situational factors such as accessibility to the collection facilities as well as designing a convenient source separation scheme are examples of opportunity in this model. The dashed lines shown in Figure 4.3 describe that belief about an activity such as source separation can change after doing the task (experience) directly or via ability. More explanation of this model is outside the scope of this chapter and the given reference is recommended.

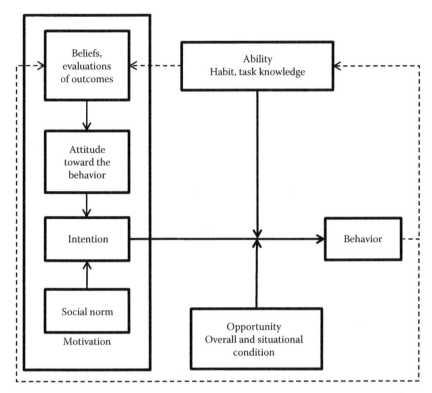

FIGURE 4.3 The motivation–ability–opportunity–behavior model. (Modified after Ölander, and Thøgersen, *J. Consum. Policy*, 18, 345–385, 1995.)

The introduction provides a general perspective to better understanding of the recycling behavior, which can lead to discovery of the different factors that influence people to participate in such source separation system. There are various studies that examined the different factors, especially in waste separation at source. For example, ascription of responsibility is one of the factors that can significantly influence recycling behavior directly, but awareness of the consequences of recycling and personal costs have indirect effects (Guagnano, Stern, and Dietz 1995). Pro-recycling attitude as an internal factor is the most significant contributor to recycling behavior, which can be influenced by knowledge of how and what to recycle and facilities and opportunity to recycle, as well as convenience, past experience, and knowledge about benefits of recycling, that is, saving energy, resources, and money (Tonglet, et al. 2004). Another study showed that situational factors such as prompts, public commitment, normative influence, goal setting, removing barriers, and providing rewards and feedback have increased recycling significantly; moreover, there is a correlation between recycling behavior and relevant specific attitudes (Schultz, et al. 1995). There are doubts as to how economic incentives can affect recycling behavior and how durable it is (Ölander and Thøgersen 1995). A meta-analyzing study showed that there are doubts as to how much sociodemographic factors such as gender, education, and income affect recycling behavior (Miafodzyeva and Brandt 2012). On the other hand, convenience, such as having access to recycling facilities and distance to the recycling stations as well as infrastructure for waste separation at source and having enough space at home, was viewed as the most important external factor (Bernstad 2014; González-Torre and Adenso-Díaz 2005; Guagnano, et al. 1995; McCarty and Shrum 1994; Tonglet, et al. 2004). In another study, Kollmuss and Agyeman (2002) pointed out that developing a model that incorporates demographic factors, internal factors, and external factors is not feasible and useful. One-dimensional models cannot figure out the different factors, which are involved in such complex systems (Kollmuss and Agyeman 2002).

Hence, it is clear that behavioral studies are complicated and it is not possible to put all the factors in a framework, especially in the case of source separation when there is confusion about the role of the householders. If the inhabitants are strongly motivated and have knowledge about recycling and the source separation is a habit for them, with an available facility nearby, there is a high possibility that the task will be done correctly. If the above-mentioned factors are not true, people may behave differently based on their understanding of the system.

According to the above-mentioned discussion, the influential factors for participation in source separation can be summarized as follows:

- *Internal factors*, such as general and specific attitude, personal beliefs and values, information and behavioral intentions, attitude toward recycling, environmental knowledge, awareness, emotion, and responsibility.
- *External factors*, such as convenience, public commitment, social and cultural norms, goal setting, removing barriers, economic forces, providing feedback, infrastructures, social institutions, information campaign, participation in decision making, storage space at home, and type of households.
- *Sociodemographic factors*, such as gender, level of education, and income.

The factors vary from case to case. For example, easy access to recycling facilities has different meaning for different people. When applying source separation schemes in a city or a part of the city, the different factors for the case should be examined in order to gain a local understanding and decision support for a durable and effective system for source separation.

4.3   COLLECTION SYSTEM FOR SEPARATED HOUSEHOLD WASTE

Household waste collection systems vary throughout the world, from no organized collection at all in some areas of developing countries to collection of 10 separated recyclable materials at the doorstep in multicompartment vehicles (Dahlen and Lagerkvist 2010). Household waste collection can be divided into property-close collection and collection at drop-off points (bring systems). Containers with different sizes and shapes are used at drop-off points. In property-close collection, combinations of bins, racks, sacks, and bags are used, which are placed in designated areas directly connected to the dwellings. Optical sorting techniques are sometimes applied, based on the use of color-coded bags for specific materials collected in the same bin. Some waste fractions such as food waste may be handled in alternative systems, for example, using food waste disposer connected to the sewage system.

Following is a presentation of collection systems typically used for separated recyclables, with examples from Swedish cities where such systems have been implemented.

4.3.1 Bring/Drop-Off System

Bring/drop-off system means that the inhabitants bring their sorted, dry, recyclables to a place outfitted with collection containers while the residual waste is usually collected close to dwellings. Depending on the goals and the strategy of the waste management in place, the facilities can be designed in different ways. This is the typical method in Sweden. There are several drop-off points in a city for collection of different packaging material as well as newsprint (usually called recycling stations). There are also a few centers set up to collect other types of materials such as worn-out household appliances, hazardous waste, and bulky waste (usually called recycling centers). The number of these stations and centers is based on the population as well as the outline of the city. For example, the average number of recycling stations per 10,000 inhabitants in Stockholm (capital city of Sweden with about one million inhabitants), Gothenburg (the second largest Swedish city), and Malmö (the third largest city) are about 3.1, 2.6, and 6.6, respectively (Nielsen 2012). The average number of citizens per recycling station in Sweden is about 1500 (FTI 2014). The distance to a recycling station varies from close in the neighborhood up to 2 km for city dwellers, while in sparsely populated areas the closest recycling station may be hundreds of kilometres away. A common location for recycling stations is in close proximity to shopping areas. Access to a recycling station is important, but not the only reason for the inhabitant to participate in the source separation schemes. If there is not enough motivation and information for the inhabitants, most of the recyclables will be left in the residual waste instead of being brought to recycling stations. For example, in Borås (the city discussed in Chapter 1), in a multicultural area that is far from any recycling station, the average ratio of missorted materials in the residual waste was about 53% (Rousta and Ekström 2013). When the drop-off point is some distance away, the inhabitants will not pass by it or visit it on a daily basis; therefore, the demands for indoor storing space primarily in the kitchen area will increase. In most cases (so far), this was not planned for when houses and apartments were outlined. Lack of indoor space and convenient kitchen equipment can be a reason for low participation rate in bring/drop-off systems. Collected materials from a drop-off point can be transferred directly to the recycling industries for processing without treatment at a MRF for mixed materials. Collection trucks are avoided in the housing areas, and the collection is more efficient from a few drop-off points instead of smaller bins at every property. Figure 4.4 shows a recycling station in the city of Borås. In a typical Swedish recycling station (2014), different containers are provided for colored glass, uncolored glass, newsprints, paper packaging, plastic packaging, and metal packaging.

FIGURE 4.4   A view of a recycling station (drop-off point) in the city of Borås, Sweden.

4.3.2   Property-Close Collection

Property-close collection can either be kerbside with waste bins placed along the street, or in waste rooms/areas close to the dwellings. Property-close collection of source-sorted waste means that the inhabitants sort many fractions at home or very close to their home and it is not necessary to bring the materials to drop-off stations. There are a number of systems in use for property-close collection of sorted waste materials; the main techniques are briefly presented below.

4.3.2.1   Two-Bin System

In Sweden, a two-bin concept is commonly used in combination with the bring/drop-off system. In the two-bin system, combustible waste is collected in one bin and sent to thermal treatment with energy recovery, while food waste is collected in another bin and sent to biological treatment with biogas and/or fertilizer production. This is combined with access to drop-off points for dry recyclables (recycling stations described above) and other collection points for hazardous waste and residual inert waste. Another two-bin system is called wet-and-dry, where the inhabitants are sup-posed to separate all the dry substances in one bin/bag and the wet fractions such as food waste in the other. Then the bags are transferred to a MRF for further mechani cal and manual processing. It is important to note that the quality of recyclables from the wet-and-dry system is generally worse than from a bring/drop-off system.

4.3.2.2   Commingled Collection of Dry Recyclables/Yellow Bin

This is the typical method that has been used in Germany for many years. In this method, all the plastic, paper, and metal packaging materials are collected in a yellow bin/bag. The glass bottles, organic waste, newsprints and cardboard, batteries, and beverage bottles are collected separately in other bins or places. The materials in yellow bags are transferred to MRFs for mechanical separation prior to further processing in the recycling industries. One of the advantages of this method as opposed to the bring/drop-off system is that the inhabitants are not confused about how to separate different packaging from each other. The need to have other bins for other materials and high investment in MRF must be taken into consideration if applying the yellow-bin concept.

4.3.2.3   Separate Bin for Each Recyclable Material

Property-close collection of several different waste fractions is commonly applied in densely populated areas, so residents in multifamily houses can leave their sorted waste in a room or designated area near the apartment or adjacent to the property. There are particular bins for each waste fraction in this room/area. Based on different designs, this system has potential to collect many sorted materials, such as different packaging, batteries, food waste, residual waste, and textiles as well as light bulbs and small electronics. The main advantage of this method is the convenience for the inhabitants when it is close to the apartment and offers a comprehensive collection system in one spot. For example, in Eskilstuna (in Sweden) about 90% of the apartments are connected to a property-close collection system (Nielsen 2012). Mölndal, a city near Gothenburg in Sweden, has provided this system for 90% of the dwelling area (Nielsen 2012). In most cities in Sweden, similar property-close collection has been implemented in apartment blocks, especially in publicly owned housing corporations (Figure 4.5).

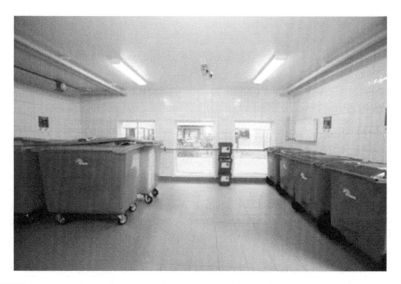

FIGURE 4.5   A view of a room for property-close collection system for apartment. (Reprinted by permission from kretslopp och vatten, Göteborgs Stad, Sweden; Courtesy of Johan Tvedberg.)

FIGURE 4.6 Two four-compartment bins per household. (Reprinted by permission from PWS Nordic AB, Perstorp, Sweden.)

4.3.2.4 Multicompartment Bins

Multicompartment bins are usually applicable for city areas with detached and semi-detached houses, not for apartment blocks. In this system, each house has two bins; each has two or four compartments. It is also possible to add small extra bins and bags for collection of materials such as batteries and small electronics and hazardous waste.

Some Swedish municipalities have implemented a standardized set of two 4-compartment bins per household (Figure 4.6). Residual waste, colored glass, food waste, and paper packaging are sorted in the first bin, and newsprints, metal packaging, plastic packaging, and uncolored glass in the second bin. The first bin, containing food waste, is emptied either every week or at least every two weeks, while the second bin may be collected less often. One major advantage of multi-compartment bins is the convenience of a comprehensive collection system in one spot close to the house. Enquiries among householders have reported high level of satisfaction, and the number of municipalities applying such collections systems is expected to increase (Nielsen 2012). However, the multicompartment bins them-selves are rather expensive and four-compartment trucks are needed, designed espe-cially for these bins in order to collect and empty the four fractions albeit separated.

4.3.2.5 Optical Sorting

In this system, the households are supposed to separate the different fractions such as paper packaging, plastic packaging, metal packaging, newsprints, food waste, and residual wastes in different color-coded plastic bags. The bags are then deposited commingled in the ordinary waste bin and transferred to a MRF with optical and mechanical sorting, which separates the bags based on their colors. However, the glass packaging cannot be included in the system and must be handled separately in the household and brought to a drop-off point. Optical sorting system is not common place in Sweden, but has been implemented, for example, since 2011 for 16,000 detached houses in the city of Eskilstuna (Nielsen 2012). The system is easy to use from the

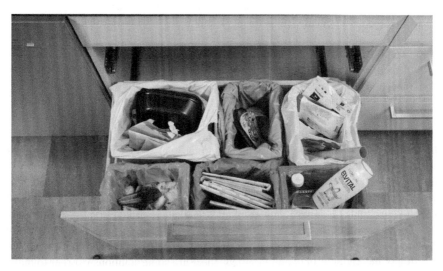

FIGURE 4.7   Place for the optical sorting method at home. (Reprinted by permission from Envac Optibag, Stockholm, Sweden.)

household perspective because after sorting the recyclables in the color-coded bags in the kitchen, everything can be placed in the same, well-known bin. However, enough space is needed in the kitchen to have different plastic bags for each fraction as it is shown in the Figure 4.7. This method needs a high investment in optical sorting facility as well as distribution of color-coded plastic bags to the households regularly.

4.3.2.6   Food Waste Separation

Food waste is a major part of household waste, representing a considerable use of energy and environmental load when the food was produced. Based on modern technology, it is possible to produce biofuel, compost and fertilizer from food waste. It is important to handle food waste separated from other wastes and thereby avoid contamination, which otherwise would possibly disturb the bioprocesses and also spoil the quality of fertilizers. Sorted food waste can be collected in separate bins or multicompartment bins as discussed above. Sorted food waste can also be collected through a completely different transport system where no vehicles are needed. Using food waste disposers, mounted in kitchen sinks, the food scraps can be ground and flushed away in the sewage system to a wastewater treatment plant where a biogas reactor is likely to be in operation. Home composting is another viable process, suitable for sparsely populated areas such as the countryside or homes with large gardens. The compost is then used for the private garden, thus, saving the cost of food waste collection and treatment facilities.

4.4   SOURCE SEPARATION: SELECTION AND EVALUATION

Solid waste management as a system, from waste generation to waste disposal, can be organized into six functional elements: (1) waste generation, (2) waste handling and separation at source, (3) collection, (4) processing and transformation of waste in MRF, (5) transfer and transport, and (6) disposal (Tchobanoglous, et al. 1993). A functional waste management system should have all these functions in order to recover resources from household waste; furthermore, the role of source separation is crucial for the entire system. The question is how should a successful source separation scheme be designed in a waste management system? First of all, to answer this question, it is necessary to establish the aim and objectives for the waste management of a city or the specific residential area. The objectives should be rational, clear, and achievable. For instance, "to reduce the landfilling by $x\%$ in the first year" or "to recover at least $y\%$ of the recyclables" can be some examples of the objectives. Based on the objectives and visions, the entire waste management system can be designed in which source separation is a part of it. To design the source separation system, it is necessary to consider both *mindware* and *hardware* (Wang and Yan 1998). Hardware refers to the technical infrastructure and facilities. The simple equation in hardware factors is: more separated fractions at source is equal to less investment in MRF. As mentioned above, source separation is a functional element between waste generation and collection, which means the separation and collection equipment must be dimensioned according to the waste amounts. Consequently, characterization of waste generation rate and content can help to better understand the choice of waste separation schemes in a specific residential arena. For instance, if the goal is to separate the plastic and paper fractions at source, it is necessary to estimate the generation rate of these fractions in order to prepare suitable volume of containers as well as appropriate collection frequency. At the same time the type of houses, access for collection vehicles, storage capacity available indoors and outdoors, and other parameters also should be considered. Moreover, the so-called mindware refers to other aspects of the system. Here, "mindware" refers to all the nontechnical factors that are related to the inhabitants in order to increase the rate of the participation as well as the correct sorting in the system. In other words, awareness of the mindware factors can help the waste management system to resolve the citizens' confusion in their roles in source separation, which was discussed in Section 4.2. For this reason, it is necessary to study the internal and external factors (as discussed in Section 4.2.1) as support to design the source separation system as well as to create a durable communication and information strategy. Right information at the right time as well as finding the appropriate communication means is the most crucial parameter that should be applied in any of the source separation methods. Economic incentives, like the choice of waste collection tariff or even how to share the income of the separated material with inhabitants, can also be a good starter for the system even though its durability is in doubt. There are two main principles for applying economic incentives in collection tariffs in Sweden: weight-based and volume-based. This means that the citizens pay for the waste management service based on the volume or on the weight of the waste that was collected.

In summary, the important point is to design a source separation, considering hardware and mindware as well as the available resources, namely, financing. Therefore, the selection of the different methods for source separation is very dependent on the local circumstances. The system that is workable for city/district "A" may not be workable for "B," and the level of ambition and means of financing the system will be decisive. For example in Sweden, when the drop-off points were sparsely located, the recyclable material was sorted less. It was also observed that there was more sorted metal, plastic, and paper packaging when dry recyclables were collected close to the property than with only drop-off points (Dahlén 2008). However, increasing the inhabitants' participation as well as reducing the missorted material in the system should be taken into account in any kind of waste collection system. Therefore, regular follow-up and evaluation of the system are crucial points to consider. Reasons for system evaluation can be summarized as follows:

- To plan/develop collection, transportation, and treatment capacity
- To follow up goals and legislation
- To monitor quality of source-sorted recyclables
- To understand recycling potential
- To monitor the effect of information/communication campaigns and incentives
- To evaluate cost-effectiveness
- To make regional and international comparisons.

In order to accomplish the evaluation, the collected waste should be characterized by composition studies, so-called *pick analysis*. Pick analysis means to estimate the different fractions in a waste stream by measuring the different materials in a sample of the waste by manual sorting and classification (Dahlén 2008). According to the aim of the evaluation, the crucial decisions for performing a household waste pick analysis by manual sorting are

- The waste category/categories to be evaluated
- Stratification, that is, the number and types of strata required with constant parameters (e.g., same type of the households)
- Level of sampling (e.g., at specific households, in containers at recycling station, from collection vehicles), the sample size, number of samples
- The choice of components in the composition characterization

Planning and performing the pick analysis precisely gives reliable and high-quality data for evaluating the system. By doing the pick analysis one can have a picture of the total amount of waste in a city/district as well as the different fractions in the waste stream. This can help one to see the feasibility of the waste management plan for a particular district. After applying the collection system, it is important to see whether inhabitants sort the fractions correctly and if the specific design works properly. Therefore, it is necessary to do the pick analysis regularly in order to find the right actions to improve the system.

4.5 SOME RECOMMENDATIONS

In summary, how to design a source separation scheme for a residential area depends on many factors and considering the local conditions is crucial.

Some recommendations worth taking into account are as follows:

1. Source separation is part of the waste management system. Therefore, view the relationships of source separation with the other functional elements, which should all be focused on reaching the goals of the waste management system.
2. Include the social factors from the beginning as parameters of the design and evaluation of the source separation system.
3. Include pick analysis as a tool to plan the facilities as well as to evaluate and improve the collection system.
4. Source separation facilities should be as simple and accessible as possible from both user and system perspectives.
5. Establish understandable, acceptable, and timely communication and information channels for citizens.
6. Keep in mind that any method applied for source separation should fulfill the citizens' requirements for good service.
7. An effective source separation system, that is, inhabitants sort the waste correctly, is economically feasible too.
8. Waste management is a complex system. It is necessary to see the opportunities, to be flexible and creative to overcome the barriers, and to include all the available resources.

REFERENCES

Ajzen, Icek. 1985. *From Intentions to Actions: A Theory of Planned Behavior*. Berlin, Germany: Springer.

Ajzen, Icek, and Martin Fishbein. 1980. *Understanding Attitudes and Predicting Social Behaviour*. Englewood Cliffs, NJ: Prentice Hall.

Bernstad, Anna. 2014. Household food waste separation behavior and the importance of convenience. *Waste Management* 34(7):1317–1323.

Dahlén, Lisa. 2008. Household waste collection: Factors and variations. Doctoral Thesis. Luleå University of Technology No 2008:33, Luleå, Sweden.

Dahlén, Lisa, and Anders Lagerkvist. 2010. Evaluation of recycling programmes in household waste collection systems. *Waste Management & Research* 28(7):577–586.

FTI. 2014. *Recycling stations* 2014 (cited June 17, 2014). Available from http://www.ftiab.se/243.html.

González-Torre, Pilar L., and B. Adenso-Díaz. 2005. Influence of distance on the motivation and frequency of household recycling. *Waste management* 25(1):15–23.

Guagnano, Gregory A., Paul C. Stern, and Thomas Dietz. 1995. Influences on attitude-behavior relationships a natural experiment with curbside recycling. *Environment and Behavior* 27(5):699 718.

Hoornweg, Daniel, and Perinaz Bhada-Tata. 2012. What a waste: A global review of solid waste management. Urban Development and Local Government Unit, World Bank.

Kollmuss, Anja, and Julian Agyeman. 2002. Mind the gap: Why do people act environmentally and what are the barriers to pro-environmental behavior? *Environmental Education Research* 8(3):239–260.

McCarty, John A., and L. J. Shrum. 1994. The recycling of solid wastes: Personal values, value orientations, and attitudes about recycling as antecedents of recycling behavior. *Journal of Business Research* 30(1):53–62.

Miafodzyeva, Sviatlana, and Nils Brandt. 2012. Recycling behaviour among householders: synthesizing determinants via a meta-analysis. *Waste and Biomass Valorization* 4:1–15.

Nielsen, Karin. 2012. Fastighetsnära insamling av förpackningar och tidningar. Kretsloppskontoret Göteborgs Stad, Sweden (in Swedish).

Rousta, Kamran, and Karin M. Ekström. 2013. Assessing incorrect household waste sorting in a medium-sized Swedish city. *Sustainability* 5(10):4349–4361.

Schultz, P., Stuart Oskamp, and Tina Mainieri. 1995. Who recycles and when? A review of personal and situational factors. *Journal of Environmental Psychology* 15(2):105–121.

Sundqvist, Jan-Olov. 2005. How should municipal solid waste be treated-a system study of incineration, material recycling, anaerobic digestion and composting. Stockholm, Sweden: IVL Swedish Environmental Research Institute.

Swedish Waste Management Association. 2014. Svensk avfallshantering 2014. Malmö, Sweden (in Swedish).

Tchobanoglous, George, Hilary Theisen, and S. A. Vigil. 1993. *Integrated Solid Waste Management: Engineering Principles and Management Issues.* McGraw-Hill series in water resources and environmental engineering. New York: McGraw-Hill.

Tonglet, Michele, Paul S. Phillips, and Adam D. Read. 2004. Using the theory of planned behaviour to investigate the determinants of recycling behaviour: A case study from Brixworth, UK. *Resources, Conservation and Recycling* 41(3):191–214.

Wang, Rusong, and Jingsong Yan. 1998. Integrating hardware, software and mindware for sustainable ecosystem development: principles and methods of ecological engineering in China. *Ecological Engineering* 11(1):277–289.

Ölander, Folke, and John Thøgersen. 1995. Understanding of consumer behaviour as a prerequisite for environmental protection. *Journal of Consumer Policy* 18(4):345–385.

5 Composting of Wastes

Antoni Sánchez, Xavier Gabarrell, Adriana Artola, Raquel Barrena, Joan Colón, Xavier Font, and Dimitrios Komilis

CONTENTS

5.1   INTRODUCTION

After many years of working on composting, we feel that the more than 20-year-old definition of composting in Haug's (1993) excellent book is more correct than ever. According to Haug, "Composting is the biological decomposition and stabilization of organic substrates, under conditions that allow development of thermophilic temperatures as a result of biologically produced heat, to produce a final product that is stable, free of pathogens and plant seeds and that can be beneficially applied to land." If one carefully reads this definition, it is easy to find the main issues related to composting and its performance. In this first introductory point, we will try to extract all these important issues to comment on them briefly. Afterward, the rest of this chapter will go deep into the crucial points of the definition and other important issues that have appeared in recent years, when composting has become a reality in modern societies.

First, composting is a biological process. Although obvious, this means that composting will only work when the conditions for the growth of several microorganisms are favored. Although a lot of excellent papers on composting microbiology have been recently published, one has to conclude that this point is still not fully understood. However, we can promote external conditions to favor the microbial growth, such as a good level of moisture, a near neutral pH, and so on. A first question arises in this point: Do all the organic substrates present these optimal conditions? The answer is clearly no. So, a new parallel definition is necessary, which is co-composting. Co-composting could be defined as the combination of wastes to have optimal conditions for microbial growth and, in consequence, for composting. In this sense, co-composting is inherent to real composting, and it will be present in practically all the composting processes.

Second, it is aerobic, so aeration is needed. Aeration can come from natural convection (chimney effect) or mechanical turning or forced convection in typical in-vessel systems. Aeration is completely related to another traditionally forgotten parameter that controls the composting process, as important as moisture or biodegradability: porosity. Porosity (often measured as air-filled space or free air space) governs the availability of oxygen inside the organic matrix, the good distribution of oxygen for microorganisms, and the absence of preferential paths for air (Ruggieri et al. 2009).

Third, it is a thermophilic process. This has important implications on the microbiology of the system, and the achievement of thermophilic temperatures is not often a simple balance among moisture, porosity, aeration, and biodegradable organic matter content. Heat comes from the oxidation of organic matter, which is used for self-heating and water evaporation. In fact, self-heating is very important as it permits the sanitation of the process, by killing the pathogens and plant seeds. As heat is spontaneously produced, we could say (using the traditional chemical reactors nomenclature) that composting is a near-to-adiabatic process.

Finally, it is beneficial to soil. It is worthwhile to notice that the definition does not mention the word "fertilizer," so compost must not be considered a substitute of chemical fertilizers, but a complement of some nutrients and mainly an organic amendment. So, its role in soils is not comparable to what mineral fertilizers do, that is, providing nutrients to plants; however, its other important benefits in such as erosion prevention, carbon sequestration, water saving, or the improvement of soil microbiology should not be forgotten. This is one of the main drawbacks when life cycle assessment (LCA) is applied to the entire composting system, since the aforementioned benefits are hard to quantify and to eventually include in the LCA with precision. LCA is inherently "from cradle to grave," so it must include these benefits, in an evident issue where research is still starting.

A transversal point in the definition are the terms "stable," related to compost, and "stabilization," related to composting. These terms are crucial to understand if compost is ready for use, and also to know the real efficiency of composting plants (Colón et al. 2012) in terms of stabilization of organic matter, which is the final objective of composting. We have the techniques and the studies to have reliable measures of stability and stabilization, so this is not an objective of research but a requirement for waste treatment plants.

5.2 COMPOSTING TO APPROACH ZERO MUNICIPAL WASTE

Our life is based on natural resources in the form of materials (e.g. metals and mineral resources, wood, and so on), water, and energy, as well as the land available to us. European economy, as well as the economy of developed countries, is characterized by a high level of resource consumption. Due to population growth at global scale we are now consuming more resources than ever. Current growth trends suggest that this consumption could increase to 140 billion tons by 2050 (UNEP 2011). The consequences of these production and consumption patterns have profound material and environmental impacts such as overexploitation, scarcity of resources, climate change, pollution, land-use change, and loss of biodiversity, which top the list of major international concerns. Traditionally waste has been considered as an inconvenient and unwanted by-product of society and industry, and although several attempts were made during the twentieth century to change this vision, most of them remained marginal. The annual total solid waste generation worldwide is approximately 17 billion tons and it is expected to reach 27 billion by 2050. About 1.3 billion tons of municipal solid waste (MSW) are currently generated by world cities, which is anticipated to increase up to 2.2 billion tons by 2025 (Laurent et al. 2014).

In 1989, Robert Frosch and Nicholas Gallopoulos wrote an article, which essentially constituted the birth of the field of industrial ecology (IE), where they compared the industrial approach to the use of materials and energy to that of nature (Harper and Graedel 2004). The main idea was to compare the industrial ecosystems with natural ones as a solution to move to more cyclic use of resources and materials, marking a shift away from thinking about waste as an unwanted burden to seeing it as a valued resource. IE incorporates a system perspective in the environmental analysis and decision making. IE not only focuses on the product and the processes, but also focuses on the current society with the system perspective, applying analogous concepts and tools at the different levels (process, firm, city, and region). The key point of the IE is the mass conservation reflected on mass balances, and the physical principles. So, the idea of converting wastes into products via system integration started to emerge.

With growing demands on the world's limited stock of resources, it is imperative that developed countries make more efficient use of both virgin materials and waste toward a zero waste society. Every European citizen throws off approximately 15–17 tons of waste per year from which 492 kg are household waste.

One definition of the zero waste concept that is widely accepted is the following:

> Zero Waste is a goal that is ethical, economical, efficient and visionary, to guide people in changing their lifestyles and practices to emulate sustainable natural cycles, where all discarded materials are designed to become resources for others to use. Zero Waste

means designing and managing products and processes to systematically avoid and eliminate the volume and toxicity of waste and materials, conserve and recover all resources, and not burn or bury them. Implementing Zero Waste will eliminate all discharges to land, water or air that are a threat to planetary, human, animal or plant health. (Zero Waste International Alliance 2004)

5.2.1 COMPOSTING WITHIN THE WASTE TREATMENT HIERARCHY

Policies for managing MSW in a sustainable manner have been key components of the European Union (EU) directives. In Europe, policies for reducing the amount of waste sent to landfills have been significantly influenced by EU Directives 1994/62/EC and 1999/31/EC. These directives aim to reduce the amount of biodegradable waste that is sent to landfills by certain percentages compared to the corresponding amounts generated in 1995. The revision brings a modernized approach to waste management, marking a shift away from thinking about waste as an unwanted burden to seeing it as a valued resource in accordance with the IE and CE (circular economy) principles. The waste directive introduced a new 50% recycling target for MSW and also introduces a five-step waste hierarchy where prevention is the first option, followed by reuse, recycling, and other forms of recovery, with disposal such as landfill as the last resort (European Commission 2008a). In addition, life cycle thinking, which seeks to identify the environmental improvement opportunities at all stages across the life cycle from raw materials to landfill, can be used to complement the waste hierarchy in order to make sure that the best overall environmental option is identified (JRC 2011).

The CE integrates the IE principles from a broader perspective not only to support resource optimization but also to move away from an accumulation society in which both IE and CE consider waste as a valuable resource that might be reintroduced into the economy. The main idea of CE concept is as follows:

> "Greater circularity in the economy has the potential to mitigate the impacts of primary extraction, processing and production as well as disposal since waste would become a valuable input to another process, and products could be repaired, reused or upgraded instead of thrown away" (Chatman House 2012).

Ideally, products made from technical nutrients are designed at the outset for advanced forms of reuse, and biotic nutrients, in any case, are nontoxic and so can be returned to the biosphere, preferably in a cascade of uses that tap as much value from them as possible. The various steps or feedback loops for technical nutrients include maintaining and repairing products, reusing and redistributing goods, refurbishing and remanufacturing goods by repairing or replacing failed parts or components, and recycling. The various steps or feedback loops for biotic nutrients include processes known as biorefining—to extract high-quality raw materials, such as fuels, power, and materials, and high-quality chemicals from biomass, but often in small volumes—and anaerobic digestion; eventually it should be possible to use all biotic nutrients as nontoxic ingredients in agricultural fertilizers (e.g., restoration or farming/collection).

Policy makers and other stakeholders are starting to appreciate the scale of the opportunities available by switching to a CE. China's leadership, inspired by

Japanese and German recycling economy laws, has formed a CE initiative that has major strategic importance worldwide and recently has adopted a new law for the CE promotion (Standing Committee of the National People's Congress 2008). In Europe, the European Commission published in 2012 a document entitled "Manifesto for a Resource Efficient Europe" in which it is clearly stated that in a world with growing pressures on resources and the environment, the EU has no choice but to go for the transition to a resource-efficient and ultimately regenerative CE (European Commission 2012). Another example is article (161), which was included in the new constitution of the Swiss Canton in June 2013, according to which the state must respect the principles of IE (Cst-GE 2014).

Food is one of the main streams in the MSW. Figure 5.1 shows the 2V "vegetable to vegetable" model: the organic matter cycle as an example of zero waste case. This model considers the entire cycle of organic matter from the generation of waste in households (and similar sources as restaurants, food stores, municipal markets, etc.) to the cultivation of vegetables. The model considers all stages of the organic fraction of MSW (OFMSW): the collection and transportation of waste, compost production, transportation from production sites to crop areas, and compost application to obtain final products (i.e., vegetables).

Furthermore, the 2V model for vegetables and home compost production avoids the transportation of waste, organic fertilizers, and vegetables to retailers. This new conception of horticultural production represents a sustainable way to treat household waste with the consequent benefits for society in accordance with the sustainable development (economic, social, and environmental; Figure 5.1).

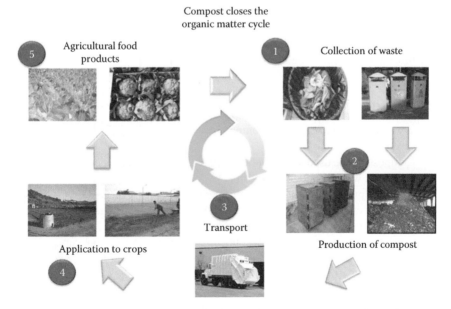

FIGURE 5.1 2V "vegetable to vegetable" model, a sustainable treatment of organic fraction from municipal solid wastes in five steps from the wastes to new agricultural products. (Data from Ecotech Sudoe, 2013.)

The production and application of compost in crops is a sustainable way to treat organic matter from MSW, which is a valuable and renewable resource for society.

Despite recent efforts to reduce the amount of solid waste sent to landfills, the MSW volume remains high. In the EU-27 countries, 37% of municipal waste was landfilled, 24% was incinerated, and 39% was recycled or composted on average in 2010 according to the recent reports by Eurostat. Therefore, it appears that the quantity of MSW landfilled nowadays is still high so that treatment of wastes needs to increase. Further, approximately 40% of biowaste from MSW still ends up in landfills in the EU countries.

European policy foresees a modern approach of waste management, where waste is no longer superfluous but becomes raw materials that end in plants instead of dumping grounds, where they are processed again into useful raw materials, compost, or fuel.

The benefits of shifting MSW management up the waste hierarchy are not limited to more efficient resource use and a reduced waste burden on the natural environment. Better waste management also offers a way to cut greenhouse gas (GHG) emissions, thus, in parallel, MSW management policies have been closely related to climate policies (Bogner et al. 2007). According to Intergovernmental Panel on Climate Change (IPCC), GHGs are defined as:

> Gases that absorb radiation at specific wavelengths within the spectrum of radiation (infrared radiation) emitted by the Earth's surface and by clouds. Gases in turn emit infrared radiation from a level where the temperature is colder than the surface. The net effect is a local trapping of part of the absorbed energy and a tendency to warm the planetary surface. Water vapour (H_2O), carbon dioxide (CO_2), nitrous oxide (N_2O), methane (CH_4) and ozone (O_3) are the primary greenhouse gases in the Earth's atmosphere. (IPCC 2014a)

Direct GHG emissions are produced in the life cycle of collection, transport, sorting, recycling, incineration, biological treatments, and landfill.

5.2.2 BENEFITS OF COMPOSTING

There are some evident benefits of composting:

- Compost can be used as a mineral fertilizer substitute in horticultural crops.
- It reduces waste dumped in landfills.
- Compost production permits a considerable reduction of GHG emissions and other contaminants to air, water, and soil; especially important is the conversion of methane to CO_2, which occurs in landfills.
- Compost provides economic benefits by saving energy for producers.
- Home compost avoids the collection of the OFMSW. This practice significantly reduces the economic, material, and energetic requirements of management and treatment.
- Furthermore, compost reduces the amount of impurities present in OFMSW by means of direct control on waste being treated.

In addition, home composting contributes to environmental awareness by involving people in the correct management of their own waste and by highlighting the importance of a number of factors influencing the treatment process.

5.3 SCIENTIFIC PRINCIPLES OF COMPOSTING

5.3.1 PHYSICOCHEMICAL PARAMETERS

Physicochemical properties can be really different, depending on the nature of the waste. Even for the same type of waste, as it would be the case of MSW, society habits and management systems implemented strongly influence its composition.

Physical properties play an important role in every stage of compost production as well as in the handling and the utilization of the end product (Agnew and Leonard 2003). The most studied physical properties of the composting materials are water content, water-holding capacity, bulk density, porosity, and air-filled porosity (also known as free air space). Water and oxygen content of an organic matrix are crucial factors for the development of biological activity. Porosity is a key parameter for the success of composting process. Ensuring the air movement through a composting solid matrix is important to maintain an optimum oxygen concentration, to remove CO_2 and excess moisture, and to limit excessive heat accumulation (Haug 1993). Usually, the wastes by themselves do not have the appropriate porosity to be composted. Air-filled porosity appears to be the best measure to determine the available porosity in an organic matrix (Ruggieri et al. 2009). Directly related with this parameter, particle size plays an important role in the maintenance of adequate porosity. Also, it is an important parameter related with the biodegradability since more surface implies more area to be degraded. A good balance between both aspects must be achieved during the composting process.

Nutrient balance is mainly defined by the C/N (carbon to nitrogen) ratio. It is important to maintain an adequate ratio (theoretically between 20 and 30) to guarantee an optimal biodegradation, both in aerobic and in anaerobic processes. Also, it is important to improve the quality of the end products (Gao et al. 2010) and to avoid ammonium emissions (de Guardia et al. 2010). There are a large number of researches where the C/N ratio has been used to formulate optimal mixtures, searching for a complementary cosubstrate. Nevertheless, this ratio should be used carefully because it is usually based on chemically measured carbon and nitrogen contents. Therefore, the biodegradable C/N ratio can be significantly different, because some organic wastes can be composed of recalcitrant carbon fractions that are not bioavailable (Puyuelo et al. 2011). In fact, some authors indicate that high C/N ratios are not a guarantee to avoid the ammonium emissions if only the biodegradable fraction is considered (Eklind and Kirchann 2000; Liang et al. 2006). De Guardia et al. (2010) evidenced that high biodegradable C/N ratio reduced the ammonia emissions. Puyuelo et al. (2011) proposed a methodology based on aerobic respiration assays to determine the biodegradable C/N ratio.

5.3.2 WASTE BIODEGRADABILITY

Waste biodegradability is a key factor for the design and successful operation of composting systems (Haug 1993). Standardized parameters such as chemical oxygen demand and biochemical oxygen demand are widely used in the design of the biological treatment for wastewater. However, despite the efforts done in research in the last years, there is no consensus for a standard parameter related with waste biodegradability for solid wastes. Still the different nature of the wastes together with social and environmental aspects makes waste biodegradability a parameter widely variable.

In the last years there have been numerous publications where the stability of organic wastes is studied in order to evaluate the effect of their disposal on landfill. The EU Directive 1999/31/EC on the landfill of waste, where the objective is to prevent or reduce as far as possible negative effects on the environment from the landfilling of waste, has had a strong impact on the research in this field. Methodologies to know the potential biodegradability of different wastes have been studied. Furthermore, the determination of stability and maturity, especially from OFMSW, has been studied by many researchers. The stability is related to the respiration activity, and methods based on oxygen consumption and CO_2 production have been developed. Nowadays all these methods are also being used to characterize wastes and predict their behavior in different treatments. In fact, the aim of many studies is to identify how the waste characteristics and the composting conditions influence both the compost's quality and the environmental impacts of composting (de Guardia et al. 2010). Clearly, the knowledge of waste biodegradability is particularly interesting from the point of view of selecting the best treatment and the use of the end product obtained.

The waste biodegradability has been measured through different parameters both physical–chemical and biological. However, the information of physical–chemical parameters does not really reflect the biological nature of the wastes. Methodologies based on biological assays appear to be more feasible. Aerobic respiration indices and biogas or methane production tests have been widely suggested in the literature as a measure of biodegradable organic matter content (Adani et al. 2004; Wagland et al. 2009; de Guardia et al. 2010; Ponsá et al. 2010a).

One of the methodologies more extended and studied to determine the potential biodegradability of wastes is based on the respirometric techniques (Barrena et al. 2006). The main methodologies differ between static and dynamics methods on the basis that oxygen uptake measurements were made in the absence (static respiration index) or in the presence (dynamic respiration index) of continuous aeration of the biomass. Although good correlations are found between them (Barrena et al. 2011), nowadays, most of the authors dealing with the biodegradability of organic wastes use a dynamic respirometric method (de Guardia et al. 2010).

There is a wide diversity of respiration methods where the main differences are the sample (quantity, size, solid or liquid state, optimal moisture) and the parameters of procedure (temperature, duration, way of results expression, etc.). In fact, one of the most controversial parameters in the assay is the temperature. This can vary from low (20°C–25°C) and high (37°C–40°C) range of mesophilic temperatures to thermophilic temperatures (58°C). Other assays are developed in isolated reactors, where the temperature profile varies according to the waste biodegradability.

Liwarska-Bizukojc and Ledakowicz (2003) investigated the influence of temperature on changes of the elemental composition of OFMSW and suggested that the optimum temperature for biodegradation process is 37°C. Anyway, different methodologies have been proposed and validated in numerous studies, and the decision to use one or the other one, whereas there is no a standard one, depends on many factors, such as the price of the equipment.

Respiration indices, especially using dynamic procedures, allow the calculation of several respirometric indices, expressed as specific rates or as cumulative consumption. Accumulated respiration activity after 4 days (AT_4) quantifies the biodegradable organic matter content while respiration index is a measure of the biodegradability rate, being high or moderate (Ponsá et al. 2010b). These parameters are correlated for stable materials and when their origin is similar. For different types of wastes, especially when the amount of biodegradable matter is high, the ratio between these parameters is variable (Ponsá et al. 2010b; Puyuelo et al. 2011). A compilation of dynamic respiration index and AT_4 values for some of the most generated wastes in the world is presented in Table 5.1. As commented, some differences with other methodologies such as temperature and dynamic conditions should be considered when comparing with other protocols. These data are the result of years of research, focused on the determination of the biological activity of organic wastes.

The respirometric activity has been proposed as a parameter to classify the biodegradability in three categories (Barrena et al. 2011):

1. Highly biodegradable wastes, respiration activity higher than 5 g O_2 kg^{-1} DM h^{-1} (which includes source-selected OFMSW, nondigested municipal wastewater sludge, and animal by-products)
2. Moderately biodegradable wastes, respiration activity within 2–5 g O_2 kg^{-1} DM h^{-1} (including mixed MSW, digested municipal wastewater sludge, and several types of manure)
3. Wastes of low biodegradability (respiration activity lower than 2 g O_2 kg^{-1} DM h^{-1})

TABLE 5.1
DRI and AT$_4$ Values for Some of the Most Generated Wastes in the World

	DRI (g O_2 kg^{-1} DM h^{-1})	AT$_4$ (g O_2 kg^{-1} DM)
OFMSW	3.7 ± 1.2	281 ± 66
MSW	2.2 ± 1.0	139 ± 54
Manure	3.8 ± 0.8	177 ± 36
Raw sewage sludge	5.4 ± 1.2	302 ± 51
Digested sewage sludge	1.9 ± 0.6	101 ± 51

Source: Data from own resources.
DRI, dynamic respiration index; AT$_4$, accumulated respiration activity after 4days.

Respiration indices have been used as a tool for optimization in composting mixtures. Barrena et al. (2011) concluded that the knowledge of the potential biological activity of a waste sample is of considerable help in formulating balanced mixtures for composting. Adhikari et al. (2013) have identified the most effective home composting formula, using respirometric tests measuring the oxygen uptake. They proposed that the most effective formula consists of a mixture of food waste and yard trimmings (wet volumetric fraction of 0.6:0.4) in supporting an active microbial activity for a fast composting process and the generation of high temperatures.

Kinetic models have been developed and tested to describe the aerobic biodegradation of wastes, obtaining the biodegradation kinetic rate constants and the different fractions in which organic matter (or organic carbon) can be classified depending on its biodegradation rate (Adani et al. 2004; Trémier et al. 2005; Komilis 2006; de Guardia et al. 2010; Ponsá et al. 2011).

5.4 COMPOSTING OPERATION

First industrial composting technologies were developed during the early twentieth century as a tool for organic farming but it was not until the second half of the twentieth century when most of the current industrial technologies were fully developed and implemented (Haug 1993).

Composting systems range from simple backyard composting to complex and highly controlled technologies. It is worth mentioning that when a composting process is properly conducted, all composting technologies can achieve a high degree of efficiency in terms of stability. Obtaining high-quality compost has been reported on several studies using a wide range of technologies, from simple home composting processes (Colón et al. 2010) to processes carried out at mechanical–biological treatment (MBT) facilities (Pognani et al. 2011). Therefore, the technology selection should be based on cost-efficiency and its associated environmental impacts rather than on the efficiency itself.

A basic distinction among composting technologies is home composting versus industrial composting. Broadly speaking, the large variety of currently available industrial composting systems can be grouped into two categories: simple industrial systems and "in-vessel" processes. Perhaps the most basic distinction is between systems in which the high-rate biodegradation phase is carried out in a reactor (in-vessel) and those in which it is not (simple industrial system). It is worth mentioning that most in-vessel processes use nonreactor systems as a maturation technology.

Another technology that will not be treated in detail in this chapter is vermicomposting. Vermicomposting involves the stabilization of organic solid wastes through earthworm consumption that converts the waste into earthworm castings (vermicast). In fact, vermicomposting is the result of combined activity of microorganisms and earthworms although earthworms are the main drivers of the process. The vermicast obtained at the end of the process is rich in plant nutrients and is free of pathogenic organisms (Singh et al. 2011). Several earthworm species have been demonstrated to be suitable for the treatment of the biowaste; *Eisenia andrei* and *Eisenia foetida* are the most commonly used. Although not widely implemented at industrial scale, vermicomposting has been reported as an adequate technique for the treatment of different organic wastes including sewage sludge, agro-industrial wastes and sludge, cattle manure, and urban solid wastes (Lleó et al. 2013).

5.4.1 HOME COMPOSTING

Home composting or backyard composting is the simplest composting technology and it can be defined as the self-composting of household biowaste as well as the compost usage in a garden belonging to a private household (European Commission 2008b). Home composting can be operated on either batch or continuous basis; for a typical family treating its organic household waste, semicontinuous feeding (twice per week) is recommended. Volume of commercial home composters ranges from 150 L to several cubic meters; minimum volume of 400 L–500 L is recommended to ensure some degree of thermal inertia if the home composter is not thermal insulated. In any case, the retention time in well-managed home composters varies between 12 and 16 weeks.

Moisture and porosity are the two key parameters to control during a backyard composting process. An excess of moisture or a lack of structure will generate compaction and poor oxygen transfer between the composter and the surrounding air. Eventually, anaerobic conditions will promote the generation of unpleasant odors and the production of unstabilized low-quality compost. Moisture and structure can be controlled by keeping a good volumetric ratio between feedstock and bulking agent; adequate ratios ranging from 1:1 to 1:2 have been reported (Colón et al. 2010). Mixing the upper layers (most oxygen-demanding fraction) on a weekly basis will enhance the aeration, prevent compaction, and also promote the homogeneity of the mixture.

Another key factor to promote home composting as an alternative to industrial systems is the sanitation control. Colón et al. (2010) reported that although temperatures did not reach the thermophilic level during a typical home composting process, hygienization occurred due to the natural decay of microorganisms.

5.4.2 SIMPLE INDUSTRIAL SYSTEMS

Industrial nonreactor processes can be subdivided based on its aeration system into "turned windrow" and "forced aeration windrow" (usually called "static pile"). It is also common to use both technologies at the same time. A full description of these systems can be found in Haug (1993).

Typically, windrows are shaped in trapezoidal form but also in triangle or round form; they usually have a height ranging from 1.5 m to 2.5 m and a width ranging from 2 m to 4 m. The length of a windrow is indeterminate as it depends on the amount of feedstock to be treated and the available area.

5.4.2.1 Turned Windrow

Feedstock is placed in rows and periodically turned. The size of a windrow is usually determined by the size of windrow turner machines, commercial machines have turning sizes ranging from 2 m to 3 m of height and 4 m to 6 m of width.

Oxygen is supplied mainly by natural ventilation produced by the chimney effect of hot gases in the windrow; turning also contributes to aeration, although to a lesser extent. In fact, turning not only contributes to aeration but also promotes the homogeneity of the mixture, prevents compaction, and ensures that all the composting material is exposed to the hotter inner zone of the pile where the sanitation of pathogens and bad seeds takes place and the biodegradation rate is higher.

Oxygen requirements as well as structural strength and moisture content of the material are some of the most important characteristics in determining the frequency of turning. The youngest windrows should be turned at least once a week during the active phase and ideally every three to four days. During the maturation phase, the turning rate is usually decreased, although it is not recommended and sometimes it is even stopped. For frequently turned, naturally ventilated windrows of biowaste, retention times of 12–20 weeks have been reported.

5.4.2.2 Static Pile

The main characteristic of static piles is the presence of a forced aeration system. Experience indicates that intermittent aeration serves to maintain aerobic conditions at an adequate level (5%–15% of oxygen in air). In order to ensure aerobic conditions along the windrow profile, it is of utmost importance to control moisture and porosity. As no turnings are performed in static piles, bulking agent must ensure enough structural strength and porosity to prevent compaction and preferential air pathways. A major drawback of static piles is the lack of homogenization between the inner and the outer part of a pile; usually temperatures reached in outer parts are not enough to ensure a proper sanitation. To overcome this drawback, the use of both technologies (forced aeration plus turning) is also frequent in many composting facilities.

Typical retention times for static piles are 3 to 4 weeks; after that period the feedstock is cured from one to several months in turned windrows.

5.4.3 IN-VESSEL PROCESSES

The term "in-vessel composting" is applied to processes in which the high-rate biodegradation phase takes place in a bioreactor. Although it is neither necessary nor specific of in-vessel composting technologies, they are usually carried out in enclosed buildings equipped with exhaust gas treatment systems.

Many systems have been designed during the last decade but only the most widespread technologies are covered in this section.

5.4.3.1 Composting Tunnels

Composting tunnel is probably the most widespread in-vessel technology. It is essentially a large-scale insulated box where a composting process is usually performed in a batch mode. Air is generally provided through a perforated floor by means of a centrifugal fan. The entire process is highly controlled via a programmable logic controller. Oxygen and temperature levels are monitored, and water (both processed or fresh water) and air (both processed or fresh air) can be added as needed. Some composting tunnels recirculate up to 80% of processed air to reduce the loss of excessive moisture and heat.

Typical dimensions of a tunnel are 4 m to 5 m width, 4 m to 5 m height, and up to 30 m length. The overall treatment process for fresh materials usually lasts about 2 to 3 weeks. Some MBT plants use composting tunnels for the curing stage of anaerobically digested sludge; in this case, the retention times are usually 1 week only.

5.4.3.2   Composting Channels/Trenches

Composting channels are reactors operated on a continuous basis; the feedstock is loaded through one end by means of a front-end loader or a conveyor belt and moved longitudinally to the other end of the reactor. A turning machine travelling above the feedstock bed carries out the longitudinal movement as well as the mixing process. Depending on the turner, material is shifted 2 m to 4 m with each turning. A forced aeration system provides the necessary air through the floor of the channel. Dimensions of individual channels vary, with depths ranging from 1.0 m to 2.5 m and widths of 2.0 m to 4.0 m. Channel length typically ranges from 50 m to 90 m, depending on the desired retention time and the properties of the turning machine.

Retention time during the high-rate composting phase is about 4 weeks; channels are also used in MBT plants as a curing stage technology, with a retention time ranging from 1 to 2 weeks.

5.4.3.3   Rotating Drum Biostabilizer

A rotating drum consists of a rotating cylinder with plug-flow conditions, so the inlet and the outlet are at the opposite ends of the reactor. The drums are approximately 45 m long and 2 m–4 m in diameter. The rotational speed is about 0.1 rpm–2 rpm. As in most in-vessel reactors, moisture and oxygen concentrations as well as temperature are monitored and maintained at optimum conditions.

Under normal operating conditions, the drum is filled from half of its capacity to about two-thirds. The retention time for the high-rate composting phase is about 1 week. Rotating drum biostabilizers are also used as a pretreatment stage with a retention time of 1–2 days for mixed MSWs; this pretreatment achieves a particle size reduction and a prefermentation of wastes and it also contributes to the disaggregation of paper and cardboard, facilitating the subsequent composting process.

5.5   COMPOSTING AND/OR ANAEROBIC DIGESTION

Anaerobic digestion and composting technologies can be applied at industrial scale to the treatment of a wide range of solid wastes such as MSW, OFMSW, waste sludge, feedstock, or manures. Readers can find a detailed description on the anaerobic digestion process in Chapter 6.

Both technologies have advantages and disadvantages, when used to treat organic waste with the final objective to its application to soil. On one side, compost is defined as an organic soil amendment with benefits when applied to soil and agriculture (see Section 5.8, Application of Compost). On the other side, the main drawback of the anaerobic digestion process, when treating solid wastes, is that the digestate is not stabilized enough to be used in agriculture. Digestate contains compounds such as fatty acids or ammonia that are known to be phytotoxic. McLachlan et al. (2004) reported phytotoxicity effect of undiluted digestate from MSW, and Pognani (2012) reported phytotoxic effect of digested pig slurry. In this last case, phytotoxicity (reduced root and hypocotyls elongation) was attributed to a possible auxin hormone-like effect. Similar results were obtained by Andruschkewitsch et al. (2013), using digestate from residual biomass. However, Komilis and Tziouvaras (2009) reported that germination tests are highly dependent on the type of seed used. Working with compost, they found that a compost that is phytotoxic to a certain seed can enhance the growth of another seed.

Besides the phytotoxic effect, digestate can produce malodors, once applied to the soil and may create difficulties during soil applications techniques (Tchobanoglous and Kreith 2002). For example, ammonia present in the digestate will partially be emitted to the atmosphere, contributing to atmospheric pollution, soil acidification, and eutrophication (Köster et al. 2014). Both, odors and phytotoxicity are related to a lack of stabilization, which can be determined with the final DRI or AT_4 values (see Section 5.7, Compost Quality). Ponsá et al. (2008), in a deep study of a MBT plant flows treating MSW, reported DRI values for digested OFMSW and MSW around 1.5 g O_2/kg^{-1} DM h^{-1}. Barrena et al. (2011) reported values for digested sewage sludge of 2.56 g O_2/kg^{-1} OM h^{-1} (ranging between 3.73 and 1.64). These studies on full-scale samples point that digestate is not stabilized enough to be directly applied to the soil.

However, clearly, anaerobic digestion takes advantages of its positive energy balance compared to composting, which is an energy-consuming system. Indeed, composting of OFMSW consumes between 65.5 kWh and 242 kWh per ton of input waste, depending on the composting technology used (Fricke et al. 2005; Blengini 2008; Colon et al. 2012), while anaerobic digestion produces 100 kWh to 150 kWh per ton of input waste (Hartmann and Ahring 2006).

Thereby, the options to consider could be composting or anaerobic digestion followed by a composting process. Focusing on municipal waste management, it seems logical that low technological composting (see Section 5.4.2, Simple Industrial Systems) plants would be installed in rural areas or in low- to medium-density population areas generating OFMSW of low to medium impurities, while in high-density areas, where probably OFMSW would have more impurities, combined anaerobic and aerobic plants with mechanical pretreatment (MBT plants) would be installed. When nonsource-selected MSW are treated, MBT plants are mandatory, based on either composting or anaerobic digestion and composting. However, as the final product could not have enough quality to be used as an organic amendment, obtaining biogas from MSW could improve the management system.

5.6 LCA AND COMPARISON WITH OTHER TECHNOLOGIES FOR WASTE TREATMENT

LCA has been proposed to support waste management decisions as this methodology provides a comprehensive view of all the processes involved and the impacts associated (Finnveden et al. 2007). Among all of the existing impact potentials those mainly used in the LCA studies of waste management and treatment are global warming potential (GWP), acidification, eutrophication, and resources consumption (energy requirements in some cases) followed by photochemical oxidation, human toxicity, abiotic depletion, and ozone depletion.

While defending the importance of LCA in environmental analysis of waste management systems, Ekvall et al. (2007) also discussed the limitations of this tool. According to these authors, the LCA's broad perspective allows considering the environmental benefits of some waste treatment processes, such as the recovery of energy in waste incineration that can reduce the need of energy obtained from other energy sources, the recovery of materials through the recycling processes, and the production of biogas and stabilized organic materials in biological treatments that can reduce the need of fertilizers and energy. Currently, the compost produced at MBT plants is not suitable for agriculture application and in most cases it ends up in landfills. The discussion around the use and application of that compost is gaining relevance as well as if this compost production should be accounted for or not within the overall recycling rates. In fact, new applications for its use are under research. However, Ekvall et al. (2007) highlight the importance of using LCA complemented with other assessment tools that permit consideration of other waste management aspects such as the economic implications. They also point that the simplifications and uncertainty associated to LCA should be kept in mind mainly when this tool is used during the decision-making process.

Taking into account the limitations stated above, Saer et al. (2013) presented composting plant gaseous emissions as one of the hotspots of the whole compost life cycle, considering waste collection, transport, processing, and compost distribution and application. Bernstand and la Cour Jansen (2011) found that the anaerobic digestion of household food waste has greater benefits with respect to GWP, to acidification potential and to ozone depletion potential than composting and incineration. The recovery of biogas derived from the anaerobic treatment was the main reason for these benefits over the other two techniques. However, these authors have also performed sensitivity analysis to conclude that the differences among the treatments studied are highly dependent on the assumptions made relating to the energy substituted by biogas.

MSW management generates GHG emissions, also referred as direct GHG emissions, the most significant being the methane (CH_4) gas, which is mostly released during the breakdown of organic matter in landfill. Collection and transport of waste also generate GHG emissions because of the use of fuel and also from the infrastructure. Biological treatments including composting and anaerobic digestion generate CO_2, CH_4, and N_2O. Relevant gases emitted during incineration include CO_2, CH_4, and N_2O but normally, emissions of CO_2 from waste incineration are more significant than CH_4 and N_2O emissions. In addition, waste management activities include upstream activities, which are needed for running waste management operations (i.e., fuel or ancillary materials), and downstream activities due to recovered materials and energy from these operations, which can be supplied back to the economic cycle, offsetting primary resources (Gentil 2011). Thus, all the GHG emissions generated over the operating activities, upstream activities (referred as indirect emissions), and downstream activities (referred as avoided emissions) may be taken into account. Figure 5.2 represents graphically the different types of emissions from waste management.

The relationship between waste management and GHG emissions has been enhanced, based on the idea that treatment and disposal of waste produce significant amounts of GHG emissions (IPCC 2006) but proper waste management can avoid GHG emissions due to controlled composting of organic waste or by waste recycling

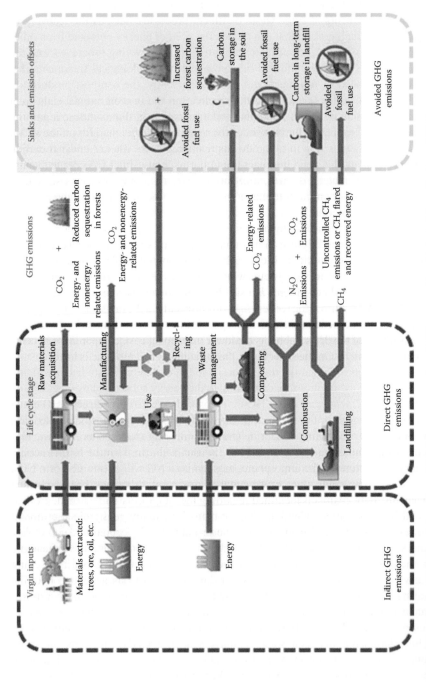

FIGURE 5.2 Types of GHG emissions from MSW management from LCA perspective. (Adapted from Sevigne, 2013.)

through the conservation of raw materials and fossil fuels (Bogner et al. 2007). This is important since the environmental consequences of waste management often depend more on the impacts of surrounding systems than on the emissions from the waste management system itself (Ekvall 2000).

The accounting, reporting, and modeling of GHG emissions began to be implemented on a global scale since the inception of the Kyoto Protocol (1997) Since then, various more protocols have been developed, such as the IPCC guideline for waste management activities (IPCC 2006; Bogner et al. 2007), which, jointly with the LCA, are currently the principal GHG quantification methods. The IPCC protocol has been widely used to report the national GHG emissions from waste management (IPCC 2006). However, with this protocol it is only possible to calculate the direct GHG emissions from disposal, biological treatment, and incineration (IPCC 2006). It is limited to the direct GHG emissions because, historically, the waste management sector broadly constituted disposal (open dumping, landfilling) and mass burn incineration, without energy recovery. Also, because upstream and downstream activities are accounted for in other sectors (i.e., energy), including them would lead to double counting (Gentil 2011). This constitutes a limitation in comparison with the LCA methodology, which allows calculating the indirect GHG emissions and the avoided GHG emissions; thus, this broad system perspective has made the LCA a powerful tool for the environmental comparison of different options for waste management.

Two important aspects regarding the GHG quantification have been widely discussed with often-conflicting interpretation: GWP values and the carbon cycle. Under the Kyoto Protocol it was decided to use the values of GWP for converting the various GHG emissions into comparable CO_2 equivalents. The GWP integrates the radiative force (how much heat is trapped by a GHG in the atmosphere) of a substance over a chosen time horizon and relative to that of CO_2 whose GWP was standardized to 1. These values are very dependent on metric type and time horizon (Myhre et al. 2013) because a gas that is quickly removed from the atmosphere may initially have a large effect but over a longer time period becomes less effective; thus, GWPs were calculated over a specific time interval, commonly 20, 100, or 500 years (Myhre et al. 2013). In addition, these values have been updated over the years due to new estimates of lifetimes, impulse response functions, and radiative efficiencies. First values appeared in the IPCC Second Assessment Report (IPCC 1995), in which the GWP for CH_4 was 21 for a time horizon of 100 years. Since then, three more assessment reports—Third (IPCC 2001), Fourth (IPCC 2007), and Fifth (IPCC 2014b)—have been published with the aim of assessing scientific, technical, and socioeconomic information concerning climate change, its potential effects, and options for adaptation and mitigation. GWP values have been updated and in the recent assessment report, the GWP for CH_4 value is 34 (Myhre et al. 2013). Thus, the GHG accounting can vary depending on GWP values.

Furthermore, for biodegradable materials (i.e., organic matter and paper), carbon will have been absorbed from the atmosphere by photosynthesis during plant growth. If this carbon is released again as CO_2 during the treatment process, then the carbon reenters the natural carbon cycle. These emissions are reported as biogenic CO_2 (Smith and Jasim 2009) and the IPCC methodology ignores the contribution of biogenic CO_2 to GHG emissions. Therefore, the IPCC methodology considers that the CO_2 emissions from biomass sources including CO_2 in landfill gas, CO_2 from composting, and CO_2 from incineration of waste biomass should not be taken into account in the GHG inventories as these are covered by changes in biomass stocks in the land use, land-use change, and forestry sectors. However, it is argued that the plant growth does not occur evenly over years and seasons, and therefore it could be several years before a flux of biogenic CO_2 emitted instantaneously from a process (i.e., combustion of biogenic carbon) is recaptured through plant growth. In addition, the atmosphere does not differentiate between a molecule of biogenic CO_2 and a molecule of fossil-derived CO_2 and the key theme is climate change and how to mitigate it, regardless of carbon sources (UNEP 2011). That is the reason why some models do not quantify biogenic CO_2 emissions, some quantify these emissions but do not consider the emission to contribute to GWP, and some quantify biogenic CO_2 emissions and consider that emissions contribute to GWP (Christensen et al. 2009).

There are numerous models based on the LCA methodology to calculate the GHG emissions of MSW management. Many of these models have been developed in north of Europe or in North America using local data (Den Boer et al. 2007; Gentil et al. 2010; Eriksson and Bisaillon 2011; Tunesi 2011). A specific tool has been developed for the Mediterranean countries, called CO2ZW®, and was later extended to other continents. CO2ZW provides a means of calculating GHG emissions (in CO_2 equivalents) for the management of MSW at the municipal, regional, or national levels with small amounts of input data (Sevigné et al. 2013). The tool is an Excel®-based calculator, which, with the input of municipality-specific waste data (or national data as a default), permits the user to obtain a municipality-level carbon footprint of waste treatments (infrastructures are not included). Stakeholders that are involved in the waste management sector (including technicians, consultants, NGOs, or politicians) are able to calculate GHG emissions. The calculator can be accessed after registration in the following web page: http://sostenipra.ecotech.cat.

5.7 COMPOST QUALITY

Benefits of composting in agriculture and positive environmental aspects imply the use of compost with adequate quality. These benefits of compost application are well studied but that compost should be a high-quality product, which should guarantee that physical, chemical, and biological characteristics have been used to evaluate the quality of composts and/or end products. "Maturity" and "stability" are terms widely used to describe compost quality. Table 5.2 presents a complete characterization of a good quality compost obtained from source-selected biowaste.

Compost quality standards of the different EU members have been established. Moreover, the wide range of limit values suggests the need for developing EU compost quality standards (Lasaridi et al. 2006). An interesting proposal is the European Quality Assurance for Compost and Digestate (European Compost Network 2011) where some parameters as quality criteria and others as precautionary requirements on the protection of environment and consumers are proposed. *Salmonella* presence;

TABLE 5.2

**Complete Characterization of a Good Quality Compost
Obtained from Source-Selected Biowaste (27 Samples)**

Property	Compost (27 Samples)
Moisture (%, wb)	44 ± 15
pH	7.7 ± 0.5
EC (dS m^{-1})	3.9 ± 2.9
OM (%, db)	57 ± 18
TNK (%, db)	2.2 ± 0.9
P (%, db)	0.8 ± 0.4
K (%, db)	1.4 ± 0.9
C/N ratio	16.0 ± 7.4
Cd (mg kg^{-1}, db)	0.3 ± 0.2
Cr (mg kg^{-1}, db)	19.3 ± 11.0
Cu (mg kg^{-1}, db)	53.1 ± 37.9
Hg (mg kg^{-1}, db)	0.1 ± 0.2
Ni (mg kg^{-1}, db)	11.3 ± 5.5
Pb (mg kg^{-1}, db)	24.2 ± 17.3
Zn (mg kg^{-1}, db)	104 ± 127
DRI (g O$_2$ kg^{-1} OM h^{-1})	0.27 ± 0.16
AT$_4$ (g O$_2$ kg^{-1} OM)	8.4 ± 8.3

wb: wet basis; db: dry basis; OM: organic matter; TNK: total nitrogen Kjeldahl; EC: electrical conductivity; DRI: dynamic respiration index; AT$_4$: cumulative respiration activity.

heavy metals; impurities like plastics, metals, and glass; and germinable seeds are parameters where limit values are proposed.

5.7.1 STABILITY AND MATURITY

"Stability" and "maturity" are the terms most cited when compost quality is evaluated. Both are quality aspects of compost that can be evaluated through some parameters, which also can be determined using several methodologies. "Compost stability" is defined as the extent to which readily biodegradable organic matter has been decomposed (Lasaridi and Stentiford 1998). Traditionally, stability has been determined by the rates of O$_2$ uptake, CO$_2$ produced, or heat released in a self-heating test. Compost requires a minimum level of biological stability to avoid problems during its storage, distribution, and final use since odor production, self-heating, and deterioration quality are consequences of the presence of biodegradable organic matter. As commented previously, respiration indices have been widely used in the literature as a measure of biodegradable organic matter and stability (Barrena et al. 2006). The term "maturity" is wider and it is associated with plant growth potential and phytotoxicity. A widely used maturity test is the germination index (Zucconi et al. 1981). Despite the extended used of this type of test, there are no standards and only some types of seeds are used. The plant response can be different depending on the time of compost application, which could lead to arriving different conclusions (Bernal et al. 2009). Whereas germination tests assess the effect of compost on the earlier stages of plant development, growth tests evaluate the effect on later stages.

In recent years, alternative composts have appeared that are derived from feed sources such as MSW and biosolids. Clearly, the final use of them is an important issue in compost quality. Stability is specifically requested for some end products when their final use is not compost agronomical application. Also, a minimum level of nutrients is required when compost is used as organic amendment. In this sense, it is important to take into account the wide variety of composting feedstocks and possibilities of end use of each compost derived. Other aspects related with composting management, such as the pretreatment process in OFMSW composting, also determine the quality of the final product. Control process and well-defined mixtures can have an important impact in the conservation of nutrients like nitrogen. Strategies for producing high-quality compost should be the objective of composting managers.

There are an important number of scientific publications on the evaluation of maturity and/or stability. Although there is a general agreement about the importance of both parameters, consensus still has not been reached about which should be the most suitable measurement. Physicochemical and biological parameters and the combination of them have also been used for evaluation.

5.7.2 PHYSICOCHEMICAL CHARACTERISTICS OF COMPOST

The physicochemical characteristics of mature composts are more in accordance with the properties of the raw feedstock rather than the process itself (Manios 2004). Physical characteristics such as odor, color, or temperature when sampling can offer information about the characteristics of compost. Particle size, inert materials, and impurities also add some information related to compost quality.

Chemical methods are widely used, including several methods to determine carbon and nitrogen forms. C/N ratio (in solid and water extract) is one of the most reported parameters (Bernal et al. 2009). A high level of NH_4-N is related to non-mature compost. Also the ratio NH_4-N/NO_3-N has been suggested as maturity index for compost of different nature (Bernal et al. 2009). Organic matter and nutrient content are indicators of the potential use of composts as soil improvers and organic fertilizers.

Heavy metal content is one of the key parameters in compost quality because it limits the compost's use as organic amendment. Heavy metal content is one of the most controversial aspects in the quality of compost from source-separated OFMSW. Compost from OFMSW is typically well below the limits of the standards of quality (Smith 2009). Nevertheless, it strongly depends on the management systems used and, in this regard, door-to-door waste collection systems have been shown as the most appropriate strategy (Colón et al. 2013). The contents of heavy metals are only slightly lower in home-made compost than in industrial compost when source-separate systems are well implemented (Barrena et al. 2014).

5.8 APPLICATION OF COMPOST

Sustainable agriculture is the production of food, fiber, or other plants or animal products using farming techniques that protect the environment, public health, human communities, and animal welfare. This form of agriculture enables us to produce healthful food without compromising future generations' ability to do the same. Organic farming can be defined as a method of production, which places the highest emphasis on environmental protection and, with regard to livestock production, on animal welfare considerations. Organic farming is considered by EU as a main driver to promote sustainability in agriculture. It avoids or largely reduces the use of synthetic chemical inputs such as fertilizers, pesticides, additives, and medicinal products. According to the European regulation, organic farming should primarily rely on renewable resources within locally organized agricultural systems. In order to minimize the use of nonrenewable resources, wastes, and by-products of plant and animal origin should be recycled to return nutrients to the land (European Union 2014). This is done through compost application to agriculture. In Europe alone, the area under organic farming for EU-27 countries was 5.8% in 2010 and 20% in 2012. Likewise, Austria, Estonia, Check Republic, and Sweden are the countries having the largest land cover as organic farming (above of 12% of total cultivated areas) in EU.

5.8.1 Role of Compost as Organic Amendment

Organic matter is essential for soils, especially when growth of plants is targeted. In the European soil map (European Commission 2005) it can be seen that there are soils of southern Europe with less than 2% organic carbon contents. The decline of soil organic matter is an emerging issue in the European policy and so measures to face that decline are promoted in a political level. In addition, it would be erroneous to undermine the value of compost as a sink of CO_2, as has been suggested by the IPCC and of several working groups on that topic (Favoino and Hogg 2008). The value of compost to sequestrate carbon is commonly accounted for in LCA studies. The term "carbon sequestration" means that the carbon contained in stable compost is not available for further degradation for some time (this time horizon is still an issue of research), though in the long term even carbon of a stable compost will eventually mineralize. Adding organic matter (e.g., via composting) to soil eventually mitigates GHG emissions, since it can offset the same amount of carbon that is released to the atmosphere by the combustion of fossil fuels (Favoino and Hogg 2008).

If we exclude the (micro) biological factor, the well-studied effects of organic matter addition into soils are listed here: Organic matter (1) could lead to a slow release of nutrients (95% of N is in organic form), (2) has a high binding capacity for organic and inorganic elements (contaminants or nutrients), (3) improves water storage due to its increased water-holding capacity, and (4) aids in the creation of soil agglomerates that can facilitate aeration of plant roots and improve water infiltration into the soil. That is, the addition of organic matter addition alone on soil alters its physicochemical properties.

Quirós (2014) and Quirós et al. (2014) presented a LCA and agronomical assessment of the following three fertilization treatments applied to horticultural cauliflower crops: industrial compost, home compost, and mineral fertilizer. This study defined the functional unit as total tons per hectare of open-field cauliflower crops. Figure 5.3 summarizes the experimental conditions for the cultivation phase. Quiros and others demonstrated that closing the organic matter flow from the 2V "vegetable to vegetable" model is better than landfilling organic resource. Furthermore, the research study also aimed at the environmental impact of the three above-mentioned fertilization treatments. The quality characteristics of samples of cauliflower (i.e., yield, diameter, and weight) from the three fertilization treatments were measured. In summary, mineral fertilizer yielded more cauliflower, which was 26% and 91% higher than home compost and industrial compost, respectively. However, other quality parameters, such the diameter and weight of cauliflower, were higher for home compost than industrial compost or mineral fertilizer. In the case of the bioactive substance content, no significant differences were found in the medians of the average values for the cauliflower harvested in any of the groups; however, mineral fertilizer had higher values for all bioactive substances except in β-carotene, for which industrial compost had the highest value. The fertilization treatment with home compost showed the best environmental performance in all categories assessed, except in eutrophication potential. In this category the best environmental performance was from industrial compost, although the difference was not so significant (<5%). The fertilization treatment with industrial compost showed the worst environmental performance. In fact, mineral fertilizer had better results in all categories assessed except in eutrophication potential and GWP. The stages that most significantly affected industrial compost were the collection of OFMSW and the transportation of the compost from the facility to the crop area. This was apparently due to the diesel requirements; therefore, considering that the collection of the OFMSW is not part of the compost process, but a consequence of the waste generation, the fertilization with industrial compost is better. Besides, if the industrial compost is used close to the production plant, as a local resource in waste generation, this impact will further be reduced.

5.8.2 Compost as Suppressor of Plant Diseases

Plant diseases are known to be primarily induced by fungi, which can form spores, rather than by bacteria. There has been a lot of research on the use of compost as a plant-disease suppressant for crops. Interestingly, some of the earliest work on the biocontrol of plant pathogens comes from the 1920s. This biocontrol is of core importance since the use of fungicides, commonly used against soil pathogens, can be reduced or totally eliminated. A research group in Ohio State University, with a long tradition on the study of the compost induced suppression of plant diseases, first implied that compost application can suppress plant diseases (Hoitink et al. 1975).

Peat (a nonrenewable material) had been used a lot in the past as a plant growth material. However, it was observed that peat favored plant diseases, and this was explained by the low microbial activity of that material. Researchers, during the 1970s, turned to compost as a substitute to peat and realized that it could do better

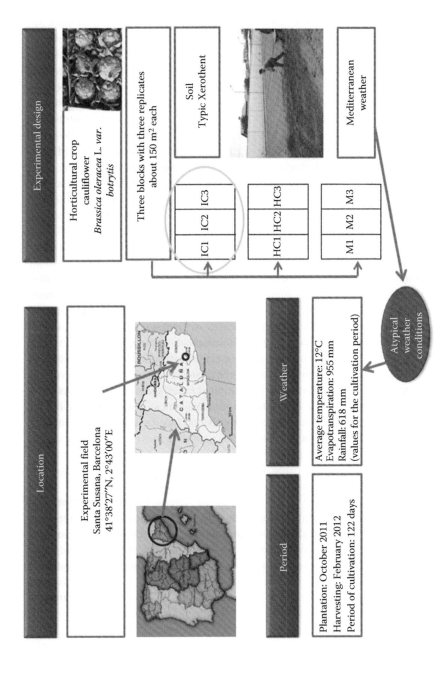

FIGURE 5.3 Experimental design and weather conditions during cultivation phase. (Data from Ecotech Sudoe, 2013.)

than peat in terms of plant diseases. Since then, the mechanisms of suppression of plant pathogens, and therefore of plant diseases, in a soil via the application of compost have been well described (Zinati 2005; Mehta et al. 2014). The suppression is generic or specific and consists of different submechanisms, which are (1) competition between pathogenic and nonpathogenic microbes, (2) antibiotic activity of certain compost microbes, (3) hyperparasitism, and (4) systemic acquired resistance (SAR) and induced systematic resistance (ISR). Generic suppression means that a wide range of microbial species are responsible for the suppression of pathogens. On the other hand, in specific suppression, only a narrow group of microbes is responsible for the suppression of a pathogen.

1. In the case of competition, the mechanism is well understood. The nonpathogenic compost microbes (bacteria, fungi, actinomycetes) compete for certain nutrients or space against the plant pathogens. This competition usually favors the nonpathogenic compost microbes, so that plant pathogens are reduced, but not eliminated. For example, iron is essential for the growth of pathogens; the increase of siderophore producers after application of compost to the soil leads to the reduction of iron, which inhibits the growth of certain pathogens (de Brito-Alvarez et al. 1995).

2. Another suppression mechanism that has been suggested comes from the production of antibiotics by compost microbes that can harm pathogens. *Pseudomonas* and *Bacillus*, for example, are known antibiotic producers and are important in the biocontrol of several plant diseases.

3. Hyperparasitism is also a suggested mechanism of plant-disease suppression. In this case, a compost microbe (such as a parasite) attacks directly a pathogenic microorganism, which is eliminated by lysis or death.

4. In addition to the above well-known mechanisms, mechanisms of SAR or ISR have been also suggested. Both SAR and ISR are considered as states of enhanced defensive capacity that a plant develops when stimulated by developing the pertinent pathogenetic genes. In both cases, the principal mechanism is that there is a prior infection or prior treatment of the plants that can result in a tolerance against pathogens or parasites. SAR can be induced by chemicals, pathogens, and beneficial soil microbes, while ISR can be developed by several bacterial and fungal microorganisms present in compost-amended substrates (Mehta et al. 2014). In SAR, an attack by a pathogen triggers local defense responses and the generation of a systemic signal in the whole plant. After the signal, all plant parts become resistant to pathogen attack via a mechanism that is probably driven by the generation of salicylic acid and several pathogenesis-related now proteins. Except for soil-borne diseases, compost application can induce resistance against foliar and root diseases (Ntougias et al. 2008). Abiotic characteristics of the compost could also trigger off pathogenesis-related genes in the roots of certain plants. Even anatomic changes in plants can be triggered by microbial communities in composts (Pharand et al. 2002).

5.9 ECONOMY OF COMPOSTING

Although extremely important, there are only a few publications on the costs of composting. In general, it is considered that the cost of composting is lower than that of other technologies used for the treatment of organic wastes, such as anaerobic digestion (Sonesson et al. 2000). However, the comparison with other final disposal systems is not so clear (e.g., incineration or landfill), due to the local restrictions and other constrains that can be found in several parts of the world where composting is applied (Meyer-Kohlstock et al. 2013). Briefly, some of the reasons that make these costs so variable can be summarized, among others, as follows:

- The need of source selection of the OFMSW, which in some parts of the world is mandatory
- The type of composting technology, which can be very simple (home composting) or very complicated (in-vessel reactors, exhaust gas treatment, etc.)
- Compost quality and its utilization
- Local waste taxes

Anyway, it is not possible to establish a general value for composting, being possible to be found values from €50 in some parts of Spain to €10 in some developing countries (Puig-Ventosa 2008). Notwithstanding this, it is evident that the economy of composting is an issue that must be coupled with the knowledge of its environmental benefits, a point that is still under research (Favoino and Hogg 2008).

5.10 CONCLUDING NOTE

Although extensively implemented and used for the management of practically all kinds of organic wastes, composting appears to be a technology where research is clearly necessary, as it is one of the most complex biotechnological processes. The presence and the interaction among several phases (gas, liquid, and solid), the microbiology, and the characterization of organic matter are to be the main focus of research in the following years. Nevertheless, composting works, regardless of lack of knowledge. This makes composting as a sustainable strategy for the management of organic solid wastes, and one of the preferred technologies to be used in any waste management scenario in developed and developing countries.

REFERENCES

F. Adani, R. Confalonieri, and F. Tambone. 2004. Dynamic respiration index as a descriptor of the biological stability of organic wastes. *Journal of Environmental Quality* 33: 1866–1876.

B.K. Adhikari, A. Trémier, S. Barrington, and J. Martinez. 2013. Biodegradability of municipal organic waste: A respirometric test. *Waste and Biomass Valorization* 4: 331–340.

J.M. Agnew and J.J. Leonard. 2003. Literature review. The physical properties of compost. *Compost Science & Utilization* 11: 238–264.

M. Andruschkewitsch, C. Wachendorf, and M. Wachendorf. 2013. Effects of digestates from different biogas production systems on above and belowground grass growth and the nitrogen status of the plant-soil-system. *Grassland Science* 59: 183–195.

R. Barrena, X. Font, X. Gabarrell, and A. Sánchez 2014. Home composting versus industrial composting: Influence of composting system on compost quality with focus on compost stability. *Waste Management* 34: 1109–1116. Available from: http://dx.doi.org/10.1016/j.wasman.2014.02.008.

R. Barrena, T. Gea, S. Ponsá, L. Ruggieri, A. Artola, X. Font, and A. Sánchez. 2011. Categorizing raw organic material biodegradability via respiration activity measurement: A review. *Compost Science & Utilization* 19: 105–113.

R. Barrena, F. Vázquez, and A. Sánchez. 2006. The use of respiration indices in the composting process: A review. *Waste Management & Research* 24: 37–47.

M.P. Bernal, J.A. Alburquerque, and R. Moral. 2009. Composting of animal manures and chemical criteria for compost maturity assessment. A review. *Bioresource Technology* 100: 5444–5453.

A. Bernstand and J. la Cour Jansen. 2011. A life cycle approach to the management of household food waste—A Swedish full-scale case study. *Waste Management* 31: 1879–1896.

G.A. Blengini. 2008. Using LCA to evaluate impacts and resources conservation potential of composting: A case study of the Asti District in Italy. *Resources, Conservation and Recycling* 52: 1373–1381.

J. Bogner, A.M. Abdelrafie, C. Diaz, A. Faaij, Q. Gao, S. Hashimoto, K. Mareckova, R. Pipatti, and T. Zhang. 2007. *Waste Management. Contribution of Working Group III to the Fourth Assessment Report of the Intergovernmental Panel on Climate Change.* Metz, B., Davidson, O.R., Bosch, P.R., Dave, R., and Meyer, L.A. (eds.). Cambridge University Press, Cambridge.

Chatman House. 2012. *A Global Redesign? Shaping the Circular Economy.* Chatman House, London. Available from: http://www.chathamhouse.org/sites/default/files/public/Research/Energy,%20Environment%20and%20Development/bp0312_preston.pdf (accessed May 2014).

T.H. Christensen, E. Gentil, A. Boldrin, A. Larsen, B. Weidema, and M. Hauschild. 2009. C balance, carbon dioxide emissions and global warming potentials. *Waste Management Research* 27: 707–715.

J. Colón, E. Cadena, M. Pognani, R. Barrena, A. Sánchez, X. Font, and A. Artola. 2012. Determination of the energy and environmental burdens associated to the biological treatment of source-separated Municipal Solid Wastes. *Energy & Environmental Science* 5: 5731–5741.

J. Colón, J. Martínez-Blanco, X. Gabarrell, A. Artola, A. Sánchez, J. Rieradevall, and X. Font. 2010. Environmental assessment of home composting. *Resources, Conservation & Recycling* 54: 893–904.

J. Colón, M. Mestre-Montserrat, I. Puig-Ventosa, and A. Sánchez. 2013. Performance of baby biodegradable used diapers in the co-composting process with the organic fraction of municipal solid waste. *Waste Management* 33: 1097–1103.

Constitution de la République et canton de Genève (Cst-GE). 2014. Available from: http://www.admin.ch/opc/fr/classified-compilation/20132788/201306010000/131.234.pdf (accessed May 2014).

M.A. de Brito-Alvarez, S. Gagne, and H. Antoun. 1995. Effect of compost on rhizosphere microflora of the tomato and on the incidence of plant growth-promoting rhizobacteria. *Applied Environmental Microbiology* 61: 194–199.

A. de Guardia, P. Mallard, C. Teglia, A. Marin, C. Le Pape, M. Launay, J.C. Benoist, and C. Petiot. 2010. Comparison of five organic wastes regarding their behaviour during composting: Part 1, biodegradability, stabilization kinetics and temperature rise. *Waste Management* 30: 402–414.

J. Den Boer, E. den Boer, and J. Jager. 2007. LCA-IWM: A decision support tool for sustainability assessment of waste management systems. *Waste Management* 27: 1032–1045.

Y. Eklind and H. Kirchmann. 2000. Composting and storage of organic household waste with different litter amendments. II: Nitrogen turnover and losses. *Bioresource Technology* 74: 125–133.

T. Ekvall. 2000. A market based approach to allocation at open-loop recycling. *Resources Conservation and Recycling* 29: 91–109.

T. Ekvall, G. Assefa, A. Björklund, O. Eriksson, and G. Finnveden. 2007. What life cycle assessment does and does not do in assessments of waste management. *Waste Management* 27: 989–996.

European Commission. 2005. Topsoil organic carbon content. Institute of Environment and Sustainability. Available from: http://eusoils.jrc.ec.europa.eu/ESDB_Archive/octop/Resources/Octop.pdf. (accessed April 2014).

European Commission. 2008a. Directive 2008/98/EC. On waste and repealing certain Directives. *Official Journal of the European Union* 312: 3–30.

European Commission. 2008b. Green paper—On the management of bio-waste in the European Union. European Commission, Brussels, Belgium.

European Commission. 2012. Manifesto for a resource-efficient Europe. European Commission, Brussels, Belgium.

European Compost Network (ECN). 2011. European Quality Assurance for Compost and Digestate. Available from: http://www.compostnetwork.info (accessed July 2014).

European Union. 2014. Council regulation (EC) No 834/2007 of 28 June 2007 on organic production and labelling of organic products. Available from: http://epp.eurostat.ec.europa.eu/ (accessed: July 2014).

E. Favoino and D. Hogg. 2008. The potential role of compost in reducing greenhouse gases. *Waste Management & Research* 26: 61–69.

G. Finnveden, A. Björklund, A. Moberg, and T. Ekvall. 2007. Environmental and economic assessment methods for waste management decision-support: Possibilities and limitations. *Waste Management Research* 25: 263–269.

K. Fricke, H. Santen, and R. Wallmann. 2005. Comparison of selected aerobic and anaerobic procedures for MSW treatment. *Waste Management* 25: 799–810.

R.A. Frosch and N. E. Gallopoulos. 1989. Strategies for Manufacturing. *Scientific American* 261: 144–152.

M. Gao, F. Liang, A. Yu, B. Li, and L. Yang. 2010. Evaluation of stability and maturity during forced-aeration composting of chicken manure and saw dust at different C/N ratios. *Chemosphere* 78: 614–619.

E.C. Gentil, A. Damgaard, M. Hauschild, G. Finnveden, O. Eriksson, S. Thorneloe et al. 2010. Models for waste life cycle assessment: Review of technical assumptions. *Waste Management* 30: 2636-2648.

E.M. Harper and T.E. Graedel. 2004. Industrial ecology: A teenager's progress. *Technology in Society* 26: 433–445.

H. Hartmann and B.K. Ahring. 2006. Strategies for the anaerobic digestion of the organic fraction of municipal solid wastes: An overview. *Water Science & Technology* 53: 7–22.

R.T. Haug. 1993. *The Practical Handbook of Compost Engineering.* Lewis Publishers, Boca Raton, FL.

H.A.J. Hoitink, A.F. Schmitthenner, and L.J. Herr. 1975. Composted bark for control of root rot in ornamentals. *Ohio Report on Research and Development* 60: 25–26.

IPCC. 1995. *Climate Change 1995: The Science of Climate Change. Contribution of Working Group I to the Second Assessment Report of the Intergovernmental Panel on Climate Change.* Houghton, J.T., Meira Filho, L.G., Callander, B.A., Harris, N., Kattenberg, A., and Maskell, K. (eds.). Cambridge University Press, Cambridge. Available from: https://www.ipcc.ch/pdf/climate-changes-1995/ipcc-2nd-assessment/2nd-assessment-en.pdf (accessed May 2014).

IPCC. 2001. *Climate Change 2001: The Scientific Basis. Contribution of Working Group I to the Third Assessment Report of the Intergovernmental Panel on Climate Change.* Houghton, J.T., Ding, Y., Griggs, D.J., Noguer, M., van der Linden, P.J., Dai, X., Maskell, K., and Johnson, C.A. (eds.). Cambridge University Press, Cambridge. Available from: http://www.grida.no/publications/other/ipcc_tar/?src=/climate/ipcc_tar/wg1/index.htm (accessed May 2014).

IPCC. 2006. *2006 IPCC Guidelines for National Greenhouse Gas Inventories.* Eggleston, H.S., Buendia, L., Miwa, K., Ngara, T., and Tanabe, K. (eds.). Hayama, Japan: Institute for Global Environmental Strategies. Available from: http://www.ipcc-nggip.iges.or.jp/public/2006gl/spanish/index.html (accessed May 2014).

IPCC. 2007. *Contribution of Working Group I to the Fourth Assessment Report of the Intergovernmental Panel on Climate Change.* Solomon, S., Qin, D., Manning, M., Chen, Z., Marquis, M., Averyt, K.B., Tignor, M., and Miller, H.L. (eds.). Cambridge University Press, Cambridge. Available from: http://www.ipcc.ch/publications_and_data/ar4/wg1/en/contents.html (accessed May 2014).

IPCC. 2014a. Glossary. Available from: http://www.ipcc.ch/pdf/glossary/ipcc-glossary.pdf (accessed May 2014).

IPCC. 2014b. Publications and data. Available from: http://www.ipcc.ch/publications_and_data/publications_and_data_reports.shtml (accessed May 2014).

Joint Research Centre (JRC). 2011. Supporting environmentally sound decisions for waste management. A technical guide to Life Cycle Thinking (LCT) and Life Cycle Assessment (LCA) for waste experts and LCA practitioners. JRC, Luxembourg. Available from: http://publications.jrc.ec.europa.eu/repository/bitstream/111111111/22582/1/reqno_jrc65850_lb-na-24916-en-n%20_pdf_.pdf (accessed May 2014).

D.P. Komilis. 2006. A kinetic analysis of solid waste composting at optimal conditions. *Waste Management* 26: 82–91.

D.P. Komilis and I.S. Tziouvaras. 2009. A statistical analysis to assess the maturity and stability of six composts. *Waste Management* 29: 1504–1513.

J.R. Köster, K. Dittert, K.-H. Mühling, H. Kage, and A. Pacholski. 2014. Cold season ammonia emissions from land spreading with anaerobic digestates from biogas production. *Atmospheric Environment* 84: 35–38.

K. Lasaridi, I. Protopapa, M. Kotsou, G. Pilidis, T. Manios, and A. Kyriacou. 2006. Quality assessment of compost in the Greek market: The need for standards and quality assurance. *Journal of Environmental Management* 80: 58–65.

K.E. Lasaridi and E.I. Stentiford. 1998. A simple respirometric technique for assessing compost stability. *Water Research* 32: 3717–3723.

A. Laurent, I. Bakas, J. Clavreul, A. Bernstad, M. Niero, E. Gentil, M.Z. Hauschild, and T.H. Christensen. 2014. Review of LCA studies of solid waste management systems—Part I: Lessons learned and perspectives. *Waste Management* 34:573–588.

Y. Liang, J.J. Leonard, J.J.R. Feddes, and W.B. McGill. 2006. Influence of carbon and buffer amendment on ammonia volatilization in composting. *Bioresource Technology* 97: 748–761.

E. Liwarska-Bizukojc and S. Ledakowicz. 2003. Stoichiometry of the aerobic biodegradation of the organic fraction of municipal solid waste (MSW). *Biodegradation* 14: 51–56.

T. Lleó, E. Albacete, R. Barrena, X. Font, A. Artola, and A. Sánchez, 2013. Home and vermicomposting as sustainable options for biowaste management. *Journal of Cleaner Production* 47: 70–76.

T. Manios. 2004. The composting potential of different organic solid wastes: Experience from the island of Crete. *Environment International* 29: 1079–1085.

K.L. McLachlan, C. Chong, and R.P. Vorony. 2004. Assessing the potential phytotoxicity of digestates during processing of municipal solid waste by anaerobic digestion: a comparison to aerobic digestion. In: *Proceedings of Sustainability of Horticultural Systems*, Acta Horticulture 638, Bertschinger, L., and Anderson, J.D. (eds.).

C.M. Mehta, U. Palni, I. Franke-Whittle, and A. Sharma. 2014. Compost: Its role, mechanism and impact on reducing soil-borne plant diseases. *Waste Management* 34: 607–622.

D. Meyer-Kohlstock, G. Hädrich, W. Bidlingmaier, and E. Kraft. 2013. The value of composting in Germany—Economy, ecology, and legislation. *Waste Management* 33: 536–539.

G. Myhre, D. Shindell, F.M. Bréon, W. Collins, J. Fuglestvedt, J. Huang, D. Koch et al. 2013. Anthropogenic and natural radiative forcing. In *Climate Change 2013: The Physical Science Basis. Contribution of Working Group I to the Fifth Assessment Report of the Intergovernmental Panel on Climate Change*, Stocker, T.F., Qin, D., Plattner, G.-K., Tignor, M., Allen, S.K., Boschung, J., Nauels, A., Xia, Y., Bex, V., and Midgley, P.M. (eds.). Cambridge University Press, Cambridge.

S. Ntougias, K.K. Papadopoulou, G.I. Zervakis, N. Kavroulakis, and C. Ehaliotis. 2008. Suppression of soil-borne pathogens of tomato by composts derived from agro-industrial wastes abundant in Mediterranean regions. *Biology and Fertility of Soils* 44: 1081–1090.

B. Pharand, O. Carisse, and N. Benhamou. 2002. Cytological aspects of compost mediated induced resistance against *Fusarium* crown and root rot in tomato. *Phythopathology* 92: 424–438.

I. Piug-Ventosa. 2008. Charging systems and PAYT experiences for waste management in Spain. *Waste Management* 28: 2767–2771.

M. Pognani. 2012. Organic matter evolution in biological solid-waste treatment plants. Raw waste and final product characterization. PhD Thesis. Chemical Engineering Department, Universitat Autònoma de Barcelona, Barcelona, Spain.

M. Pognani, R. Barrena, X. Font, F. Adani, B. Scaglia, and A. Sánchez. 2011. Evolution of organic matter in a full-scale composting plant for the treatment of sewage sludge and biowaste by respiration techniques and pyrolysis-GC/MS. *Bioresource Technology* 102: 4536–4543.

S. Ponsá, T. Gea, L. Alerm, J. Cerezo, and A. Sánchez. 2008. Comparison of aerobic and anaerobic stability indices through a MSW biological treatment process. *Waste Management* 28: 2735–2742.

S. Ponsá, T. Gea, and A. Sánchez. 2010a. The effect of storage and mechanical pretreatment on the biological stability of municipal solid wastes. *Waste Management* 30: 441–445.

S. Ponsá, T. Gea, and A. Sánchez. 2010b. Different indices to express biodegradability in organic solid wastes. *Journal of Environmental Quality* 39: 706–712.

S. Ponsá, B. Puyuelo, T. Gea, and A. Sánchez. 2011. Modelling the aerobic degradation of organic wastes based on slowly and rapidly degradable fractions. *Waste Management* 31: 1472–1479.

B. Puyuelo, S. Ponsá, T. Gea, and A. Sánchez. 2011. Determining C/N ratios for typical organic wastes using biodegradable fractions. *Chemopshere* 85: 653–659.

R. Quirós. 2014. Environmental assessment of technologies to treat unsorted municipal solid waste and the organic matter to produce compost and its application in horticultural crops. PhD thesis. Universitat Autònoma de Barcelona, Barcelona, Spain.

R. Quirós, G. Villalba, P. Múñoz, X. Font, and X. Gabarrell. 2014. Environmental and agronomical assessment of three fertilization treatments applied in horticultural open field crops. *Journal of Cleaner Production* 67: 147–158.

L. Ruggieri, T. Gea, A. Artola, and A. Sánchez. 2009. Air filled porosity measurements by air pycnometry in the composting process: a review and a correlation analysis. *Bioresource Technology* 100: 2655–2666.

A. Saer, S. Lansing, N.H. Davitt, and R.E. Graves. 2013. Life cycle assessment of a food waste composting system: Environmental impact hotspots. *Journal of Cleaner Production* 52: 234–244.

E. Sevigné, C.M. Gasol, R. Farreny, X. Gabarrell, and J. Rieradevall. 2013. CO2ZW®: Carbon footprint tool for municipal solid waste management for policy options in Europe. Inventory of Mediterranean countries. *Energy Policy* 56: 623–632.

R.P. Singh, P. Singh, A.S.F. Araujo, M.H. Ibrahim, and O. Sulaiman. 2011. Management of urban solid waste: Vermicomposting as a sustainable option. *Resources, Conservation and Recycling* 55: 719–729.

S.R. Smith. 2009. A critical review of the bioavailability and impacts of heavy metals in municipal solid waste composts compared to sewages ludge. *Environment International* 35: 142–156.

S.R. Smith and S. Jasim. 2009. Small-scale home composting of biodegradable household waste: Overview of key results from a 3-year research programme in West London. *Waste Management & Research* 27: 941–950.

U. Sonesson, A. Björklund, M. Carlsson, and M. Dalemo. 2000. Environmental and economic analysis of management systems for biodegradable waste. *Resources, Conservation and Recycling* 28: 29–53.

Standing Committee of the National People's Congress (NPSCS). 2008. *Circular Economy Promotion Law of the People's Republic of China.* Order of the President of the People's Republic of China No. 4. Beijing, China.

G. Tchobanoglous and F. Kreith. 2002. *Handbook of Solid Waste Management.* McGraw-Hill, New York.

A. Trémier, A. de Guardia, C. Massiani, E. Paul, and J.L. Martel. 2005. A respirometric method for characterising the organic composition and biodegradation kinetics and the temperature influence on the biodegradation kinetics, for a mixture of sludge and bulking agent to be cocomposted. *Bioresource Technology* 96: 169–180.

United Nations Environmental Program (UNEP). 2011. *Decoupling Natural Resource Use and Environmental Impacts from Economic Growth.* Fischer-Kowalski, M., Swilling, M., von Weizsäcker, E.U., Ren, Y., Moriguchi, Y., Crane, W., Krausmann, F., Eisenmenger, N., Giljum, S., Hennicke, P., Romero Lankao, P., Siriban Manalang, A., and Sewerin, S. (eds.). Working Group on Decoupling to the International Resource Panel. Available from: http://www.unep.org/resourceefficiency/Home/Policy/SCPPolicies/National ActionPlansPovertyAlleviation/NationalActionPlansIntroduction/CircularEconomy/ tabid/78389/Default.aspx (accessed July 2014).

S.T. Wagland, S.F. Tyrrel, A.R. Godley, and R. Smith. 2009. Test methods to aid in the evaluation of the diversion of biodegradable municipal waste (BMW) from landfill. *Waste Management* 29: 1218–1226.

Zero Waste International Alliance. Available from: http://zwia.org/ (accessed July 2014).

G. Zinati. 2005. Compost in the 20th century: A tool to control plant diseases in nursery and vegetable crops. *Hortechnology* 15: 61–66.

F. Zucconi, A. Pera, M. Forte, and M. de Bertoldi. 1981. Evaluating toxicity of immature compost. *Biocycle* 22: 54–57.

6 Biogas from Wastes
Processes and Applications

Maryam M. Kabir, Gergely Forgács, Mohammad J. Taherzadeh, and Ilona Sárvári Horváth

CONTENTS

6.1   INTRODUCTION

Municipal solid waste (MSW) is the waste generated in a community; hence, MSW includes residential (e.g., households), commercial (e.g., from stores, markets, shops, hotels, and restaurants), and institutional (e.g., schools and hospitals) waste. According to a World Bank report, 1.3 billion tons of MSW were generated by 3 billion urban residents all over the word in 2012, which is expected to increase to 2.2 billion tons by 2025 (Hoornweg and Bhada-Tata 2012). Typically, 50%–70% of the MSW is classified in a broad category known as organic or biodegradable waste, including paper, cardboard, garden waste, and food waste. Commonly, this fraction of MSW is referred to as the organic fraction of the MSW (OFMSW) (Hoornweg and Bhada-Tata 2012).

If the OFMSW is not collected and treated, it will biologically decompose in the environment, leading to contamination of land, water, and air. According to current environmental regulations, traditional disposal of organic wastes in landfills is restricted. The legislation in the European Union restricts the amount of biodegradable waste entering landfills (National Archives 1996). Additionally, the uncontrolled degradation of the organic matter through microbial activities at dumping sites results in the emission of methane (CH_4) and carbon dioxide (CO_2) into the atmosphere, contributing to global warming (Baldasano and Soriano 2000; Chynoweth et al. 2001). Hence, the controlled production of biogas from these kinds of wastes simultaneously reduces the demand for fossil fuels and lowers the emission of greenhouse gases, while it treats and stabilizes the waste.

Anaerobic digestion (AD) is a biological process where the organic material is converted into mainly methane and carbon dioxide in an oxygen-free environment. The main advantage of AD compared to other biological processes is that many types of organic materials can be utilized as a substrate, which is mainly composed of proteins, fats, and carbohydrates (Gunaseelan 1997).

Within the last few decades, AD has become an attractive technology for treating the OFMSW (De Baere 2000). Since the OFMSW composition varies greatly by season, geographical area, and the socioeconomic level of the community, various technological solutions have been developed (Kayhanian et al. 2007). In developing countries, biogas is produced in simple, small-scale, "home-made" reactors and it is used for cooking, while in developed countries centralized industrial-scale digesters are developed, which generate heat and electricity for the community, or the produced biogas is upgraded to biomethane and utilized as a substitute for natural gas.

6.2 AD PROCESS

Biogas formation is a result of several interdependent, complex, sequential, and parallel biological reactions, in which the products generated by one group of microorganisms serve as substrates for the next group (Noykova et al. 2002). The degradation of the organic substrate in the AD process is typically divided into four steps: hydrolysis, acidogenesis, acetogenesis, and methanogenesis (Figure 6.1), where the initial point as well as the degradation pathways depend on the nature of the organic matter utilized. H_2, CO_2, and acetate formed during the degradation steps are converted into methane and carbon dioxide in the final step, making up the biogas (Ziemiński and Frąc 2012). The microorganisms involved in this process are bacteria and archaea, which partially have a syntrophic relation to each other with different requirements on the environment (Deublein and Steinhauser 2008).

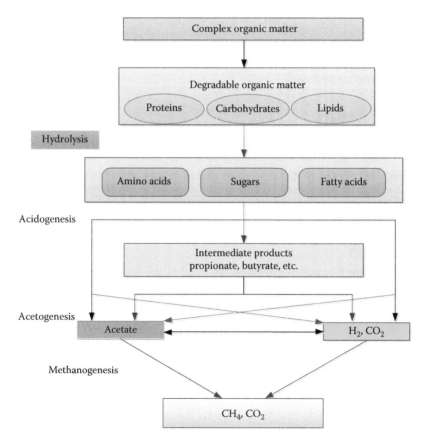

FIGURE 6.1 The main degradation pathways in anaerobic digestion. (Data from Batstone, et al., *Water Sci Technol*, 45, 65–73, 2002.)

Furthermore, it should be noted that the biochemistry and microbiology of the AD of the complex organic substrates are still not fully understood (Van Lier et al. 2001; Michaud et al. 2002). Therefore, anaerobic degradation is still an interesting and challenging subject, specifically relating to the biochemical and microbiological aspects.

6.2.1 HYDROLYSIS

Hydrolysis is the first stage in the AD process, which includes extracellular biological processes mediating the transformation of the particulate organic matter into solubilized monomers (Batstone et al. 2002). In this step, carbohydrates, proteins, and lipids are disintegrated into macromolecules, namely, monosaccharides, amino acids, and fatty acids, respectively, and then, hydrolyzed to soluble compounds. The enzymes involved in this process are extracellular enzymes (Table 6.1), including cellulases, proteases, and lipases excreted by the facultative anaerobes (Parawira et al. 2005; Taherzadeh and Karimi 2008). The hydrolysis step can act as the rate-limiting step in the AD process, if the substrate, which is fed to the digester, is difficult to degrade (Vavilin et al. 1996). However, if it is a readily degradable substrate, the rate-limiting step will be acetogenesis and methanogenesis (Björnsson et al. 2001). When the substrate is hydrolyzed, it becomes available for cell transport and can be degraded by the fermentative bacteria in the following acidogenesis step.

6.2.2 ACIDOGENESIS

The monomers and the soluble oligomers formed in the hydrolytic phase are utilized as substrates by the obligate and facultative anaerobes, and will mainly be converted into short chain organic acids, C1–C5 molecules, such as valeric acid, butyric acid, propionic acid, acetic acid, formic acid, hydrogen, and alcohols.

TABLE 6.1

Important Groups of Enzymes Involved in the Hydrolysis Step and Their Functions

Enzymes	Substrate	Degradation Products
Cellulase	Cellulose	Cellobiose and glucose
Hemicellulase	Hemicelluloses	Sugars, such as glucose, xylose, and mammose arabinose
Pectinase	Pectin	Sugars, such as galactose and arabinose
Proteinase	Protein	Amino acids
Lipase	Fat	Fatty acids, glycerol, and polygalacturonic acid

Source: Schnürer, A., and Jarvis, Å. Mikrobiologisk handbok för biogas anläggningar. *Rapport U2009* 3, 2009.

Acidogenesis of sugars and amino acids is performed without involving an electron acceptor or donor, while long chain fatty acids (LCFA) are oxidized using hydrogen ions as electron acceptors. It should be mentioned that the concentration of the intermediately formed hydrogen ions regulates the expected products in this step (Deublein and Steinhauser 2008). In a stable anaerobic digester, with low partial pressure of hydrogen, the main degradation pathway results in the formation of acetate, carbon dioxide, and hydrogen. This degradation pathway is more favorable since it gives a higher energy yield for the microorganisms, and the methanogenic microorganisms can directly utilize the products. However, when the concentration of the hydrogen and the formate is high, intermediates, such as volatile fatty acids (VFA), and alcohols are formed (Batstone et al. 2002). In this case, the products from the acidogenesis step will contain about 51% acetate, 19% H_2/CO_2, and 30% reduced products (VFA, alcohols, or lactate; Batstone et al. 2002). The acidogenesis step is usually considered as the fastest step in the AD of complex organic matter (Vavilin et al. 1996).

6.2.3 ACETOGENESIS

The intermediate products of acidogenesis (i.e., VFA, alcohols, or lactate), excluding acetate, are further oxidized to acetate and hydrogen in the acetogenic step by the action of the obligate hydrogen-producing acetogens. Acetate, hydrogen, and carbon dioxide are appropriate substrates for the methanogenic biomass (Deublein and Steinhauser 2008). It is known that acetogenesis reactions are thermodynamically feasible only when the concentration of the hydrogen in the digester medium is low (lower than 10^{-5} bar). Therefore, the syntrophic association between acetogens (hydrogen producers) and hydrogenotrophic methanogens (hydrogen utilizers) is of vital importance to regulate the hydrogen concentration and, hence, to the whole digestion process (Schink 1997; Chandra et al. 2012).

6.2.4 METHANOGENESIS

The products of acidogenic and acetogenic reactions are converted into biogas by the methanogenic archaea in the final methanogenesis step. The majority (two-thirds) of the methane producers perform biogas production through the acetoclastic pathway. There, the two carbons of the acetate will be split; one will be oxidized to carbon dioxide and the other will be reduced to methane ($CH_3COOH \rightarrow CH_4 + CO_2$) (Klass 1984; Zinder 1984; Garcia et al. 2000). These groups of methanogens are known to have the slowest growth rate and are the most sensitive organisms working in the digester. The acetate oxidation pathway becomes more favorable at high temperatures and low partial pressure of hydrogen (Schink 1997). The rest (one-third) of the methane is produced through the hydrogenotrophic pathway using H_2 and CO_2 (Pavlostathis and Giraldo-Gomez 1991; Vogels et al. 1988). The hydrogenotrophic pathway is vital for the whole digestion process, since it is in charge of maintaining the low hydrogen pressure by removing hydrogen, which is required for the acetate production (Harper and Pohland 1986). On the other hand, the partial pressure of hydrogen has to be above a minimum level (higher than 10^{-6} bar) for the reaction to be exergonic (Bruni 2010). Lastly, very small groups of the methanogens use the methyl group as a precursor for the methane production ($CH_3OH + H_2 \rightarrow CH_4 + H_2O$) (Astals et al. 2013).

In the anaerobic digestion (AD) system, methanogens, compared to bacteria, are known to be the most sensitive group of microorganisms affected by the environmental and operational changes. They are strictly anaerobes; hence, the presence of molecular oxygen is highly toxic for them. Even the presence of an inorganic source of oxygen, such as nitrates, might inhibit their growth, due to an increase in the oxidation–reduction potential in the reactor. Methanogenic archaea also show a high sensitivity to ammonia; however, ammonia inhibition for acetoclastic methanogens is stronger than that for the hydrogenotrophic microorganisms (Van Velsen 1979; Hansen et al. 1998). It has been suggested that free ammonia (NH_3) is an active component causing ammonia inhibition (Hashimoto 1986a; Angelidaki and Ahring 1993), since free ammonia can diffuse into the cell, leading to a proton imbalance or causing a potassium loss (Chen et al. 2008). In addition, they have the longest generation time, between 2 and 25 days, compared to the other microorganisms in the digester, which is the reason that this step is considered to be the rate-limiting step for the easily hydrolyzed materials.

6.3 OPERATIONAL AND ENVIRONMENTAL FACTORS AFFECTING AD PERFORMANCE

AD is a complex biological process, which is highly influenced by various factors including feedstock characteristics, reactor design, and operational conditions. Among the operational conditions such as pH, temperature, organic loading rate (OLR), and retention time, the micro- and macronutrient availability are vital parameters. Thus, to reach high efficiency within the process, these parameters should be efficiently controlled and maintained within the optimum range for all the microorganisms involved in the AD (Ward et al. 2008).

6.3.1 TEMPERATURE

Temperature is one of the physical factors that has a significant effect on the performance of the digestion process. Changes in the temperature directly affect the kinetics and the thermodynamics of the reactions as well as the growth rate and the metabolism of the microorganisms. Therefore, it determines the degradation pathway and biomass dynamics throughout the digestion process. Among the microorganisms present in the AD systems, methanogenic archaea are the ones that most strongly rely on temperature. Temperature variations might be positive for certain groups and harmful to other groups. AD can be applied, using a wide range of temperatures from psychrophilic (<20°C) to extreme thermophilic conditions (>60°C) (Van Lier et al. 1997; Lepistö and Rintala 1999; Kashyap et al. 2003). However, the most common AD digesters operate under mesophilic (around 35°C) or thermophilic (around 55°C) conditions. Currently, about 60% of the MSW digesters run at mesophilic temperatures due to the higher process stability and the lower energy requirements achieved during these conditions (Astals et al. 2013). Increasing the temperature can increase the solubility of the organic compounds; decrease the viscosity, reducing the energy requirements for mixing; improve the diffusivity of the soluble substrate; increase the chemical and biological reaction rates; increase the death rate of the pathogenic bacteria, especially under thermophilic conditions; and increase the degradation of the LCFA, VFA, and other intermediates (Boe 2006). The disadvantages of the high temperatures are that it decreases the pKa of ammonia, thus increasing the fraction of the free ammonia, which is inhibitory to microorganisms; it increases the pKa of VFA, thus increasing its un-dissociated fraction, especially at low pH (4–5) such as in the acidogenic reactor (Boe 2006). Moreover, the diversity of the microorganism under thermophilic conditions is much lower than the mesophilic temperature (Schnürer and Jarvis 2009). These are the reasons why the thermophilic process is generally more sensitive to inhibition.

Anaerobic digesters are assumed to operate at constant temperatures. However, there are situations when the digester is subjected to changes in the temperature. A farm bioreactor might be subjected to temperature fluctuations due to the variation in the outdoor temperature, especially in northern climates (Masse et al. 2003; Alvarez et al. 2006). Bioreactors working during daily temperature fluctuations have been investigated in thermophilic (El-Mashad et al. 2003), mesophilic (Chayovan et al. 1988), and psychrophilic (Masse et al. 2003; Alvarez and Lidén 2008a) ranges. The results of these studies showed that the digestion performance decreases significantly as the operating temperature is reduced; however, the digesters were able to recover and remain stable as the temperature returned to the optimal higher range.

6.3.2 NUTRIENTS

There are many organic and inorganic substances, which are indispensable for the synthesis and growth of the enzymes and cofactors involved in the biochemical and metabolic pathways of the AD microorganisms. Nutrients are classified into two groups: the macronutrients and the micronutrients. In order to have a well-balanced system, both macro- and micronutrients need to be present in the digester in the right portions and concentrations (Mara and Horan 2003).

The concentration of the nutrients is of vital importance since even a small shortage of the nutrient may inhibit the metabolism of the microorganisms. Therefore, in case of feedstock nutrient deficiencies, additional nutrients must be provided to stimulate the digestion process. Additionally, if the concentration surpasses a certain limit, it might slow down the growth or even cause severe inhibition in the system (Chen et al. 2008).

Important micronutrients known to be essential for the growth and metabolism are iron (Fe), nickel (Ni), cobalt (Co), molybdenum (Mo), and tungsten (W; Zandvoort et al. 2006). Several functions of the anaerobic microorganisms are dependent on the presence of these micronutrients (Oleszkiewicz and Sharma 1990), since they play an important role in the formation of the active sites for several key enzymes involved in the digestion process. The optimum micronutrient requirements in the digester have to be optimized based on the inherent micronutrient concentrations of the substrate, inocula, and the general process conditions within the digester (Jagadabhi 2011). Fundamental macronutrients such as carbon (C), nitrogen (N), phosphorus (P), and sulfur (S) are essential for the growth and multiplication of the microorganisms. Furthermore, the nitrogen content of the feedstock has an important role in this process, since it results in a neutral pH stability by releasing ammonium ions (Speece 1983; Gunnerson et al. 1986).

6.3.3   C/N Ratio

The composition of the feedstock is also an important factor in the AD process in terms of the process stability and gas production. Apart from the nutritional elements and vitamins that are responsible for regulating the activity of the microbial enzyme systems, the ratio of carbon to nitrogen (C/N ratio) is also considered to be of vital importance. If the C/N ratio is too low, the process can easily suffer from ammonia inhibition and consequently leads to process failure (Alvarez and Lidén 2008b; Shanmugam and Horan 2009). On the other hand, since nitrogen is necessary for the growth of the microorganisms, if the C/N ratio is too high, the microbial community of the AD process may experience nitrogen deficiency, which can lead to insufficient consumption of the carbon source and, consequently, process termination (Yen and Brune 2007; Resch et al. 2011). The optimal C/N ratio in the different AD systems varies depending on the process conditions and feedstock characteristics. The values of C/N ratio reported in the literature vary between 10 and 30, with an optimum range between 15 and 25 (Yadvika et al. 2004; Hansson and Christensson 2005; Yen and Brune 2007; Liu et al. 2008). There is a tight correlation between the formation of fatty acids and the C/N ratio (Liu et al. 2008). An increase in the C/N ratio within the range 10–30 stimulates the production of fatty acids during the process (Callaghan et al. 2002; Yen and Brune 2007; Liu et al. 2008). If the level of the VFA is not too high, it can also increase the methanogenic activity (Callaghan et al. 2002; Sosnowski et al. 2003; Yen and Brune 2007).

6.3.4   pH and Alkalinity

The microorganisms in the anaerobic digester are sensitive to the pH, and different microorganisms in different steps of the AD process have widely varying requirements on pH for best growth (Yadvika et al. 2004). Fermentative bacteria can live within a wide range of pH, that is, between 4 and 9, although the optimum pH range for this group of microorganisms is reported to be around 5–6. In contrast, the methanogenic archaea present a narrow survival pH range, which is stated to be between 6.0 and 8.5, with an optimum around a neutral pH (Zehnder and Wuhrmann 1977; Huser et al. 1982; Boe 2006; Appels et al. 2008). Methanogens are defined as the key organisms of the process; therefore, the digesters are designed and operated to maintain a neutral pH in favor of this sensitive group, since a pH outside of the neutral range can severely affect the methanogens and, consequently, result in imbalances in the system (Dague 1968; Schnürer and Jarvis 2009). The changes in the pH during the digestion depend on the amount of intermediates, such as VFAs present in the digester. As the pH decreases, the methanogens' activity is affected;

however, the acid-forming bacteria are still able to grow and degrade more organic matter. Therefore, VFAs will accumulate, which will result in further inhibition of methanogens and, finally, failure of the process (Mosey and Fernandes 1989). The acetoclastic methanogens are more sensitive to a decrease in the pH than the hydrogenotrophic methogens (Brummeler 1993).

In order to maintain an optimum pH in the digester, it is important to have a high stable alkalinity. The alkalinity, or buffer capacity, is the capacity of the digester medium to moderate the pH changes by neutralizing the acids formed during the process. In the AD process, the alkalinity is mainly provided by a few acid-based pairs, including carbon dioxide–bicarbonate, ammonium–ammonia, and dihydrogen phosphate–hydrogen phosphate. The major buffer produced in the anaerobic digesters is bicarbonate (HCO_3^-), with a pKa of 6.3, and the main acid is VFA, with an aggregate pKa around 4.8 (Boe 2006). In addition, the protein-rich substrate releases ammonia when degraded, which contributes to the alkalinity as well. For the process stability, the recommended VFA: alkalinity ratio should be maintained at less than 0.3 (Gerardi 2003; Schnürer and Jarvis 2009).

6.3.5 HYDRAULIC RETENTION TIME AND OLR

The OLR is regarded as the daily added organic material, expressed as the volatile solids (VS) per reactor volume of the digester (kg VS $m^{-3}day^{-1}$). However, for liquid feedstock, this is measured based on the chemical oxygen demands (COD); thus, the OLR is expressed as kg COD $m^{-3}day^{-1}$ (Vandevivere et al. 2003). Generally, during the start-up period, the process needs a lower OLR, while a well-balanced process can survive on a higher OLR. Using a too high OLR normally results in the accumulation of VFAs or other inhibitors in the digester, which may finally terminate the methane production (Mata-Alvarez et al. 2000; Bouallagui et al. 2004). On the other hand, feeding the digester with a too low organic loading (i.e., underloaded system) is not economical, since the capacity of the digester is not entirely used. The biological performance of the AD system is very sensitive to the OLR and waste composition (Sharma et al. 1999; Zuo et al. 2013).

The rate of the bioconversion of the substrate to biogas is highly dependent on the retention time. The retention time is normally expressed as the hydraulic retention time (HRT), which refers to the average time spent by the input slurry inside the digester before it comes out. According to the chemostat theory, at steady-state conditions, the specific growth rate of a microorganism and the residence time are inversely proportional (Doran 1995; Villadsen et al. 2011). A disturbance in the system occurs if the inverse of the residence time exceeds the maximum specific growth rate of the microorganism. In this situation, there is a risk for washout of the active bacterial population by decreasing the retention time (Yadvika et al. 2004). Apparently, shorter retention times are favorable to minimize the system cost, by reducing the size of the digester and increasing the productivity of the process (Chandra et al. 2012). However, a decrease in the residence time must be carried out without stressing the fermentation process. The minimal HRT is directly linked to the type of substrate to be digested. In general, HRT becomes a more significant factor when the feedstock is complex and difficult to digest (Hashimoto 1982; Forgács 2012).

6.4 PROCESS IMPROVEMENT

6.4.1 SUBSTRATE

The characteristics of the solid waste have a significant effect on the AD process, in terms of biogas production potential and biodegradability. The characteristics of the organic food wastes may differ depending on the collection method, weather, season, and cultural habits of the community. The biodegradability and bioconversion of the organic waste to biogas have a high correlation to the amount of the main components in the waste, such as lipids, proteins, and carbohydrates, including cellulose, hemicelluloses, and lignin (Hartmann and Ahring 2006). The expected theoretical methane potential can be determined using the general chemical formulas for carbohydrates ($C_6H_{10}O_5$), proteins ($C_5H_7O_2N$), and lipids ($C_{57}H_{104}O_6$), which are 0.42, 0.50, and 1.01 m^3 CH_4 kg^{-1} VS, respectively (Møller et al. 2004). Although the methane yield of lipids is higher than that of the other components, studies have reported that the rate of hydrolysis for kitchen waste with an excess amount of lipids is slow compared to less fatty kitchen wastes. Since lipids can be adsorbed onto the solid surfaces of other components, this will interrupt the hydrolysis process by reducing the accessibility for the hydrolytic enzymes (Neves et al. 2008).

The composition of the food waste also determines the C/N ratio present in the waste. As mentioned in Section 6.3.3, a high C/N ratio is an indication of nitrogen deficiency, which is not suitable for bacterial growth. Therefore, the methane production rate and the degradability of solids will be low. On the other hand, the degradation of waste with a very low C/N ratio results in ammonia accumulation, which is toxic to the AD microorganisms (Hartmann and Ahring 2006).

Apart from the substrate composition, the substrate particle size also has a significant effect on the AD process of solid wastes, particularly during the hydrolysis step. This is because a smaller particle size offers a larger accessible surface area for enzymatic attacks (Hartmann and Ahring 2006; Hajji and Rhachi 2013). Table 6.2 summarizes the methane potential of the organic wastes and by-products in the AD.

6.4.2 PRETREATMENT

MSWs are mostly composed of a wide variety of objects of different sizes and shapes. The received wastes are subjected to size reduction, in order to obtain a reasonably uniform final product, which has a greater density than the original form. Inert materials, for example, sand, clay, and glass, and floating materials, such as plastics, also need to be eliminated prior to the AD process. The size reduction of the solid waste is carried out by shredding machines that are capable of converting the waste into a form that is easier to handle for processing. The machines that are used for size reduction of MSWs are called shredders, crushers, or millers. The reduction of the particle size has shown a significant improvement in the AD of solid waste (Hartmann and Ahring 2006; Hajji and Rhachi 2013). A study on the effect of the particle size on the AD of food wastes showed that the bigger particle size leads to a decrease in the digestion rate (Kim et al. 2000).

TABLE 6.2
Methane Potential of Some Organic Waste

Waste	Methane Yield ($m^3 CH_4 kg^{-1} VS$)	References
Industrial and commercial waste		
Expired food	0.47–1.10	Braun et al. (2003)
Sludge from distilleries	0.40–0.47	Braun et al. (2003)
Potato waste	0.69–0.89	Braun et al. (2003)
Molasses	0.31	Angelidaki and Ellegaard (2003)
Edible oil sludge	1.10	Braun et al. (2003)
Municipal waste		
Household waste	0.40–0.50	Angelidaki and Ellegaard (2003)
Garden waste	0.10–0.20	Angelidaki and Ellegaard (2003)
Paper	0.08–0.37	Owens and Chynoweth (1993)
Market waste	0.90	Braun et al. (2003)
Municipal solid waste	0.20–0.22	Chynoweth et al. (1993)
Banana peel	0.27–0.32	Gunaseelan (2004)
Citrus waste	0.43–0.73	Gunaseelan (2004)
Vegetable waste	0.19–0.40	Gunaseelan (2004)
Animal and slaughterhouse waste		
Animal fat	1.00	Braun et al. (2003)
Stomach and gut contents	0.4–0.46	Ahring et al. (1992)
Rumen content	0.35	Braun et al. (2003)
Blood	0.65	Braun et al. (2003)
Agricultural waste		
Cow manure	0.15–0.30	Angelidaki and Ellegaard (2003)
Swine manure	0.30–0.51	Ahring et al. (1992); Møller et al. (2004)
Poultry manure	0.30	Ahring et al. (1992)
Straw and other plant residues	0.15–0.36	Angelidaki and Ellegaard (2003); Møller et al. (2004)
Green plant, crops, grain	0.18–0.28	Angelidaki and Ellegaard (2003)
Sugarcane	0.23–0.30	Chynoweth et al. (1993)
Sorghum	0.26–0.39	Chynoweth et al. (1993)

If the OFMSW is blended with lignocellulosic feedstock such as yard waste, tree trimmings, woods, and cardboards, it may need a further pretreatment process prior to the AD. Since the hydrolysis is considered as a rate-limiting step in the biogas production, the recalcitrance structure of these types of biomass prevents the adequate physical contact between the hydrolytic enzymes and the substrate (Taherzadeh and Karimi 2008). Therefore, developing a suitable pretreatment method to enhance the accessibility of the enzymes to the biomass is still an interesting field for further investigations. According to literature, a proper pretreatment can increase the methane yield by improving the hydrolysis. Pretreatments for lignocelluloses, such as woody biomass, straw, grasses, and OFMSW, can be mechanical, chemical, biological, or physiochemical methods (McMillan 1994; Sun and Cheng 2002; Taherzadeh and Karimi 2008; Zheng et al. 2009; Cheng and Hu 2010; Ibrahim et al. 2011).

6.4.3 CODIGESTION

Codigestion is another interesting alternative to improve the biogas production. Codigestion consists of AD of two or more waste streams, which are simultaneously digested together. The improvement in the biogas production is a consequence of the positive interaction between the substrates. Advances in agricultural waste digestion have resulted in the concept of centralized AD, where many farms collaborate to feed a single, large-scale digestion plant (Mata-Alvarez et al. 2000). The system principally aims to minimize the generated waste, together with energy production for local utilization for the agricultural sectors. In addition, codigestion may improve the biogas yield due to the synergistic effect between the feedstock and also because of giving a better nutritional balance in the digester. For instance, the digestion of manure usually results in a relatively low methane yield. However, codigestion of manure with different substrates significantly enhances the biogas production (Callaghan et al. 2002; Fernández et al. 2005; Kaparaju and Rintala 2005). The methane production of manure was $0.13–0.15$ m^3 CH_4 kg^{-1} VS, while the addition of a cosubstrate (potato waste) with a feed ratio of 80–20 (pig manure to potato waste) clearly increased the methane production to $0.30–0.33$ m^3 CH_4 kg^{-1} VS.

Generally, codigestion takes place in a wet, single-stage process using CSTR or continuously stirred tank reactor (Bouallagui et al. 2005). Recently, the digestion of animal by-product and slaughterhouse waste has been considered as an attractive waste management alternative (Ahring et al. 1992; Banks and Wang 1999). The use of rumen, stomach, blood waste, and sludge from the slaughterhouse wastewater treatment as substrates is becoming common in the codigestion plants in Denmark and Sweden (Hedegaard and Jaensch 1999; Edström et al. 2003; Murto et al. 2004). Digestion of slaughterhouse waste without the addition of a cosubstrate is a sensitive process and tends to fail (Edström et al. 2003; Wang and Banks 2003). This is due to the accumulation of the nonprotonized ammonia, which results from the degradation of nitrogen-rich protein components of the slaughterhouse waste (Hashimoto 1986b; Koster and Lettinga 1988; Hansen et al. 1998). In order to solve the ammonium inhibition challenges, codigestion with other substrates was proposed. The results of different experimental studies with a focus on the codigestion of slaughterhouse waste with manure and agricultural waste (Murto et al. 2004; Pages-Diaz et al. 2014) as well as food waste (Edström et al. 2003) proved that the codigestion led to a more stable process and higher biogas yield.

Fruit and vegetable wastes compose a great proportion of MSW, which showed a high potential to be used as feedstock for the AD. This feedstock is also reported to be suitable for codigestion with the nitrogen-rich slaughterhouse waste, due to its low nitrogen and phosphorus content (Callaghan et al. 1999; Sharma et al. 2000; Misi and Forster 2001).

6.5 TYPES OF AD REACTORS FOR ORGANIC SOLID WASTES

Anaerobic digesters are designed from small-scale and simple reactors for household purposes and for laboratory investigations to large-scale and industrial systems for energy generation. AD systems can be classified and compared based on the biological and technical performance and characteristics:

- Reactor configuration (batch vs. continuous)
- Solid content of the feedstock (wet vs. dry)
- Number of reactors (single-, two-, or multistage systems)

6.5.1 BATCH AND CONTINUOUS SYSTEMS

Batch reactors are fed once at the start-up, and all the feedstock is digested at once. In batch digestion, the material stays in the digester during the entire digestion period; no new material is added, and no material is taken out during the process. The methane production is generally the highest at the beginning and decreases towards the end of the process. When the material is completely digested, the digester is emptied before it is fed with a new batch of substrates again.

In continuous systems, the feed is charged and the digestate is discharged continuously, allowing a steady state to be reached in the reactor with a constant gas yield. This kind of system is possible when the substrate is a fluid or at least can be pumped for continuous feeding. Otherwise, a semicontinuous process is applied with a discrete amount of feed several times a day.

Two commonly used continuously-fed reactors are CSTR (in which the contents of the reactor are mixed by mechanical agitation or effluent or biogas recirculation) and plug-flow reactors (PFR), where the contents of the reactor are pushed along a horizontal reactor. Although continuous reactors have higher operating costs than the batch reactors due to the pumping requirements, these reactors are able to maintain microorganisms within the system at a steady-state level, thus, preventing a lag phase associated with the adaptation and growth of the microorganisms in the batch reactors (Chaudhary 2008).

6.5.2 WET AND DRY AD

The choice of the digestion technology in the treatment of solid waste is directly connected to the total solid (TS) content of the treated material. Wet systems are designed for processing dilute organic slurry with a TS content of 10%–15% or less. For the feedstock with TS higher than 15%, the slurry will be diluted with fresh, recirculated process water, or will be codigested with other organic matter containing a lower amount of TS content. The most common reactor configuration used for wet AD of organic slurry is CSTR (De Baere 2000). Wet AD systems have been successfully applied to treat the various ranges of low solid materials, including sewage sludge and food industry effluents; nevertheless, there are still a number of challenges that need to be overcome when using these systems to treat the OFMSW. The major challenges are as follows:

- The generation of the wet slurry from the OFMSW might result in the loss of volatile organics.
- The requirement of a mechanical stirrer, because of the fact that, the wet slurry in the digester tends to separate into layers. Then, the top layer, which is the floating layer of the scum cannot be digested since the physical contact between the enzymes and the substrate is not provided.
- It requires a relatively large digester volume, higher capacity of wastewater treatment facilities, and higher heat and energy demand.

Another possible technology to treat the OFMSW is solid-state fermentation, also called dry digestion. However, due to the high solid content (25%–40% TS) of the waste in these kinds of systems, the technical approach regarding the waste handling and treatment is fundamentally different from those applied in the wet systems (Verma 2002). The digester used in the dry digestion processes may be designed as PFRs and run without a mechanical stirrer. Nevertheless, due to the high viscosity in the dry digestion systems, heat and nutrient transfer is not as efficient as it is in the wet AD systems. Therefore, mixing is of vital importance since it guarantees adequate inoculation, and also prevents local overloading and acidifications (Wellinger et al. 1993; Luning et al. 2003). However, the conventional mechanical mixers are not used in the dry digesters; instead, mixing is performed by recirculation of the waste or biogas injection to the reactor (Luning et al. 2003).

6.5.3 NUMBER OF STAGES

Digesters are also categorized based on the number of reactors in operation. Digesters can be operated as single-, two-, or multistage reactors.

6.5.3.1 Single-Stage Process

The single-stage reactors operate in favor of both acidogenic and methanogenic microorganisms. This type of reactors can be applied for both wet and high solid digestion. Their simplicity and the lower capital and operating costs make them attractive and allow them to operate for several decades (Vandevivere et al. 2003; Kelleher 2007). The very first single-stage anaerobic digester was applied in the treatment of the MSW and industrial organic waste, which has been running over a century, using a conventional reactor from a wastewater treatment plant. The industrialization of the AD commenced in 1859, with the first digestion plant in Mumbai (previously Bombay), India, utilizing sewage sludge (McCarty 1981). The most common reactor design for the single-stage process is the CSTR. The process requires the digestate to be continuously stirred, hence, completely mixed throughout the digestion process. The OLR is proportional to the rate of effluent removed. The retention time varies between 14 and 28 days depending on the type of the substrate and the operating temperature.

6.5.3.2   Two- or Multistage Process

In general, the logic behind the two- and multistage systems is that the feedstock conversion to biogas is mediated by a sequence of biochemical reactions, which do not essentially have the same optimal environmental conditions. Providing optimal conditions for each group of microorganisms by phase separation may lead to a larger reaction rate and, consequently, a higher biogas yield (Ghosh et al. 2000).

In the two-stage reactors, the hydrolysis and acido/acetogenesis processes are separated from methanogenesis. Therefore, the first step operates under the most favorable conditions for the growth of hydrolytic and acidogenic microorganisms, whereas the second step is optimized to operate in favor of acetate- and methane-forming microorganisms (Ince 1998). In the case of difficult-to-degrade substrates, such as lignocellulosic waste, the digestion is limited by the rate of hydrolysis carried out in the first stage. However, the rate-limiting factor in the second stage is typically the rate of microbial growth (Chaudhary 2008). Phase separation offers an optimal condition in terms of nutritional needs, growth kinetics, and sensitivity to the environment for both groups of microorganisms. This allows for longer biomass retention times in the methanogenesis phase, which enhances the biogas yield (Verma 2002). This kind of digesters usually have a more stable performance than single-stage digesters, since the latter ones may suffer from process disturbances caused by the changes in the pH or ammonia accumulation, which may consequently lead to process failure (Chaudhary 2008; Rapport et al. 2008).

Multistage reactors will provide an even better phase separation for hydrolysis, acido-, aceto-, and methanogenesis using three reactors, which offers a better process control and optimization over each step, thus, increasing the methane production (Griffin 2012).

6.6   GLOBAL OVERVIEW OF THE AD DIGESTERS

6.6.1   Small-Scale AD Digesters

AD technology has been introduced to households, since it was a rather simple and cheap method for rural energy production. As a result, the number of small-scale digesters has significantly increased during the last few years; for example, in China today more than 30 million household digesters are available (Jiang et al. 2011; Thien Thu et al. 2012). Apparently, China has rapidly increased its investments in biogas infrastructure, and it is expected that the number of household biogas digesters will increase to 80 million by the year 2020, and the produced biogas will be utilized as a main source of fuel by 300 million people (NDRC 2007; Rajendran et al. 2012). India is the second leading country in terms of utilization of household digesters for biogas production. The number of household digesters operating in India is 3.8 million, followed by Nepal with 0.2 million, and Bangladesh with 60,000 digesters (Jiang et al. 2011; Thien Thu et al. 2012). It has been reported that in Europe (2012–2013), more than 7500 farm-scale biogas digesters were in operation in Germany, about 300 in Austria, 89 in Switzerland, and 82 in Denmark, followed by 53 in the United Kingdom and Sweden with 26 farm-scale plants (Raven and Gregersen 2007; Wilkinson 2011; IEA Bioenergy 2014). This shows a significant improvement in the number of digesters even in Europe compared to those that were under operation in 2005 (IEA Bioenergy 2005). In the United States, 162 farm-scale plants were in operation in 2010, supplying energy for 41,000 homes, while in Canada, 17 small-scale plants were in operation (AgSTAR 2010). However, the progress in the biogas technology for household purposes is still low in many African countries. In addition, among the existing plants, only a few small-scale biogas plants are currently in operation. The reason for this failure is the poor technical quality of construction as well as lack of knowledge on AD systems in practice (Omer and Fadalla 2003; Amigun et al. 2008).

6.6.1.1 Designs of Household Digesters

In the design of the digesters, various factors should be considered such as the type of feedstock, availability of substrate, geographical location, and climate conditions, as well as local circumstances.

For instance, a digester used in mountainous regions is designed to have less gas volume in order to avoid gas loss. For tropical countries, it is preferred to have digesters underground due to the possible utilization of geothermal energy. Out of all the different digesters developed, the fixed dome model developed in China and the floating drum model developed in India are still commonly used today. Recently, plug-flow digesters have been gaining attention due to their portability and easy operation (Rajendran et al. 2012).

6.6.2 INDUSTRIAL-SCALE DIGESTERS

As a result of the increasing demand for AD of organic wastes, several commercial anaerobic digesters have been developed over the past few years. There are several processes/technologies that have been developed and that are available on the market, especially in the European countries. These processes are patented based on the basic characteristic of the digesters such as batch or continuous mode of feeding, number of stages, TS content of waste, and so on. Other factors such as mixer characteristic and reactor configuration (vertical or horizontal, rectangular, and cylindrical) and process flow (completely mixed or plug flow) are also important for the design of the digesters.

Figure 6.2 presents the most common AD technologies for solid waste treatment, especially in Europe. The high solid (dry digestion) technology has been developed since 1980. The most successfully proven reactor design in the dry digestion modes are processes developed by Dranco (OWS, Belgium), Kompogas (Axpo Kompogas AG, Germany), and Valorga (Valorga Internationals S.A.S, France) for one-stage continuous process and by Biocel (Arcadis Heidemij Realisatie Bv, the Netherlands) for batch operation.

The very first design for dry AD was the Valorga system (Figure 6.3a) with a unique process design. The reactor is cylindrical with a vertical median inner wall on approximately two-thirds of its diameter. The feed enters through an inlet close to the bottom of the reactor and slowly moves around the vertical plate until it is discharged through an outlet that is located diametrically opposite to the inlet. Discontinuous injection of recirculated biogas is applied at a high pressure through a network of injectors at the bottom of the reactor for pneumatic mixing, every 15 minutes (Fruteau

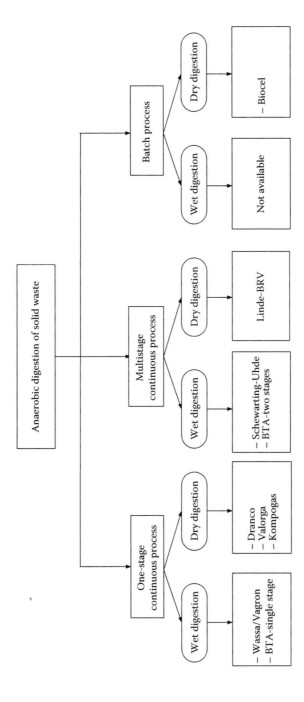

FIGURE 6.2 Available anaerobic digestion methods. (Modified from Nayono, S. E. Anaerobic digestion of organic solid waste for energy production. PhD diss., Fakultät für Bauingenieur-, Geo- und Umweltwissenschaften Universität Fridericiana zu Karlsruhe, Karlsruhe, Germany, 2010.)

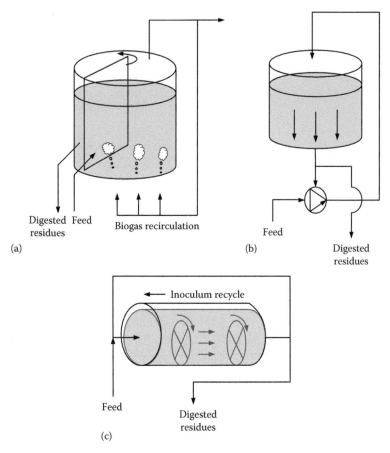

FIGURE 6.3 Different digester designs used in one-stage continuous dry digestion. (a) Valorga design, (b) Kompogas, and (c) Dranco. (Modified from Vandevivere, P. et al., Types of anaerobic digester for solid wastes, 2003.)

de Laclos et al. 1997). Although the mixing process seems promising, the injectors need regular maintenance, as they are prone to clogging (Verma 2002).

In the Dranco reactor (Figure 6.3b), the feed enters from the top and the digestate residue is discharged through an outlet that is located at the bottom of the reactor. The mixing in the Dranco process is performed only by downward plug flow of the waste. Some parts of the processed waste are extracted from the bottom of the reactor and mixed with the fresh feed, and introduced to the reactor (Figure 6.3b). The Kompogas process (Figure 6.3c) functions almost in the same manner, except that the mixing is done by the slowly rotating impellers inside a horizontal PFR. The configuration and mixing facilitate a homogenization of the waste inside the reactor and result in an accumulation of the heavier particles such as sand and glass at the bottom of the reactor (Wellinger et al. 1993; Vandevivere et al. 2003).

The Biocel system is based on the batchwise dry digestion. In the Biocel process, the TSs of the organic wastes are maintained at 30%–40%. Several reactors, of a rectangular shape made of concrete, are coupled in series and operate mostly at mesophilic conditions. The digesters are equipped with chambers at the floor of the digesters for collecting the leachate. The digestate from the previous reactor is mixed with the fresh biowaste and fed to the next reactor by shovels. After loading, the reactors are closed with airtight doors. In order to maintain the temperature at 35°C–40°C, the preheated leachate is sprayed from nozzles on the top of the reactors. The retention time of the process is commonly between 15 and 21 days (Brummeler 2000). It is reported that a full-scale Biocel plant is capable of successfully treating vegetable, garden, and fruit wastes with the capacity of 35,000 tons year^{-1}.

Examples of available reactor designs for the wet digestion include Wassa, BTA single stage for continuous processes, and the Schwarting-Uhde together with the BTA two stage for multistage continuous process. The Wassa process is a one-stage, wet anaerobic system. The process is performed in a vertical, completely mixed reactor, which is divided internally to form a predigestion chamber. The mixing in the digester is done by a mechanical stirrer, and gas is injected from the bottom of the digester (Williams et al. 2003). The process can be operated at both thermophilic and mesophilic conditions. The retention time for the thermophilic condition is normally about 10 days and for the mesophilic conditions is 20 days (Egigian Nichols 2004). The feedstock preparation involves a separation process that removes the inert material and sand from the substrate. Process water is also added to the fresh biowaste to provide the desired TS concentration of 10%–15%.

The BTA process can be run as both single- and multistage continuous anaerobic process; however, the single-stage system is more commonly used. The single-stage systems are relatively small, decentralized waste management units whereas the multistage systems are mainly used for plants with capacities of more than 50,000 tons per year. The BTA process involves two major stages: the hydromechanical pretreatment and the AD process; both are performed in completely mixed reactors where the mixing is carried out by biogas injection. Throughout the hydromechanical pretreatment, the solids are diluted with recirculated process water to achieve a maximum TS content of 10%. Furthermore, the impurities such as plastics, textiles, wood, stones, batteries, and metals are removed. The temperature in the BTA process is normally kept in the mesophilic range. The digestate is dewatered using a decanter centrifuge and sent further to an aerobic posttreatment. The gas yield varies between 80 m^3 ton^{-1} and 120 m^3 ton^{-1} of biowaste, depending on the composition of the waste utilized (Kübler et al. 2000; Egigian Nichols 2004).

The Schwarting-Uhde reactors are used in a series of two vertical PFRs, which adopt a two-stage wet AD process. The phase separation in this type of digester provides the proper environment in favor of different groups of microorganisms involved in the process. The temperature range for the first reactor is mesophilic, where the hydrolysis and acidification reactions take place. The second reactor operates under thermophilic range for the methanogens. Prior to the AD process, the biowaste is shredded into a smaller particle size and then diluted to a TS content of around 12%. The slurry is preheated to the right temperature (around 35°C) by heat exchangers and then pumped to the first reactor through a number of perforated plates set within the reactor. Although mechanical mixing is not applied in this process the column of liquid in the tank is raised and lowered to create turbulence at the perforated plates using time-controlled impulse pumps to provide proper mixing. The overall retention time for both reactors is about 10–12 days, with an individual retention time of 5–6 days. The heavy solids, which settle at the bottom of the digester, are then frequently removed by the screw pumps. Biogas is collected from the top of the reactor. The design of the Schwarting-Uhde process has an advantage in minimizing the formation of a thick floating scum layer that is a common problem in the wet AD. Nevertheless, the high risk of clogging at the perforated plates makes this process only suitable for treating clean, highly biodegradable biowastes (Lissens et al. 2001; Vandevivere et al. 2003).

6.7 UTILIZATION OF BIOGAS

Biogas is mainly composed of 50%–70% methane (CH_4) and 30%–50% carbon dioxide (CO_2). Moreover, it might contain impurities such as nitrogen (0%–5%), oxygen (1%), hydrocarbons (1%), hydrogen sulfide (0.5%), ammonia (0.05%), water vapor (1%–5%), and siloxanes (0–50 mg m^{-3}) (IEA Bioenergy 1999). Biogas can be utilized in a variety of ways; however, in most cases, a part or all of the impurities should be removed prior to any application (Persson et al. 2006). Figure 6.4 shows the different utilization forms of the biogas and the required cleanup.

6.7.1 Heating/Cooking

Heating or cooking is one of the simplest utilization methods for biogas. In developing countries, including China and India, biogas is produced in small-scale facilities, and it is mainly used for cooking (Surendra et al. 2014). For this purpose, the biogas produced is distributed directly through a pipe from the household digester (1–3 m³) to the kitchen, where the gas is burned for cooking on a gas stove. On the other hand, in industrial countries, high-efficiency boilers are present for the utilization of biogas. Typically, the efficiency rate varies between 75% and 90%. These boilers do not have a high gas quality requirement. Typically, gas pressure has to be around 8 mbar to 25 mbar and hydrogen sulfide concentration should be low (<1000 ppm), since hydrogen sulfide converts to sulfurous acid in the boiler, which can cause serious corrosion (Kaparaju et al. 2013). Heat produced by these boilers is used for district heating; however, there is no constant demand for heat, so this type of utilization is not common, except in Finland, where around 60% of the produced biogas is utilized in the form of heat (IEA Bioenergy 2014).

6.7.2 Producing Heat and Electricity

Burning biogas in combined heat and power (CHP) plants is probably the most common utilization form of biogas in industrial countries (Makaruk et al. 2010). Most European countries support electricity generation from biogas because it is considered as green electricity. In a CHP unit, heat and electricity are simultaneously generated from biogas. The size of the engines is in the range from 45 kW on the small farms up to several MW in the large-scale facilities. Currently, most CHP plants employ gas and diesel engines. The electrical efficiency of the CHPs varies between

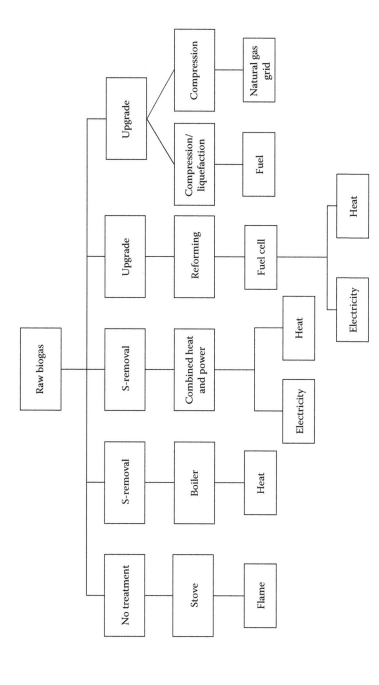

FIGURE 6.4 Biogas utilization purposes and required upgrading methods. (Data from IEA Bioenergy, *Biogas Upgrading and Utilisation*, Task 24, 1999.)

30% and 40%, while its thermal efficiency is 40%–50%, depending on the size. The production of heat and electricity, applying gas turbines, is a promising technology. Modern gas turbines are equally efficient as internal combustion engines and are very robust. They allow for the recovery of the heat in the form of valuable steam. Unfortunately, the efficient turbines are available only at scales greater than 800 kW. The requirements for the quality of the gas in a CHP unit are comparable to those for boilers; however, the H_2S level should be lower and the solixanes should be removed to guarantee a long operation of the CHP plant.

6.7.3 UPGRADING

When biogas is upgraded, which means that carbon dioxide, hydrogen sulfide, ammonia, particles, and water (and sometimes other trace components) are removed to get a product gas with a methane content above 97%, it can be used as a substitute for natural gas (Pöschl et al. 2010). Several upgrading techniques exist, such as water scrubbing, pressure swing adsorption, and chemical absorption, as well as cryogenic and membrane separation (Makaruk et al. 2010; Ryckebosch et al. 2011). Currently, the high-pressure water scrubbing is the most widespread technology used, because of its low cost. Table 6.3 summarizes the current biogas upgrading technologies.

Upgraded biogas, usually referred to as biomethane, can be used as a vehicle fuel or injected directly to the natural gas grid. In different countries, different quality specifications are applied for both cases. Biomethane is considered as one of the cleanest vehicle fuels with regard to the environment, climate, and human health. Currently, the utilization of methane as a vehicle fuel is only widespread in Sweden, due to Swedish financial support (IEA Bioenergy 2014).

TABLE 6.3
Common Biogas Upgrading Technologies

Separation Method	Process	Principle	Final CH_4 Content
Physical absorption	Pressurized water wash	Dissolution of CO_2 in water at high pressure	>96%
	Wash in solvent	Dissolution of CO_2 in solvent	>96%
Adsorption	Pressure swing adsorption	Adsorption of CO_2 on a molecular sieve	>96%
Chemical absorption	Monoethanolamine (MEA) wash	Chemical reaction between CO_2 and MEA	>99%
Cryogenic process	Low temperature process	CO_2 goes to liquid phase, while CH_4 remains gaseous	>99.9%
Membrane separation	Polymer membrane gas separation (dry)		>80%
	Membrane gas separation (wet)	Membrane permeability of CH_4 is lower than H_2S and CO_2	>96%

Source: IEA Bioenergy, *Biogas Country Overview (Country Reports)*, Task 37, 2014.

6.7.4 Fuel Cells

Fuel cells (FCs) are power-generating systems that produce electricity without combustion by combining fuel (in this case biomethane) and oxygen (from the air) in an electrochemical reaction (Alves et al. 2013). Currently, the commercial FCs, which utilize H_2 and FCs that are made for direct utilization of biomethane, are still under development (Lanzini and Leone 2010; Shiratori et al. 2010). Therefore, to be able to utilize methane in commercial FCs, it should first be converted into hydrogen, either by a catalytic steam reforming conversion or by using a (platinum) catalyst. This process is usually referred to as a methane-reforming process. The most common methane-reforming processes for hydrogen production are known as steam reforming (Maluf and Assaf 2009), partial oxidation reforming (Dantas et al. 2012), autothermal reforming (Cai et al. 2006), dry reforming (Fidalgo et al. 2008), and dry oxidation reforming (Izquierdo et al. 2012). In the second stage, hydrogen is converted into electricity. The by-products of the reaction are water and clean CO_2. The efficiency of the conversion into electricity is expected to exceed 50%. FCs demonstrate relatively constant efficiencies over a wide range of loads. There are five types of FCs, classified by the type of electrolyte used: alkaline (AFC), phosphoric acid (PAFC), molten carbonate (MCFC), solid oxide (SOFC), and proton exchange membrane (PEM) (IEA Bioenergy 1999; Sharaf and Orhan 2014). Table 6.4 summarizes the main characteristics of these FCs.

6.8 ECONOMICS OF AD

AD is a steadily growing industry in Europe and all over the world (Bond and Templeton 2011; van Foreest 2012). The economics of AD is affected by various factors, including (1) capital cost, which depends on the type, the size, and the location

TABLE 6.4
Fuel Cells and Their Characteristics

Fuel Cell and Characteristic	AFC	PAFC	MCFC	SOFC	PEM
Electrolyte	Alkaline solution KOH	Phosphoric acid	Molten carbonate LiKCO$_3$	Solid oxide Y$_2$O$_3$ and ZrO$_2$	Proton exchange membranes
Operating temperature	150°C–200°C	200°C	650°C	1000°C	50°C–120°C
Efficiency	60%–70%	40%–45%	50%–57%	45%–50%	40%–50%
Module size	0.3–5 kW	0.2–2 MW	2 MW	3–100 kW	50–250 kW

Source: Sharaf, O. Z., and Orhan, M. F., *Renew Sust Energ Rev*, 32, 810–853, 2014; IEA Bioenergy, *Biogas Upgrading and Utilisation*, Task 24, 1999.
AFC: Alkaline fuel cell; PAFC: phosphoric acid fuel cell; MCFC: molten carbonate fuel cell; SOFC: solid oxide fuel cell; PEM: proton exchange membrane.

of the plant; (2) the operating costs; (3) the cost of the biomass feedstock; (4) the products' revenues; and (5) the financial support systems (Forgács 2012).

6.8.1 CAPITAL COST

The capital cost depends on several factors, for example, plant size, location, engineering, and the composition of the waste treated. The characteristics of the incoming feedstock are important to consider, because they also define the essential units required for the preprocessing steps prior to the digestion. Generally, a larger plant size requires less investment per production unit, and a vertical digester is considered to be more expensive compared to the horizontal design (Shafiei et al. 2011; Pasangulapati et al. 2012). The majority of the capital cost relates to the construction, followed by the costs of the grid connection and infrastructure (Ove Arup & Partners 2011). Jones and Salter (2013) estimated the capital cost for the farm-scale AD plants to be £4000–£8000 per kW installed electricity capacity. The capital cost of an AD plant utilizing municipal waste is higher since additional preprocessing steps, such as the removal of plastic, glass, and metal, are required, which increase the cost.

6.8.2 OPERATING COST

Operating cost is related to the operation of the biogas plant; it includes the costs associated with the operating staff (salaries, insurances, etc.), transportation, licenses, price of the feedstock, and maintenance (Turton et al. 2008). These costs are either direct, which means that they are dependent on the amount of feedstock used, number of workers employed at the plant, and so on, or indirect, which means that they are independent regardless of whether the plant is in operation or not. Indirect costs include insurance, taxes, and salaries (Turton et al. 2008). Typically, the insurance costs are 1% of the capital cost, while the maintenance costs are approximately 2.0%–2.5% of the capital, and other costs, including advice and laboratory testing, add up to 1% of the capital costs (Jones and Salter 2013; Forgács et al. 2014).

6.8.3 COST OF THE BIOMASS FEEDSTOCK

The cost or price of the feedstock shows significant variations, depending on the type of the waste and the location of the AD plant. In some countries, including England, the AD plant charges a waste management gate fee per ton waste treated by the plant. In other countries, such as Germany, the waste management sites, including the AD plants, compete for wastes/feedstocks (Dolan et al. 2011). In these countries, the gate fee for this kind of wastes is zero, and the biogas plants must sometimes even pay to get the waste. In that case, the income can be calculated and compared to the cost of the feedstock (Redman 2008). The income from the substrate can be calculated for a biogas plant generating heat and electricity as follows:

$$\text{Income} = \text{YCH}_4 * \text{ECH}_4 * \left(\eta_e * P_e + \eta_t * P_h \right) + \text{IFD} + \text{ID}$$

where:
YCH$_4$ is the methane yield per ton of substrate
ECH$_4$ is the energy content of methane
η_e is the electrical conversion efficiency
P$_e$ is the price of the electricity
η_t is the thermal conversion efficiency
P$_h$ is the price of the heat
IFD is the income from feed tariff
ID is the income from digestate

6.8.4 PRODUCT REVENUE

The two main sources of income from AD come from the sale of biogas and digester-related products; however, in many cases, the AD plants have other income sources as well. The most common revenues are

- Sales of biogas (or a product from its processing such as heat and electricity)
- Sales of liquid digestate as fertilizer
- Income from financial support system (e.g., Renewables Obligation Certificates, feed tariffs)
- Gate fees

6.8.5 FINANCIAL SUPPORT SYSTEMS

The European Union and many other countries support AD as an alternative energy production. However, each country has its own "anaerobic digestion strategy" and support system. In most cases, the extent of the support depends on the size of the plant. Table 6.5 summarizes the financial support system in some countries (IEA Bioenergy 2014).

6.9 CONCLUDING REMARKS

The current chapter clearly indicates that the AD technology is probably one of the most effective biological processes to treat a wide range of organic solid wastes. However, different factors such as the composition and quality of the substrate and cosubstrate and process parameters, including temperature, pH, OLR, and microbial dynamics, influence the efficiency of the AD process and must be maintained in an optimum range to achieve the maximum benefit from the process in terms of organic waste management and energy production.

Developments of the AD technology over the past few decades have led to considerable progress on digester designs and cost reduction as well as optimization of the process especially for biogas from the MSW. Different political schemes have been attempted at the national level and various kinds of partnerships have been established in several countries. Therefore, this technology could benefit from more accurate data and information that already have been gathered. Therefore, it can be concluded that AD is a rather known technology to deal with

TABLE 6.5

Financial Support System for AD in Some Countries

Country	Main Financial Support System	Additional Support/Requirement
The United States	A renewable portfolio standard legislation that creates a market in tradable renewable or green electricity certificates	Many states also offer programs to offset the costs of digester
	There are many federal and state funding sources: U.S. Treasury grants cover 30% of the cost of new biogas recovery systems The USDA Rural Energy for America Program (REAP) offers grants and loans	The Environmental Quality Incentives Program (EQIP) is a program that gives landowner financial and technical assistance for installation of a methane digester
China		Tax incentives: VAT reduction and exemption, preferential income tax Special funds: for biogas construction in rural areas, for biomass utilization
Brazil	There are no tariffs or subsidies for biogas	Low Carbon Agriculture program, which gives access to rural technical assistance to promote improvement of infrastructure. Additional credit to producers to develop new technologies
Germany	<150 kW$_e$ 0.143 € kWh^{-1} <500 kW$_e$ 0.123 € kWh^{-1} <750 kW$_e$ 0.110 € kWh^{-1} <5 MW$_e$ 0.110 € kWh^{-1} <20 MW$_e$ 0.060€ kWh^{-1}	Additional support after specific substrates including crops and organic fraction of municipal solid waste
The United Kingdom	<250 kW 0.1213 £ kWh^{-1} <500 kW 0.1122 £ kWh^{-1} >500 kW 0.0949 £ kWh^{-1}	Renewable heat tariff of 0.068 £ kWh^{-1} for biomethane injected into the natural gas grid and combusted downstream. Renewable Transport Fuel Obligation Certificates 0.1115 £ L^{-1}
	Double Renewable Obligation Certificates (ROCS) apply to AD	Investment grants are available for small scale units from On-Farm AD
Austria	Feed-in tariff for electricity: 0.1950 € kWh^{-1} < 250 kW$_e$, 0.1693 € kWh^{-1} 250–500 kW$_e$, 0.1334 € kWh^{-1} 500–750 kW$_e$, 0.1293 € kWh^{-1} > 750 250 kW$_e$ + 0.02 EUR kWh^{-1} if biogas is upgraded + 0.02 EUR kWh^{-1} if heat is used efficiently	Minimum 30% of manure should be used as a substrate If organic waste is used, the feed-in tariff is reduced by 20%

(Continued)

TABLE 6.5 (*Continued*)
Financial Support System for AD in Some Countries

Country	Main Financial Support System	Additional Support/Requirement
Denmark	0.056 EUR kWh^{-1} for biogas used in a CHP unit or injected into the grid (115 DKK GJ^{-1}) 0.037 EUR kWh^{-1} for direct usage for transport or industrial purposes (75 DKK GJ^{-1})	Investment grants are available for plants digestion of mainly manures
Sweden	Sweden has no feed-in tariffs Support systems promoting biomethane as automotive fuel including: No carbon dioxide or energy tax on biogas 40% reduction of the fringe benefit taxation for the use of company NGVs until 2016	Investment grants for marketing of new technologies A joint electricity certificate market in Norway and Sweden 0.2 SEK kWh^{-1} (~0.02 EUR kWh^{-1}) for manure based biogas production to reduce methane emissions from manure
India	Central Financial Assistance (CFA) system Support for the installation of biogas based power generation units: <20 kW 40,000 INR kW^{-1} <100 kW 35,000 INR kW^{-1} <250 kW 30,000 INR kW^{-1}	

Source: IEA Bioenergy, *Biogas Country Overview (Country Reports)*, Task 37, 2014.

the biodegradable waste streams that could allow us to get one step closer to a zero waste society.

REFERENCES

AgSTAR. 2010. *U.S. Farm Anaerobic Digestion Systems: A 2010 Snapshot.* Available from http://www.epa.gov/agstar/documents/2010_digester_update.pdf (Accessed June 12, 2014).

Ahring, B. K., Angelidaki, I., and Johansen, K. 1992. Anaerobic treatment of manure together with industrial waste. *Water Sci Technol* 25 (7):311–318.

Alvarez, R., and Lidén, G. 2008a. The effect of temperature variation on biomethanation at high altitude. *Bioresour Technol* 99 (15):7278–7284.

Alvarez, R., and Lidén, G. 2008b. Semi-continuous co-digestion of solid slaughterhouse waste, manure, and fruit and vegetable waste. *Renew Energy* 33 (4):726–734.

Alvarez, R., Villca, S., and Liden, G. 2006. Biogas production from llama and cow manure at high altitude. *Biomass Bioenergy* 30 (1):66–75.

Alves, H. J., Bley Junior, C., Niklevicz, R. R. et al. 2013. Overview of hydrogen production technologies from biogas and the applications in fuel cells. *Int J Hydrogen Energy* 38 (13):5215–5225.

Amigun, B., Sigamoney, R., and Von Blottnitz, H. 2008. Commercialisation of biofuel industry in Africa: A review. *Renew Sust Energ Rev* 12:690–711.

Angelidaki, I., and Ahring, B. K. 1993. Thermophilic anaerobic digestion of livestock waste: The effect of ammonia. *Appl Microbiol Biotechnol* 38 (4):560–564.

Angelidaki, I., and Ellegaard, L. 2003. Codigestion of manure and organic wastes in centralized biogas plants: status and future trends. *Appl Biochem Biotechnol* (109):95–99.

Appels, L., Baeyens, J., Degrève, J., and Dewil, R. 2008. Principles and potential of the anaerobic digestion of waste-activated sludge. *Prog Energy Combust Sci* 34 (6):755–781.

Astals, S., Romero-Güiza, M., and Mata-Alvarez, J. 2013. Municipal solid waste. In *Alternative Energies, Advanced Structured Materials*, edited by G. Ferreira. Berlin, Germany: Springer.

Baldasano, J., and Soriano, C. 2000. Emission of greenhouse gases from anaerobic digestion processes: comparison with other municipal solid waste treatments. *Water Sci Technol* 41 (3):275–282.

Banks, C. J., and Wang, Z. 1999. Development of a two phase anaerobic digester for the treatment of mixed abattoir wastes. *Water Sci Technol* 40 (1):69–76.

Batstone, D. J., Keller, J., Angelidaki, I. et al. 2002. The IWA anaerobic digestion model no 1(ADM 1). *Water Sci Technol* 45 (10):65–73.

Björnsson, L., Murto, M., Jantsch, T. G., and Mattiasson, B. 2001. Evaluation of new methods for the monitoring of alkalinity, dissolved hydrogen and the microbial community in anaerobic digestion. *Water Res* 35 (12):2833–2840.

Boe, K. 2006. Online monitoring and control of the biogas process. PhD diss., Institute of Environment & Resources Technical University of Denmark, Lyngby, Denmark.

Bond, T., and Templeton, M. R. 2011. History and future of domestic biogas plants in the developing world. *Energy Sustain Dev* 15 (4):347–354.

Bouallagui, H., Torrijos, M., Godon, J. J. et al. 2004. Two-phase anaerobic digestion of fruit and vegetable wastes: Bioreactors performance. *Biochem Eng J* 21 (2):193–197.

Bouallagui, H., Touhami, Y., Ben Cheikh, R., and Hamdi, M. 2005. Bioreactor performance in anaerobic digestion of fruit and vegetable wastes. *Process Biochem* 40 (3):989–995.

Braun, R., Brachtl, E., and Grasmug, M. 2003. Codigestion of proteinaceous industrial waste. *Appl Biochem Biotechnol* 109 (1–3):139–153.

Brummeler, E. T. 1993. Dry anaerobic digestion of the organic fraction of municipal solid waste. PhD diss., Landbouwuniversiteit te Wageningen, the Netherlands.

Brummeler, E. T. 2000. Full scale experience with the BIOCEL process. *Water Sci Technol* 41 (3):299–304.

Bruni, E. 2010. Improved anaerobic digestion of energy crops and agricultural residues. PhD diss., Department of Environmental Engineering, Technical University of Denmark, Lyngby, Denmark.

Cai, X., Dong, X., and Lin, W. 2006. Autothermal Reforming of Methane over Ni Catalysts Supported on $CuO-ZrO_2-CeO_2-Al_2O_3$. *J Nat Gas Chem* 15 (2):122–126.

Callaghan, F. J., Wase, D. A. J., Thayanithy, K., and Forster, C. F. 1999. Co-digestion of waste organic solids: Batch studies. *Bioresour Technol* 67 (2):117–122.

Callaghan, F. J., Wase, D. A. J., Thayanithy, K., and Forster, C. F. 2002. Continuous co-digestion of cattle slurry with fruit and vegetable wastes and chicken manure. *Biomass Bioenerg* 22 (1):71–77.

Chandra, R., Takeuchi, H., and Hasegawa, T. 2012. Methane production from lignocellulosic agricultural crop wastes: A review in context to second generation of biofuel production. *Renew Sust Energ Rev* 16 (3):1462–1476.

Chaudhary, B. K. 2008. Dry continuous anaerobic digestion of municipal solid waste in thermophilic conditions. PhD diss., Asian Institute of Technology, Bangkok, Thailand.

Chayovan, S., Gerrish, J. B., and Eastman, J. A. 1988. Biogas production from dairy manure: the effects of temperature perturbations. *Biol Waste* 25 (1):1–16.

Chen, Y., Cheng, J. J., and Creamer, K. S. 2008. Inhibition of anaerobic digestion process: A review. *Bioresour Technol* 99 (10):4044–4064.

Cheng, H., and Hu, Y. 2010. Municipal solid waste (MSW) as a renewable source of energy: Current and future practices in China. *Bioresour Technol* 101 (11):3816–3824.

Chynoweth, D. P., Owens, J. M., and Legrand, R. 2001. Renewable methane from anaerobic digestion of biomass. *Renew Energ* 22 (1):1–8.

Chynoweth, D. P., Turick, C. E., Owens, J. M., Jerger, D. E., and Peck, M. W. 1993. Biochemical methane potential of biomass and waste feedstocks. *Biomass Bioenerg* 5 (1):95–111.

Dague, R. R. 1968. Application of digestion theory to digester control. *J Water Pollut Control Fed*:2021–2032.

Dantas, S. C., Resende, K. A., Rossi, R. L., Assis, A. J., and Hori, C. E. 2012. Hydrogen production from oxidative reforming of methane on supported nickel catalysts: An experimental and modeling study. *Chem Eng J* 197:407–413.

De Baere, L. 2000. Anaerobic digestion of solid waste: State-of-the-art. *Water Sci Technol* 41:283–290.

Deublein, D., and Steinhauser, A. 2008. *Biogas from Waste and Renewable Resources: An Introduction.* Weinheim, Germany: Wiley.

Dolan, T., Cook, M. B., and Angus, A. J. 2011. Financial appraisal of wet mesophilic AD technology as a renewable energy and waste management technology. *Sci Total Environ* 409 (13):2460–2466.

Doran, P. M. 1995. *Bioprocess Engineering Principles.* London: Academic Press.

Edström, M., Nordberg, Å., and Thyselius, L. 2003. Anaerobic treatment of animal byproducts from slaughterhouses at laboratory and pilot scale. *Appl Biochem Biotechnol* 109 (1–3):127–138.

Egigian Nichols, C. 2004. Overview of anaerobic digestion technologies in Europe. *Biocycle* 45 (1):47–53.

El-Mashad, H. M., van Loon, W. K. P., and Zeeman, G. 2003. A model of solar energy utilisation in the anaerobic digestion of cattle manure. *Biosys Eng* 84 (2):231–238.

Fernández, A., Sanchez, A., and Font, X. 2005. Anaerobic co-digestion of a simulated organic fraction of municipal solid wastes and fats of animal and vegetable origin. *Biochem Eng J* 26 (1):22–28.

Fidalgo, B., Domínguez, A., Pis, J. J., and Menéndez, J. A. 2008. Microwave-assisted dry reforming of methane. *Int J Hydrogen Energy* 33 (16):4337–4344.

Forgács, G. 2012. Biogas production from citrus wastes and chicken feather: Pretreatment and co-digestion. PhD diss., Chemical and Biological Engineering, Chalmers University of Technology, Göteborg, Sweden.

Forgács, G., Niklasson, C., Sárvári Horváth, I., and Taherzadeh, M. 2014. Methane production from feather waste pretreated with $Ca(OH)_2$: Process development and economical analysis. *Waste Biomass Valorization* 5 (1):65–73.

Fruteau de Laclos, H., Desbois, S., and Saint-Joly, C. 1997. Anaerobic digestion of municipal solid organic waste: Valorga full-scale plant in Tilburg, the Netherlands. *Water Sci Technol* 36 (6):457–462.

Garcia, J.-L., Patel, B. K. C., and Ollivier, B. 2000. Taxonomic, phylogenetic, and ecological diversity of methanogenic Archaea. *Anaerobe* 6 (4):205–226.

Gerardi, M. H. 2003. *The Microbiology of Anaerobic Digesters.* Hoboken, NJ: Wiley.

Ghosh, S., Henry, M., Sajjad, A., Mensinger, M., and Arora, J. 2000. Pilot-scale gasification of municipal solid wastes by high-rate and two-phase anaerobic digestion (TPAD). *Water Sci Technol* 41 (3):101–110.

Griffin, L. P. 2012. Anaerobic digestion of organic wastes: The impact of operating conditions on hydrolysis efficiency and microbial community composition. MSc diss., Colorado State University, Fort Collins, CO.

Gunaseelan, V. N. 1997. Anaerobic digestion of biomass for methane production: A review. *Biomass Bioenergy* 13 (1):83–114.

Gunaseelan, V. N. 2004. Biochemical methane potential of fruits and vegetable solid waste feedstocks. *Biomass Bioenergy* 26 (4):389–399.

Gunnerson, C. G., Stuckey, D. C., Greeley, M. et al. 1986. Anaerobic digestion: Principles and practices for biogas systems. In *Integrated Resource Recovery*. World Bank Technical Paper no 49. Washington, DC: World Bank.

Hajji, A., and Rhachi, M. 2013. The influence of particle size on the performance of anaerobic digestion of municipal solid waste. *Energy Procedia* 36 (0):515–520.

Hansen, K. H., Angelidaki, I., and Ahring, B. K. 1998. Anaerobic digestion of swine manure: Inhibition by ammonia. *Water Res* 32 (1):5–12.

Hansson, A., and Christensson, K. 2005. *Biogas ger energi till ekologiskt lantbruk*: Jordbruksverket.

Harper, S. R., and Pohland, F. G. 1986. Recent developments in hydrogen management during anaerobic biological wastewater treatment. *Biotechnol Bioeng* 28 (4):585–602.

Hartmann, H., and Ahring, B. 2006. Strategies for the anaerobic digestion of the organic fraction of municipal solid waste: An overview. *Water Sci Technol* 53 (8):7–22.

Hashimoto, A. G. 1982. Methane from cattle waste: effects of temperature, hydraulic retention time, and influent substrate concentration on kinetic parameter (K). *Biotechnol Bioeng* 24 (9):2039–2052.

Hashimoto, A. G. 1986a. Ammonia inhibition of methanogenesis from cattle wastes. *Agr Wastes* 17 (4):241–261.

Hashimoto, A. G. 1986b. Pretreatment of wheat straw for fermentation to methane. *Biotechnol Bioeng* 28 (12):1857–1866.

Hedegaard, M., and Jaensch, V. 1999. Anaerobic co-digestion of urban and rural wastes. *Renew Energy* 16 (1):1064–1069.

Hoornweg, D., and Bhada-Tata, P. 2012. *What a Waste: A Global Review of Solid Waste Management*. World Bank, Washington DC.

Huser, B. A., Wuhrmann, K., and Zehnder, A. J. B. 1982. Methanothrix soehngenii gen. nov. sp. nov., a new acetotrophic non-hydrogen-oxidizing methane bacterium. *Arch Microbiol* 132 (1):1–9.

Ibrahim, M. M., El-Zawawy, W. K., Abdel-Fattah, Y. R., Soliman, N. A., and Agblevor, F. A. 2011. Comparison of alkaline pulping with steam explosion for glucose production from rice straw. *Carbohydr Polym* 83 (2):720–726.

IEA Bioenergy. 2001. *Biogas Upgrading and Utilisation*. Task 24. Available from http://www.biogasmax.eu/media/biogas_upgrading_and_utilisation__018031200_1011_24042007.pdf (Accessed March 23, 2015).

IEA Bioenergy. 2005. *Biogas Production and Utilization*. Task 37. Available from http://www.biogasmax.eu/media/2_biogas_production_utilisation__068966400_1207_19042007.pdf (Accessed March 23, 2015).

IEA Bioenergy. 2014. *Biogas Country Overview (Country Reports)*. Task 37. Available from http://www.iea-biogas.net/country-reports.html?file=files/daten-redaktion/download/publications/country-reports/Summary/Countryreport2013.pdf (Accessed March 23, 2015).

Ince, O. 1998. Performance of a two-phase anaerobic digestion system when treating dairy wastewater. *Water Res* 32 (9):2707–2713.

Izquierdo, U., Barrio, V. L., Lago, N. et al. 2012. Biogas steam and oxidative reforming processes for synthesis gas and hydrogen production in conventional and microreactor reaction systems. *Int J Hydrogen Energy* 37 (18):13829–13842.

Jagadabhi, P. S. 2011. Methods to enhance hydrolysis during one and two-stage anaerobic digestion of energy crops and crop residues. PhD diss., Department of Biological and Environmental Science, University of Jyväskylä, Jyväskylä, Finland.

Jiang, X., Sommer, S. G., and Christensen, K. V. 2011. A review of the biogas industry in China. *Energy Policy* 39 (10):6073–6081.

Jones, P., and Salter, A. 2013. Modelling the economics of farm-based anaerobic digestion in a UK whole-farm context. *Energy Policy* 62 (0):215–225.

Kaparaju, P., and Rintala, J. 2005. Anaerobic co-digestion of potato tuber and its industrial by-products with pig manure. *Resour Conserv Recy* 43 (2):175–188.

Kaparaju, P., Rintala, J., and Rasi, S. 2013. Biogas upgrading and compression. In *Bioenergy Production by Anearobic Digestion*. Hoboken, NJ: Taylor & Francis.

Kashyap, D. R., Dadhich, K. S., and Sharma, S. K. 2003. Biomethanation under psychrophilic conditions: a review. *Bioresour Technol* 87 (2):147–153.

Kayhanian, M., Tchobanoglous, G., and Brown, R. C. 2007. Biomass conversion processes for energy recovery. In *Handbook of Energy Efficiency and Renewable Energy*. Boca Raton, FL: CRC Press.

Kelleher, M. 2007. Anaerobic digestion outlook for MSW streams. *Biocycle* 48 (8):51.

Kim, I., Kim, D., and Hyun, S. 2000. Effect of particle size and sodium ion concentration on anaerobic thermophilic food waste digestion. *Water Sci Technol* 41 (3):67–73.

Klass, D. L. 1984. Methane from anaerobic fermentation. *Science* 223 (4640):1021–1028.

Koster, I. W., and Lettinga, G. 1988. Anaerobic digestion at extreme ammonia concentrations. *Biol Waste* 25 (1):51–59.

Kübler, H., Hoppenheidt, K., Hirsch, P. et al. 2000. Full scale co-digestion of organic waste. *Water Sci Technol* 41 (3):195–202.

Lanzini, A., and Leone, P. 2010. Experimental investigation of direct internal reforming of biogas in solid oxide fuel cells. *Int J Hydrogen Energ* 35 (6):2463–2476.

Lepistö, R., and Rintala, J. 1999. Kinetics and characteristics of 70°C, VFA-grown, UASB granular sludge. *Appl Microbiol Biotechnol* 52 (5):730–736.

Lissens, G., Vandevivere, P., De Baere, L., Biey, E., and Verstraete, W. 2001. Solid waste digestors: Process performance and practice for municipal solid waste digestion. *Water Sci Technol* 44 (8):91–102.

Liu, X., Liu, H., Chen, Y., Du, G., and Chen, J. 2008. Effects of organic matter and initial carbon–nitrogen ratio on the bioconversion of volatile fatty acids from sewage sludge. *J Chem Technol Biotechnol* 83 (7):1049–1055.

Luning, L., Van Zundert, E., and Brinkmann, A. 2003. Comparison of dry and wet digestion for solid waste. *Water Sci Technol* 48 (4):15–20.

Makaruk, A., Miltner, M., and Harasek, M. 2010. Membrane biogas upgrading processes for the production of natural gas substitute. *Sep Purif Technol* 74 (1):83–92.

Maluf, S. S., and Assaf, E. M. 2009. Ni catalysts with Mo promoter for methane steam reforming. *Fuel* 88 (9):1547–1553.

Mara, D., and Horan, N. J. 2003. *Handbook of Water and Wastewater Microbiology*. London: Academic Press.

Masse, D. I., Masse, L., and Croteau, F. 2003. The effect of temperature fluctuations on psychrophilic anaerobic sequencing batch reactors treating swine manure. *Bioresour Technol* 89 (1):57–62.

Mata-Alvarez, J., Macé, S., and Llabrés, P. 2000. Anaerobic digestion of organic solid wastes. An overview of research achievements and perspectives. *Bioresour Technol* 74 (1):3–16.

McCarty, P. L. 1981. One hundred years of anaerobic treatment. *Paper read at the 2nd International Symposium on Anaerobic Digestion*, Travemünde, Germany.

McMillan, J. D. 1994. Pretreatment of Lignocellulosic Biomass. In *Enzymatic Conversion of Biomass for Fuels Production*. Washington, DC: American Chemical Society.

Michaud, S., Bernet, N., Buffière, P., Roustan, M., and Moletta, R. 2002. Methane yield as a monitoring parameter for the start-up of anaerobic fixed film reactors. *Water Res* 36 (5):1385–1391.

Misi, S. N., and Forster, C. F. 2001. Batch co-digestion of multi-component agro-wastes. *Bioresour Technol* 80 (1):19–28.

Møller, H. B., Sommer, S. G., and Ahring, B. K. 2004. Methane productivity of manure, straw and solid fractions of manure. *Biomass Bioenergy* 26 (5):485–495.

Mosey, F. E., and Fernandes, X. A. 1989. Patterns of hydrogen in biogas from the anaerobic digestion of milk-sugars. *Water Sci Technol* 21 (4–5):187–196.

Murto, M., Björnsson, L., and Mattiasson, B. 2004. Impact of food industrial waste on anaerobic co-digestion of sewage sludge and pig manure. *J Environ Manage* 70 (2):101–107.

The National Archives. 1996. *The Landfill Tax Regulations*. Available from http://www.legislation.gov.uk/uksi/1996/1527/contents/made (Accessed June 30, 2014).

National Development and Reform Commission (NDRC). 2007. *Medium and Long-Term Development Plan for Renewable Energy in China*. Beijing, China.

Nayono, S. E. 2010. Anaerobic digestion of organic solid waste for energy production. PhD diss., Fakultät für Bauingenieur-, Geo- und Umweltwissenschaften Universität Fridericiana zu Karlsruhe (TH), Karlsruhe, Germany.

Neves, L., Gonçalo, E., Oliveira, R., and Alves, M. 2008. Influence of composition on the biomethanation potential of restaurant waste at mesophilic temperatures. *Waste Manage* 28 (6):965–972.

Noykova, N., Muèller, T. G., Gyllenberg, M., and Timmer, J. 2002. Quantitative analyses of anaerobic wastewater treatment processes: identifiability and parameter estimation. *Biotechnol Bioeng* 78 (1):89–103.

Oleszkiewicz, J. A., and Sharma, V. K. 1990. Stimulation and inhibition of anaerobic processes by heavy metals—A review. *Biol Waste* 31 (1):45–67.

Omer, A. M., and Fadalla, Y. 2003. Biogas energy technology in Sudan. *Renew Energ* 28 (3):499–507.

Ove Arup & Partners. 2011. *Review of the generation costs and deployment potential of renewable electricity technologies in the UK*. Department of Energy and Climate Change, London.

Owens, J. M., and Chynoweth, D. P. 1993. Biochemical methane potential of municipal solid waste (MSW) components. *Water Sci Technol* 27 (2):1–14.

Pages-Diaz, J., Pereda-Reyes, I., Taherzadeh, M. J., Sárvári Horváth, I., and Lundin, M. 2014. Anaerobic co-digestion of solid slaughterhouse wastes with agro-residues: Synergistic and antagonistic interactions determined in batch digestion assays. *Chem Eng J* 245:89–98.

Parawira, W., Murto, M., Read, J. S., and Mattiasson, B. 2005. Profile of hydrolases and biogas production during two-stage mesophilic anaerobic digestion of solid potato waste. *Process Biochem* 40 (9):2945–2952.

Pasangulapati, V., Ramachandriya, K. D., Kumar, A. et al. 2012. Effects of cellulose, hemicellulose and lignin on thermochemical conversion characteristics of the selected biomass. *Bioresour Technol* 114 (0):663–669.

Pavlostathis, S. G., and Giraldo-Gomez, E. 1991. Kinetics of anaerobic treatment. *Water Sci Technol* 24 (8):35–59.

Persson, M., Jönsson, O., and Wellinger, A. 2006. *Biogas upgrading to vehicle fuel standards and grid injection*. Paris, France: IEA Bioenergy.

Pöschl, M., Ward, S., and Owende, P. 2010. Evaluation of energy efficiency of various biogas production and utilization pathways. *Appl Energ* 87 (11):3305–3321.

Rajendran, K., Aslanzadeh, S., and Taherzadeh, M. J. 2012. Household biogas digesters—A review. *Energies* 5 (8):2911–2942.

Rapport, J., Zhang, R., Jenkins, B. M., and Williams, R. B. 2008. *Current Anaerobic Digestion Technologies Used for Treatment of Municipal Organic Solid Waste*. Contractor Report to the California Integrated Waste Management Board, University of California, Davis, CA.

Raven, R., and Gregersen, K. H. 2007. Biogas plants in Denmark: Successes and setbacks. *Renew Sust Energ Rev* 11 (1):116–132.

Redman, G. 2008. *A Detailed Economic Assessment of Anaerobic Digestion Technology and Its Suitability to UK Farming and Waste Systems*. Melton Mowbray: The Andersons Centre for the National Non-Food Crops Centre.

Resch, C., Wörl, A., Waltenberger, R., Braun, R., and Kirchmayr, R. 2011. Enhancement options for the utilisation of nitrogen rich animal by-products in anaerobic digestion. *Bioresour Technol* 102 (3):2503–2510.

Ryckebosch, E., Drouillon, M., and Vervaeren, H. 2011. Techniques for transformation of biogas to biomethane. *Biomass Bioenerg* 35 (5):1633–1645.

Schink, B. 1997. Energetics of syntrophic cooperation in methanogenic degradation. *Microbiol Mol Biol Rev* 61 (2):262–280.

Schnürer, A., and Jarvis, Å. 2009. Mikrobiologisk handbok för biogas anläggningar. *Swed Gas Technol Cent, Rapport* 3.

Shafiei, M., Karimi, K., and Taherzadeh, M. J. 2011. Techno-economical study of ethanol and biogas from spruce wood by NMMO-pretreatment and rapid fermentation and digestion. *Bioresour Technol* 102 (17):7879–7886.

Shanmugam, P., and Horan, N. J. 2009. Optimising the biogas production from leather fleshing waste by co-digestion with MSW. *Bioresour Technol* 100 (18):4117–4120.

Sharaf, O. Z., and Orhan, M. F. 2014. An overview of fuel cell technology: Fundamentals and applications. *Renew Sust Energ Rev* 32 (0):810–853.

Sharma, A., Unni, B. G., and Singh, H. D. 1999. A novel fed-batch digestion system for bio-methanation of plant biomasses. *J. Biosci. Bioeng.* 87 (5):678–682.

Sharma, V. K., Testa, C., Lastella, G., Cornacchia, G., and Comparato, M. P. 2000. Inclined-plug-flow type reactor for anaerobic digestion of semi-solid waste. *Appl Energ* 65 (1):173–185.

Shiratori, Y., Ijichi, T., Oshima, T., and Sasaki, K. 2010. Internal reforming SOFC running on biogas. *Int J Hydrogen Energy* 35 (15):7905–7912.

Sosnowski, P., Wieczorek, A., and Ledakowicz, S. 2003. Anaerobic co-digestion of sewage sludge and organic fraction of municipal solid wastes. *Adv Environ Res* 7 (3):609–616.

Speece, R. E. 1983. Anaerobic biotechnology for industrial wastewater treatment. *Environ Sci Technol* 17 (9):416A-427A.

Sun, Y., and Cheng, J. 2002. Hydrolysis of lignocellulosic materials for ethanol production: A review. *Bioresour Technol* 83 (1):1–11.

Surendra, K. C., Takara, D., Hashimoto, A. G., and Khanal, S. K. 2014. Biogas as a sustainable energy source for developing countries: Opportunities and challenges. *Renew Sust Energ Rev* 31 (0):846–859.

Taherzadeh, M., and Karimi, K. 2008. Pretreatment of lignocellulosic wastes to improve ethanol and biogas production: A review. *Int J Mol Sci* 9 (9):1621–1651.

Thien Thu, C. T., Cuong, P. H., Hang, L. T. et al. 2012. Manure management practices on biogas and non-biogas pig farms in developing countries using livestock farms in Vietnam as an example. *J Clean Prod* 27:64–71.

Turton, R., Bailie, R. C., Whiting, W. B., and Shaeiwitz, J. A. 2008. *Analysis, Synthesis and Design of Chemical Processes*. Upper Saddle River, NJ: Pearson Education.

van Foreest, F. 2012. *Perspectives for Biogas in Europe*. Oxford: Oxford Institute for Energy Studies.

Van Lier, J. B., Rebac, S., and Lettinga, G. 1997. High-rate anaerobic wastewater treatment under psychrophilic and thermophilic conditions. *Water Sci Technol* 35 (10):199–206.

Van Lier, J. B., Van der Zee, F., Tan, N., Rebac, S., and Kleerebezem, R. 2001. Advances in high rate anaerobic treatment: Staging of reactor systems. *Water Sci Technol* 44 (8):15–25.

Van Velsen, A. F. M. 1979. Adaptation of methanogenic sludge to high ammonia-nitrogen concentrations. *Water Res* 13 (10):995–999.

Vandevivere, P., De Baere, L., & Verstraete, W. (2003). Types of anaerobic digester for solid wastes. In J. Mata-Alvarez (Ed.), Biomethanization of the organic fraction of municipal solid wastes (pp. 111–140). IWA Publishing.

Vavilin, V. A., Rytov, S. V., and Lokshina, L. Y. 1996. A description of hydrolysis kinetics in anaerobic degradation of particulate organic matter. *Bioresour Technol* 56 (2):229–237.

Verma, S. 2002. Anaerobic digestion of biodegradable organics in municipal solid wastes. MSc diss., School of Engineering & Applied Science, Columbia University, New York.

Villadsen, J., Nielsen, J., and Lidén, G. 2011. *Bioreaction Engineering Principles*. Berlin, Germany: Springer.

Vogels, G. D., Keltjens, J. T., and Van Der Drift, C. 1988. Biochemistry of methane production. In *Biology of anaerobic microorganisms*. New York: Wiley.

Wang, Z., and Banks, C. J. 2003. Evaluation of a two stage anaerobic digester for the treatment of mixed abattoir wastes. *Process Biochem* 38 (9):1267–1273.

Ward, A. J., Hobbs, P. J., Holliman, P. J., and Jones, D. L. 2008. Optimisation of the anaerobic digestion of agricultural resources. *Bioresour Technol* 99 (17):7928–7940.

Wellinger, A., Wyder, K., and Metzler, A. E. 1993. Kompogas-a new system for the anaerobic treatment of source separated waste. *Water Sci Technol* 27 (2):153–158.

Wilkinson, K. G. 2011. A comparison of the drivers influencing adoption of on-farm anaerobic digestion in Germany and Australia. *Biomass Bioenerg* 35 (5):1613–1622.

Williams, R. B., Jenkins, B. M., and Nguyen, D. 2003. Solid waste conversion: A review and database of current and emerging technologies. *Final Report. CIWMB Interagency Agreement IWM-C0172.*

Yadvika, Santosh, Sreekrishnan, T. R., Kohli, S., and Rana, V. 2004. Enhancement of biogas production from solid substrates using different techniques—A review. *Bioresour Technol* 95 (1):1–10.

Yen, H.-W., and Brune, D. E. 2007. Anaerobic co-digestion of algal sludge and waste paper to produce methane. *Bioresour Technol* 98 (1):130–134.

Zandvoort, M. H., Van Hullebusch, E. D., Fermoso, F. G., and Lens, P. N. L. 2006. Trace metals in anaerobic granular sludge reactors: bioavailability and dosing strategies. *Eng Life Sci* 6 (3):293–301.

Zehnder, A. J. B., and Wuhrmann, K. 1977. Physiology of a Methanobacterium strain AZ. *Arch Microbiol* 111 (3):199–205.

Zheng, Y., Pan, Z., and Zhang, R. 2009. Overview of biomass pretreatment for cellulosic ethanol production. *Int J Agr Biol eng* 2 (3):51–68.

Ziemiński, K., and Frąc, M. 2012. Methane fermentation process as anaerobic digestion of biomass: Transformations, stages and microorganisms. *Afr J Biotechnol* 11 (18):4127–4139.

Zinder, S. 1984. Microbiology of anaerobic conversion of organic wastes to methane: Recent developments. *Am Soc Microbiol News* 50 (7): 294–298.

Zuo, Z., Wu, S., Zhang, W., and Dong, R. 2013. Effects of organic loading rate and effluent recirculation on the performance of two-stage anaerobic digestion of vegetable waste. *Bioresour Technol* 146:556–561.

7 Combustion of Wastes in Combined Heat and Power Plants

*Anita Pettersson, Fredrik Niklasson,
and Tobias Richards*

CONTENTS

7.1   INTRODUCTION

Initially, when incineration of waste started, the main purpose was to reduce the volume of the solid waste, and in many cases it was handled as open dump fires. However, it became clear that there were toxic emissions from the flue gases and that leaching occurred from the residual solid material and by necessity a more controlled combustion started. Thereafter, the released energy was utilized to produce heat and power. From a global warming perspective, waste contains partly renewable material and partly fossil-based compounds and will thus have an impact on the net CO_2 emissions. However, it should be compared to the scenario where the waste is dumped in a landfill and where landfill gases have formed from the organic substances. This gas consists roughly of 50% CH_4 (methane) and 50% CO_2. Due to the higher impact of CH_4 (21 times higher than CO_2), as a greenhouse gas, and the difficulty of collecting all landfill gases from the large areas, the overall net CO_2 (CO_2 equivalents) emission decreases when waste combustion is used, compared to landfilling. Regardless of global warming, landfilling does not treat other harmful substances; moreover, the leached water must be controlled and treated for many decades before the levels are acceptable.

One main objective of waste combustion is to dispose of waste fractions that cannot be recycled or reused. Combustion is a better alternative than landfill from an environmental point of view (presuming that the waste fraction in question is combustible). However, landfill may be a cheaper alternative if not penalized by a legislated deposition fee. Waste combustion should be conducted in a plant where the waste is completely burnt; furthermore, the plant should be equipped with a flue gas–cleaning system to keep emission levels low (regulated by standards). Newer plants for waste combustion show a considerable improvement in the handling of emissions. When it comes to emissions of heavy metals, particulates, and dioxins (toxic equivalent [TEQ] basis) in the flue gases, the reduction for an average plant in the United Kingdom from 1991 to 2001 was more than 99% (Porteous 2001). The reduction of SO_2 was 89%. Corresponding numbers in the United States are the same for dioxins (TEQ basis), 96% or more for heavy metals and particulates, and the SO_2 reduction was 88% (Michaels 2010; Stantec 2011).

In many cases, the heat released is recovered to produce either electric power or heat for residential/industrial use. The combustion of waste reduces its volume radically and also results in a sterile residue where the principal residue is bottom ash. The light ashes (fly ash) are separated from the flue gas stream by dust filters.

Compared to conventional fuels (e.g., coal and wood), waste fuels are generally more inhomogeneous. Properties of the waste vary, for example, by local sources, season, and business activity. Therefore, to burn waste successfully, the boiler must have rather wide fuel flexibility.

In 2011, Europe had more than 470 waste-to-energy (WtE) plants handling municipal solid waste (MSW) in 20 countries (Haukohl 2012). In 2010, the United States had 86 waste incinerators. Today, Japan has more than 300 plants for the thermal treatment of waste and handles around 40 million tons of waste every year (Themelis 2012). Taking Sweden as an example, of the 4.4 million tons of MSW that were collected in 2012, 51.6% or 2,270,400 tons were energy recovered. (Swedish Waste Management 2013).

7.1.1 General Considerations about WtE Plants

The WtE facility consists of several parts. Initially, there is a presorting and size-reduction step where the incoming waste fuel is conditioned to fit the restrictions of the boiler, and a magnet is used to remove any magnetic metals. The next stage is the combustion where the fuel is introduced into a hot environment and air is added. This effectively reacts with the organic materials and forms mainly CO_2 and H_2O and at the same time releases energy due to the exothermic reactions.

When combusted, the fuel undergoes several steps: First, it is dried and degassed, which means that the water is vaporized and volatile hydrocarbons are removed. The temperature of the solid material is around 100°C to 300°C. Next is the pyrolysis and gasification process, where different gases, such as H_2 and CO, and different hydrocarbons are released into the gas phase, leaving a solid consisting of mainly carbon from the organic material together with the inorganic materials. The temperature during this part is mainly around 500°C–1000°C but can be higher depending on conditions. The last step is the oxidation, where oxygen comes into contact with the solid carbon and completes the solid reactions.

The most versatile form of energy that can be delivered from a WtE plant is electricity, as it can easily be distributed over long distances and can be used for many purposes. A steam cycle is utilized to produce electricity from part of the heat released by combustion. The heat from the combustion is transferred from the hot flue gas to a closed water/steam loop by heat transfer surfaces located in the walls of the combustion chamber and through the tube bundles in the path of the flue gas next to the combustion chamber. However, the fraction of the heat released during combustion that can be converted into electricity is theoretically limited by the steam temperature, which in turn is practically limited by material temperatures. For WtE plants, the limiting temperature is usually that of the final super heater (the last heat exchanger in the steam loop, raising the steam temperature before the steam turbine). The flue gas, especially the fly ash, in the waste-fired boilers is very aggressive on the steel used in the heat transfer surfaces. The problems of corrosion and deposit formation generally increase with material temperature. The corrosion problem can be counteracted, to some extent, by using high alloy steels, but that increases the cost of installation. It is also possible to limit the corrosion by changing the fuel properties, for example, by introducing an additive or by using a beneficial fuel mixture. An example of an additive that reduces the super heater corrosion is kaolin. Mixing a waste fuel with sewage sludge also reduces the corrosion. A superheated steam temperature of 400°C is commonly used in WtE plants. This steam temperature is chosen with some safety margin and corresponds to a moderate electricity conversion in order of 25%. The rest of the heat released from the combustion can be used for other purposes, such as industrial processes or residential heating, but the heat can only be utilized close to the plant, as the heat loss in the long water/steam tubing is considerable.

A number of key environmental issues must be considered by the WtE plant (European Commission 2006). These include

- Overall process emissions into air and water (including odor)
- Overall process residue production
- Process noise and vibration
- Energy usage and transformation
- Raw material (reagent) consumption
- Fugitive emissions—mainly from waste storage
- Handling and treatment of hazardous wastes

7.1.2 Waste as a Fuel

As already stated in the waste hierarchy, Figure 1.2, WtE is found at the bottom of the inversed pyramid only followed by landfilling. This means that the wastes intended for use as recovered energy are wastes that are not suitable for reuse or recycling. The waste classification in Chapter 1 (Section 1.4.1) shows that the main part of the wastes that should be recovered for energy are found in the group residual waste, meaning combustible wastes with a low material recycling value because of the low quality of the material, high contamination, and mixed materials (e.g., diapers and tissues). However, today, a lot of waste that could be reused or material recovered is combusted. This depends on many different factors such as legislations, infrastructure of waste management, collecting and sorting systems, and awareness among the waste producers in the system, namely, households and industries.

Regarding WtE, the waste is usually not referred to as residual waste but is instead divided into MSW, refuse-derived waste, industrial waste, and/or combustible waste. However, there is no clear definition of these classifications, and it is therefore important to clearly describe the waste considered in every case (Kara 2012; Reza et al. 2013; Rajendran et al. 2014; Krüger et al. 2014; Zheng et al. 2014). The largest amount of wastes treated by energy recovery are by-products from various industries and the properties of the waste they produce have a wide span, that is, demolition wood, food from pet-food industry, fiber sludge, and mixed polymers (see Figure 7.1).

All WtE plants have their own unique waste fuel mix depending mostly on which industries that are delivering waste to the plants. However, what is common for these fuel mixes is the high alkali (especially Na), S, and Cl content compared to most other fuels. These elements react and cause deposits and corrosion on the heat transfer surfaces in the boiler, leading to high maintenance cost, reduced heat transfer, and material losses. In the worst case, it might also cause unplanned production stops. In addition, there are also other elements in the waste fuels that make them difficult to handle, such as heavy metals. Some of these have a low melting point and are vaporized (Zn, Pb, and Hg) and react with available Cl and form chlorides that are carried and distributed in the hot parts of the boiler, increasing deposition and corrosion. Other materials, for example, Fe and Cu, act as catalysts for dioxin formation. Dioxins are formed when the flue gases are cooled in the convection path in the temperature range of 400°C to 250°C. This means that fast cooling of the flue gases in this range decreases the dioxin formation.

FIGURE 7.1   Industrial waste waiting for pretreatment and mixing with household waste in Borås, Sweden. (Courtesy of Anita Pettersson.)

7.2   BOILER DESIGN

For municipal WtE installations, there are two main combustion methods used: stoker (grate) and fluidized bed (also commonly denoted FB). Other alternatives exist in some applications, for instance, incineration of hazardous or medical wastes (Brunner 2002), but those will not be described here. The selection of furnace type depends mainly on the expected fuel supply and the waste fuel characteristics.

Generally, complete combustion requires that the combustion device provides sufficient temperature, time, and turbulence. The turbulence is needed to mix combustible gases with the oxygen supplied by air. A high temperature is necessary to fire up the reaction kinetics while the residence time has to be sufficient for the combustion reactions to reach completion before the gases leave the combustion chamber. Some standards prescribe minimum values of residence time and combustion temperature, for example, the European Commission Directive 2000/76/EC that states that the residence time should be at least 2 seconds at a temperature over 850°C (European Commission 2000). Turbulent mixing is induced by high speed over fire air jets. The total amount of air supplied to the combustion chamber has to be in excess of what is theoretically needed, because it is not possible to achieve perfect mixing of the oxygen and fuel. It is, however, desired to keep the excess air as low as possible in order to optimize the boiler efficiency.

7.2.1 Grate Furnace Boiler

There are many different grate furnace designs on the market, developed by various manufacturers for a range of fuels and boiler sizes. The grate concept (Figure 7.2) means keeping a fuel bed on top of a grate while letting primary air pass through the grate from beneath. While passing through the fuel bed, generally, almost all the oxygen supplied in the air will be consumed by the combustion process. The grate also has to convey the burning fuel from the fuel-feed side of the furnace toward the ash discharge at the other end of the grate. The speed of the movement of the fuel bed should be appropriate for the fuel to reach burnout just before the end of the grate. For waste fuels, the most common design is a sloping reciprocating grate in which every second row of grate elements are moving to drive the fuel, while the other second rows of grate elements remain fixed. Each row of grate elements overlaps the row beneath it. The moving elements push the refuse over the stationary elements. The moving grate elements are supported by a frame. Such a design accomplishes a fuel migration along the grate even for an inhomogeneous fuel, such as unsorted

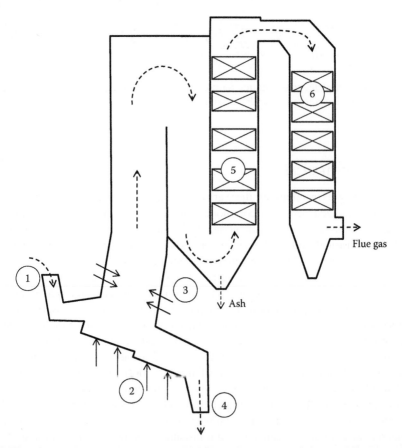

FIGURE 7.2 Example of waste-fired grate boiler: (1) fuel chute, (2) primary air, (3) overfire air injection, (4) bottom ash, (5) super heaters, and (6) economizer.

waste. If the waste fuel has a high heating value, it may be necessary to use water cooling, at least in the first part of the grate where the combustion is the most intense.

Larger grates are often designed as a number of modules with individual drives and air plenums. Such modules can be preassembled at the factory. Typically, a grate can consist of 2–4 modules in a series and 1–2 modules in parallel. It is possible for the boiler operator to control the air flow and fuel migration rate of each individual module.

A criterion for the sizing of the grate is that it must be wide enough to distribute the incoming waste fuel to a bed of appropriate height The specific fuel load on the grate area is another design parameter. The specific heat release rate from the grate is usually in the order of 1 MW/m^2, but it can be lower for high-moisture fuels (Kitto and Stultz 2005). The depth of the grate depends on the fuel's heating value; a highly calorific (dry) fuel burns rapidly and needs only a relatively short grate compared to a low calorific (wet) fuel that requires a longer grate (longer residence time) for complete burnout. There is also a minimum amount for the boiler load so that it can be operated without complications on a specific grate.

The primary combustion air is supplied by the air plenums beneath the grate and has to be distributed evenly across the fuel bed of each module. This is usually accomplished by small air holes or slits in the grate bars. As a guideline, about 3% open area usually provides a high enough pressure drop over the grate to distribute the air sufficiently. Under normal operation, about 60% of the total combustion air is supplied as primary air through the grate. The rest of the combustion air is supplied as overfire air, through the front and rear walls in the combustion chamber above the bed. The overfire air system is designed to mix the air and the combustible gases emitted from the solid fuel. Sometimes, recirculated flue gas is added to the overfire air system in order to enhance the gas mixing and to limit the NO_x formation.

When the fuel burns, the ash formed is divided into either light (fly ash) or coarse bottom ash. The bottom ash (sometimes called *clinker*) remains on the grate until it falls into an ash quench, where it is cooled before being transported to the ash storage. It is common to use a water quench to rapidly cool the bottom ash. The other ash fraction, the light fly ash, is entrained into the flue gas and follows the gas path until it is deposited in the boiler system or collected in the flue gas–cleaning system.

7.2.2 Fluidized Bed Boiler

In a fluidized bed boiler, there is a bed of sand (or similar inert material) at the bottom of the combustion chamber (Figure 7.3). The bed rests upon an air distributor, which allows the primary air to pass into the bed from below but prevents the bed particles from pouring down through the air distributor. This is usually accomplished by a large number of specially designed nozzles distributed over the bottom of the reactor. The nozzles are designed to give a sufficiently high pressure drop to distribute the primary air evenly over the bed cross section. The primary air then passes through the sand bed, some interstitially between the particles but most of the air forms rising bubbles. As a result of the air acting on the sand particles, the bed becomes fluidized. There are two main categories of fluidized bed boilers:

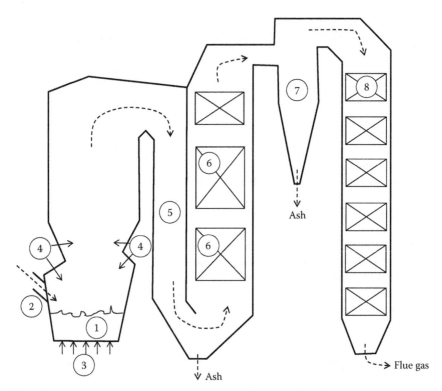

FIGURE 7.3 Example of a waste-fired fluidized bed boiler: (1) fluidized bed, (2) fuel feed, (3) primary air, (4) overfire air, (5) empty gas pass, (6) superheaters, (7) cyclone, and (8) economizer.

(1) bubbling fluidized bed (or simply fluidized bed) in which the primary air velocity is too low to carry the bed particles out of the reactor and (2) circulating fluidized bed where the gas velocity through the bed is high enough to carry a substantial fraction of the bed material out from the reactor. In a circulating fluidized bed, the bed material leaving the reactor has to be captured and recirculated back to the reactor in order to maintain the bed material in the reactor. For that purpose, circulating fluidized bed boilers are equipped with a relatively large cyclone to capture the bed material and a return leg to feed it back to the reactor via a particle seal.

Of these two fluidized bed designs, the fluidized bed with a bubbling bed is the preferred choice for moderately sized boilers, in order of 20 MW_{th}, while the somewhat more complex circulating fluidized bed design is a more attractive alternative for the larger boilers because the power produced per cross sectional area unit is higher in a circulating fluidized bed.

The fuel is fed into the bed by one or more fuel feeders. The number of fuel feed points depends on the cross-sectional area of the combustion chamber because it is important to avoid severe maldistribution of the fuel. The fuel used in a fluidized bed has to be somewhat more finely crushed than what is needed for a grate-fired boiler. When the fuel reaches the bed, it is rapidly heated before the combustion starts. During this initial heating and drying, heat is transferred to the fuel from the bed material. When the fuel starts to burn, heat is transferred to the bed material. The high thermal capacity of the fluidized bed material and the continuous stirring caused by erupting bubbles keep the temperature relatively uniform within the bed. The primary air serves two purposes: to fluidize the bed and to supply oxygen needed for the combustion. To control the bed temperature, recirculated flue gas can be added to the primary air flow. The height of the fluidized bed within the boiler is usually kept to about 1 m, in order to avoid excessive pressure losses.

As mentioned, the temperature in the fluidized bed is well controlled, but for a highly volatile fuel, such as waste, a major part of the heat released comes from the volatiles that tend to burn above the bed, where the overfire air is injected from ports in the vertical boiler walls. The overfire air should enhance the gas mixing and supply sufficient oxygen, in the same way as in a grate-fired furnace.

In a fluidized bed boiler, the bottom ash from the fuel mixes with the bed material, leading to an accumulation of material in the bottom bed, if not counteracted by bottom ash discharges through the ash discharge chutes at the bottom of the fluidized bed. The removed bottom ash is mixed with the original bed sand. These bottom ash discharges are necessary, not only to prevent the accumulation of bottom ash but also to renew the bed material to avoid bed agglomeration, because particles in the fluidized bed are prone to getting a slowly growing coating of ash constituents, for example, alkali silicates. The ash chemistry is complex, but as a result of the chemical reactions, the coatings on the bed material particles may become sticky. In the worst case, this may lead to bed sintering, which would mean that the boiler has to be taken out of operation. To counteract this potential problem, fresh sand is intermittently fed into the combustion chamber, replacing the old bed material that is discharged with the bottom ash. The amount of fresh sand needed depends on the fuel properties as well as the sand quality and bed temperature.

In contrast to the grate-fired boiler, a fluidized bed boiler does not involve any moving parts, which reduces the investment cost of the boiler. On the other hand, a fluidized bed boiler requires fresh sand and a more homogeneously crushed fuel, which may increase the running operational cost.

7.3   FLUE GAS–CLEANING SYSTEM

In the early 1980s, waste-fired boilers were usually equipped with only an electrostatic precipitator (ESP) for particulate capture. Thereafter, after growing health and environmental concerns, air pollution agencies began setting regulations for many other substances, such as hydrochlorides, dioxins, furans, and various heavy metals, implying that an ESP alone was no longer a sufficient solution for the flue gas cleaning.

The flue gas–cleaning system should remove potentially harmful substances from the flue gas before being released to the atmosphere. In general, due to the heterogeneous source of most waste fuels, the untreated flue gas contains a number of substances that are harmful for both health and the environment. Different gas compounds and substances require different cleaning devices for efficient capture. Thus, the cleaning of the flue gas demands a number of devices in series, with each device designed to remove certain substances. Different cleaning devices also have specific temperature ranges at which they are most efficient. There are a number of technologies available on the market, each with a specific set of requirements that should be fulfilled for optimum performance. Each WtE plant has a unique set of specific requirements that has to be fulfilled, for instance, regarding available area, budget, and emission standards. Accordingly, the designs of the existing WtE plants, including the flue gas–cleaning system, vary considerably from case to case.

7.3.1 Particle Precipitation

The concentration of the particulate matter (PM) in the flue gas depends on the fuel as well as on the design of the combustion system. The fly ash concentrations are generally higher in the flue gas from the fluidized bed boilers than from the grate boilers. The particle sizes vary from submicron particles up to a few hundred microns. Particles smaller than 2.5 μm (PM 2.5) are of particular health concern, because they can easily be inhaled and reach far into the lungs. Furthermore, toxic organics tend to adsorb on these small particles. Hence, the particle precipitator is a very important part in a flue gas–cleaning system.

The cyclone separators are probably the most used particle-collection devices in the world. A cyclone consists of a vertical cylinder body with a conical bottom from which precipitated particles are discharged (Figure 7.4). The gas enters the circular body tangentially to create a swirl-type flow along the walls of the cylinder. Particles suspended in the gas are driven by the centrifugal force toward the walls, where they lose their momentum and slide down toward the bottom due to gravity. It is a simple device that is efficient for larger particles, but it is not sufficiently efficient for small particles to meet most legislated emission standards for power plants.

In an ESP, the particles in the flue gas are driven toward the hanging steel curtains by electrostatic force. ESPs are much more efficient for small particles than cyclones. The basic concept of an ESP is to charge the particles and then collect them in an electrostatic field. Although it is a two-step process, it is generally carried out simultaneously in one chamber. The charging process is quick, while the collecting process is relatively slow and therefore size limiting. Dust collected on the collection plates form solid dust cakes, which are intermittently removed by rapping the plates. Most of the dust cake falls down in a hopper below, but a small fraction become re-entrained into the flue gas. To minimize this effect, an ESP is designed with a number of systems in a series, each with an individual power supply and rapping system. Most of the particles are captured in the first system, which accordingly needs rapping most frequently. The re-entrained particles from the first system are then collected in the next downstream system. The last system of the ESP does not need rapping very often, but when it does, the re-entrained particles leave through the outlet of the ESP. This is one of the reasons why it is difficult to reach ultralow emissions with an ordinary ESP. In general, the collecting efficiency of an ESP is proportional to the total area of the collecting surface area of the ESP. However, for very high collecting efficiencies this proportionality is no longer valid because the last fraction of the particles is harder to capture. However, there is an alternative design, the wet ESP, which can provide very low particulate emissions because the

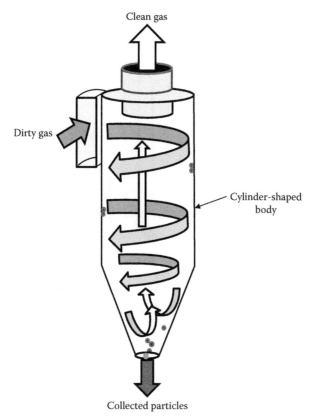

Clean gas

Dirty gas

Cylinder-shaped
body

Collected particles

FIGURE 7.4 Illustration of principle of cyclone separator.

collecting electrodes are flushed with water to remove the collected particles. The re-entrainment from rapping is hence avoided, but the wet ESP consumes water and provides a wet rest product.

For modern WtE plants, the most used particle-collection device is the fabric filter. In such a device, the particle-laden flue gas passes through a cloth-like filter medium. Solid particles on the windward side are deposited on the filter where they agglomerate and form a filter cake. The flue gas passes through the filter cake and the filter medium while particles are precipitated. The presence of a filter cake increases the collecting efficiency of the particles as well as improves the sorbent utilization for the cases when lime is added upstream to capture HCl and SO_2. However, the filter and the filter cake cause a substantial pressure drop in the flue gas that has to be overcome by the fan work. Over a period of time, the filter cake becomes thicker and the pressure drop increases. To prevent very high pressure losses, the filter is regularly cleaned from its filter cake, for example, by a pulse of pressurized air from the clean side of the filter or by a mechanical shaking mechanism. The fabric filters can collect submicron particles more efficiently than is possible by an ordinary ESP, as long as it is operated with a low gas velocity through the filter material, requiring a relatively large total filter surface area.

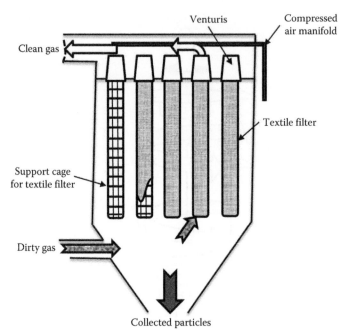

FIGURE 7.5 Illustration of an example of a baghouse filter, which is cleaned by compressed air pulses.

A typical industrial fabric filter module is a baghouse filter (Figure 7.5), in which a number of filter tubes hang vertically; the direction of the flow of the flue gas is from the outside of the tubes and going inwards. The openings of the tubes, for the exit of the clean gas, are located at the top. The filter bags are internally supported by wire cages to avoid deflation due to the pressure drop over the filter surface. The regular cleaning stresses the filter material, and as a result, a bag rupture may occasionally occur. A modular designed baghouse filter allows for the replacement of filter bags while still running the plant by taking only one of several parallel filter modules out of service. Of course, a prerequisite is that the total filter area in the baghouse must be large enough to allow for the outage of one module without getting too high gas velocity through the other filter modules.

7.3.2 CO CONTROL

Continuous monitoring of the CO emission can be used as an indicator of the combustion quality. A high concentration of CO in the flue gas indicates poor combustion, which could be accompanied by the presence of other, more harmful, hydrocarbons in the flue gas. For that reason, the reduction of CO is preferably accomplished by modifications in the combustor and not by secondary flue gas–cleaning reactors. To fully oxidize CO and other combustible gases that are released from the burning solid fuels, the combustible gases have to be mixed with oxygen by the overfire air system at sufficient high temperature and enough residence time for complete burnout.

7.3.3 SCRUBBERS FOR HCL AND SO₂ REMOVAL

The gas emissions of HCl and SO_2 are generally controlled by scrubbers, of which three main categories can be distinguished: wet scrubbers, wet–dry scrubbers, and dry scrubbers. The main parameter of performance for a gas-cleaning device is its collecting (or removal) efficiency, which is simply the fraction of a substance (or gas compound) that is captured by the device. Another measure of performance for the scrubbers is the stoichiometric ratio—the ratio of reagents fed into the reactor divided by the theoretical needed amount to capture all the HCl and SO_2 in the flue gas. This ratio is usually higher than unity due to the mass transfer limitations and incomplete mixing. In general, the stoichiometric ratio used is lower in a wet scrubber than in a dry scrubber. As an example, the chemical reactions for the capture of HCl and SO_2 when using hydrated lime are

$$Ca(OH)_2 + 2HCl \rightarrow CaCl_2 + 2H_2O$$

$$Ca(OH)_2 + SO_2 \rightarrow CaSO_3 + H_2O$$

A wet scrubber can be divided into two consecutive stages, either combined in a single two-stage scrubber (Figure 7.6) or as two separate vessels in a series. The first stage uses acidic water to capture HCl while the second stage uses either lime slurry or another reagent such as caustic soda for SO_2 absorption. In a wet scrubber, the direction of the flue gas flow is generally upwards, meeting the falling droplets of injected water

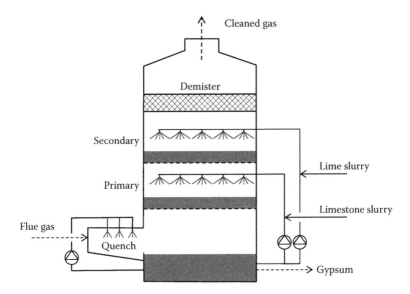

FIGURE 7.6 Example of the outline of a wet scrubber system.

and reagents. The removal efficiencies of HCl and SO$_2$ can be >99% in wet scrubbers, and the operating cost is relatively low as the lime (or other soda) utilization is high. However, the captured rest product is wet and has to be further treated before it can be disposed.

In a wet–dry scrubber (i.e., "semidry scrubber" or "spray dryer"), both the flue gas and the lime slurry usually enter from the top of the reactor (Figure 7.7). Contrary to the wet scrubber, the amount of water added is not sufficient to saturate the flue gas, resulting in a dry rest product. The acid gases in the flue gas are neutralized by the hydrated lime (or equivalent) when the gas comes into contact with the finely atomized droplets of the slurry. Heavy metals in the flue gas are to some extent absorbed on the particles and droplets within the reactor. The reactor is designed to give a sufficient long residence time of the droplets to evaporate inside the vessel. Furthermore, the outlet temperature should be high enough to ensure a dry product and to avoid agglomeration of the PM by calcium chloride. A typical outlet temperature of the flue gas from the scrubber is 140°C. The outlet temperature is controlled

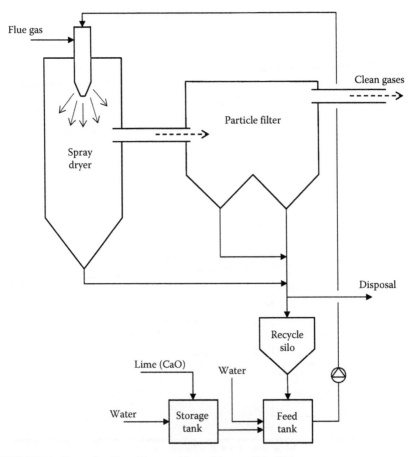

FIGURE 7.7　Example of semidry scrubber with a particle filter.

by the amount of water sprayed into the reactor, and the concentrations of the acid gases are controlled by the injected amount of reagent (e.g., lime). A fraction of the solid particles is trapped at the bottom of the reactor while the majority of the solids are carried by the gas flow into the downstream baghouse filter. Activated carbon is typically injected between the reactor and the baghouse filter in order to further collect the heavy metals, dioxins, and furans. The absorbing reactions started in the scrubber continue in the filter cake on the fabric filters, enhancing the removal efficiency of the system.

A dry scrubber is based on the injection of the powdered hydrated lime, activated carbon, and recirculated fly ash in a venturi reactor (or simply in the flue gas duct) upstream of the baghouse filter. The fly ash recirculation allows for a large amount of reactants to be present in the system, which improves the collecting efficiency. The chemical reactions of the sorbents are similar to the previously described scrubbers. The flue gas temperature at the sorbent injection is usually in the range of 150°C–320°C, depending on the sorbent used and other aspects of the process (Hasselriis 2002). The main advantage of the dry scrubber is the lower investment cost, compared to the other alternative scrubbers. On the other hand, the utilization of added lime is lower.

7.3.4 NO_x Removal

Some of the nitrogen in the fuel and in the combustion air can be oxidized to form nitric oxide (NO) and nitrogen dioxide (NO_2), commonly clustered as NO_x. In combustion plants, more than 90% of the NO_x is usually in the form of NO. However, after a period of time in the atmosphere, most emitted NO will eventually form NO_2. The two gas compounds, thus, cause equal environmental damage. There are three main chemical mechanisms identified regarding the NO_x formation: thermal, prompt, and fuel NO_x formation. Thermal NO_x formation is usually negligible at temperatures below 1300°C (de Nevers 1995). The prompt NO_x is formed under fuel-rich conditions during the first part of the combustion by hydrocarbon radicals reacting with nitrogen; furthermore, it is hard to counteract this mechanism, but it is usually a relatively small source of the total NO_x formation when burning fuels containing nitrogen (Kitto and Stultz 2005). The fuel NO_x formation originates from the nitrogen in the fuel, which via a number of chemical reactions leads to NO production following a reaction with oxygen. This is usually the major source of NO_x formation when burning fuels containing nitrogen, such as waste. To limit the fuel NO_x formation, it is beneficial to keep the oxygen concentration low in the high-temperature zone of the combustion chamber. The NO_x emissions can be significantly reduced by an optimized staged air technology. However, there is a risk of increased CO emissions by reducing the available oxygen in the combustion zone.

Reduction of NO_x can be achieved by both primary and secondary measures. Primary measures are taken in the combustor, for instance, by limiting the combustion temperature to avoid the formation of thermal NO_x and by keeping the air factor low in the primary combustion zone.

Secondary measures include the two reduction methods: SNCR (selective noncatalytic reduction) and SCR (selective catalytic reduction). In the SNCR method, ammonia, or urea, is injected using nozzles inserted in the upper part of the combustion chamber in a region where the flue gas has a temperature between 850°C and 950°C (Tabasová et al. 2012). The added ammonia, or urea, displaces the chemical equilibrium of the NO_x forming chemical reactions, leading to reduced NO_x concentrations in the flue gas. A common figure for NO_x reduction by SNCR is about 60% (Tabasová et al. 2012). To reach a high NO_x reduction, somewhat more reagent (ammonia or urea) has to be injected than is actually consumed, which leads to unreacted ammonia in the flue gas leaving the boiler, which is called ammonia slip. Generally, the more reagent that is injected, the better the NO_x reduction at the cost of increased ammonia slip, but it is also important to distribute the reagent evenly over the flue gas cross section to maximize the use of the reagent added. It can be mentioned that if the flue gas–cleaning system contains wet scrubbers, they can be used to remove most of the ammonia slip from the flue gas.

The alternative method, SCR, can achieve more than 90% NO_x reduction (Kitto and Stultz 2005; Tabasová et al. 2012). In the SCR process, a catalyst is used to promote the reaction between ammonia and NO_x to form nitrogen and water. Upstream of the catalyst, ammonia or urea is injected into the flue gas. The catalyst enhances the reactions and the ammonia slip is negligible for an optimized injection system (Kitto and Stultz 2005). However, the catalyst is expensive and can easily become poisoned by the constituents in waste fuels. Therefore, to avoid damaging the catalyst when used after waste-fueled boilers, the SCR reactor is preferably placed at the end of the flue gas–cleaning system, after scrubbers and particulate filters. Depending on the layout of the system, the flue gas may have to be reheated to reach the temperature window of about 250°C–450°C, where the catalyst is active (Tabasová et al. 2012). There are catalysts that can operate at lower temperatures if periodically reactivated by heating (Hitachi Zosen Inova AG 2014).

In conclusion, an SCR can reduce the NO_x emissions to lower levels and give a lower ammonia slip than SNCR, but at a higher investment cost.

7.4   ASHES

Almost all combustion results in by-products such as ash and flue gases, so also WtE. The ash content or the incombustible fraction of an average waste fuel is around 20% but can be anything between 10% and 50%, depending on the material and origin. The ash (incombustibles) consists of the inorganic elements of the fuel such as: Si, Ca, S, P, K, Na, Al, Mg, Ti, and Fe. Additional impurities like glass, ceramics, soil, and gravel may also be present in waste not properly sorted. These materials can be called ash-forming matter. Most of the ash-forming matter is inert, that is, not reacting during combustion, but some are very reactive and/or form compounds with low melting and evaporation temperatures, for example, alkali silicates, chlorides, and sulfates.

The ashes formed are collected and removed from the different locations in the system. Ashes are divided into two main groups: bed ash (bottom slag) and fly ash. The bed ash is collected in the bottom of the boiler and contains most of the impurities (glass, ceramics, gravel, metal pieces, etc.) coming in with the fuel. Furthermore, in the case of fluidized bed technology, also sand (bed material) is collected. The fly ash can be divided into several fractions, which are decided by the specific type of flue gas cleaning–system that is used. Very often, the first cleaning step is directly when the flue gas leaves the boiler and is led through an empty pass with a sharp bend where the particles with the highest density and size are collected. This fly ash fraction is usually called boiler ash. The next step could be a cyclone and/or an ESP.

7.5 ASH TREATMENT/DISPOSAL

Ash handling is an important issue today. Stricter regulations, decreasing demand for material for landfill construction and landfill coverage, and current processes that are time and/or space limited are enforcing a development in this area. Bed and fly ash are most often treated separately because of the differences in the properties of these ashes.

Some of the ash problem could be solved by the minimization of the ash produced by improved pretreatment of the waste before combustion to sort out glass, gravel, sand, soil, metals, and others that otherwise end up as bed ash. This pretreatment is more or less efficient depending on the grade of contamination of the combustible waste by these materials.

A thorough presorting of the waste could also minimize the ash production. A lot of material that could be classified as biofuel or could be material-recycled ends up in the combustible waste and could be sorted out at the waste source or at a pretreatment plant.

7.5.1 BED ASH

Bed or bottom ash/slag is the coarsest ash in the system. It is also most often the largest part of the formed ash, almost 1 million tons/year in Sweden, corresponding to 15%–20% of the total weight of the waste fuel burned. The ash is crushed and as much as possible the metal pieces are separated and material recovered. The remaining bed ash fraction now separated from most of the metals is most often classified as inert or nonhazardous waste and is sometimes used instead of gravel in road constructions but most often as construction material at landfills (Avfall Sverige 2013). However, the amount of active landfills are decreasing, and the bed ash must also compete with other inert or nonhazardous waste fractions such as contaminated sand and soil as construction material; thus, the demand is decreasing. This means that it is of utmost importance to find new ways to deal with the ash. Some suggested methods are

- Minimization of the bed ash production by
  - Waste pretreatment
  - Waste presorting
- Posttreatment of the bed ash to
  - Improve the mechanical metal separation
  - Minimize oxidation
  - Improve the metal recovery of reacted metals
- Selective waste combustion to improve ash quality

7.5.2 Fly Ash

In 2012, Sweden alone handled about 220,000 tons of industrial and MSW fly ash (Avfall Sverige 2013). Fly ash is the most contaminated ash and is often classified as hazardous waste. Most of this ash is used as a neutralizer for waste acids and fillers in an old mine on the island of Langøya in Norway (Avfall Sverige 2013; NOAH AS 2012). A small portion of it is also used in Germany in a similar way, and the remaining amount is deposited in Sweden. However, none of these three options are economically or environmentally sustainable. In addition, in about 10 years, the Norwegian mine will be completely filled; yet, no other options have been found. However, many of these ashes are high in both metals—for example, Zn, and Cu— and nutrients—namely, K, Na, P, and S—which could be recovered by some means.

Suggested methods for reducing the fly ash cost and the environmental impact:

- Minimization of the fly ash production by
  - Waste pretreatment
  - Waste presorting
- Posttreatment of the bed ash to
  - Improve the metal recovery of metals
  - Nutrient recovery
  - Ash cleaning/stabilization
- Selective waste incineration to improve the ash quality
- Selective ash collection

7.5.3 Posttreatment of Bed Ash

The required posttreatment of bed and fly ash is somewhat different because of the difference in the properties between the two. Concerning bed ash, mechanical metal sorting is an important treatment, which is impossible to use on most fly ash not containing metal pieces. Mechanical sorting is mostly conducted using several different separation techniques in a series. Examples of techniques used are sieving with different techniques and size distribution, magnetic sorting, Eddy-current separation, shaking table, and density separations; the manner in which the sequence is put together varies between the plants.

The resulting bed ashes from the different boilers also have different properties due to the combustion technique used (grate boiler/fluidized bed), temperature in the boiler, elemental composition of the fuel, and so on.

7.5.4 Posttreatment of Fly Ash

Fly ash from the waste incineration contains high levels of metals and often also dioxins and furans. Additives used in the flue gas–cleaning system to capture the chlorines, dioxins, and heavy metals, for example, slaked lime and active carbon, dilute the fly ash. In particular, the added slaked lime could be a problem when posttreating the ash. By using acid leaching, metals such as zinc and lead can be extracted from the fly ash (Karlfeldt et al. 2014; Schlumberger and Bühler 2012;

Schlumberger et al. 2007), substituting virgin metals in the industry. Some fly ashes are high on phosphorus and thus interesting as a P source.

7.5.5   Selective Waste Incineration

One solution might be to decide which waste incineration plants should take care of which type of waste fractions. Today, most plants receive waste from a whole range of companies having very different element compositions in their waste. If plants specialized in a certain type of waste, it might result in ashes that could be high in metal or nutrient concentrations and thus be more attractive to the industry. The most critical substances could be concentrated in one or two plants avoiding that they contaminates cleaner wastes. It would also result in a smaller fraction of highly hazardous ash allowing the ash treatment cost for this small amount to be high compared to the treatment cost of other ashes produced.

7.5.6   Selective Ash Collection

To decrease the amount of hazardous ash or to have ash fractions with higher concentrations of different compounds, a special selection has been suggested. All ashes collected in different parts of the flue gas–cleaning system have their own composition because of the temperature, cleaning technique, additives, and position in the system. Instead of mixing all the fly ash in one or two silos, all the fractions could be collected separately. However, to get an economic sustainability, a certain amount of each type of ash has to be produced within an area to provide an upgrading facility with the materials, which can be a problem.

7.6   COST AND REVENUES

7.6.1   Cost of WtE Plants

The cost of waste for power-producing units (including WtE units) varies between different countries and even within large countries (IEA 2010). In a Danish report (Energinet.dk 2012), the investment cost for a 25–35 ton/h fuel input WtE plant (CHP) was estimated to be between €4.7/ton/h and €6.8/ton/h (corresponding to €7–10 million/MW$_{electricity}$). Compared with other fuels, the investment cost is about three times higher than a woodchip CHP using a Rankine cycle (€2.6 million/MW$_{electricity}$) and four times higher than a pulverized coal power plant (€2.04 million/MW$_{electricity}$). However, the actual numbers vary with the site location. In another study (Hogg 2001), the price for Sweden was given as €7 million/MW$_{electricity}$ for a 15 MW$_{fuel}$ plant and €4 million/MW$_{electricity}$ for a 114 MW$_{fuel}$ size. A case in Germany, with an annual waste input of 200,000 tons (fuel input 25 ton/h), was estimated to cost €122 million (equals €4.9/ton/h). Furthermore, the cost depends heavily on size, where a larger plant renders a lower specific cost (Hogg 2010; see Figure 7.8).

In addition to the investment cost, there is the running cost—operation and maintenance (O&M) cost. They can be divided into two parts: fixed and variable. The fixed O&M costs include all costs that are present regardless of whether the plant is

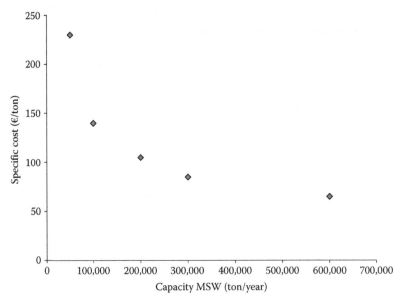

FIGURE 7.8   Specific cost of a WtE plant. (Data from Hogg, D., Costs for municipal waste management in the EU, Final Report to Directorate General Environment, European Commission, 2010.)

running or not, such as administration, planned maintenance, service agreements, property tax, and reinvestments within the expected lifetime. The variable O&M costs include the actual running costs (fuel costs were excluded because they are a matter of negotiation and dependent on specific conditions) and consist of, for example, treatment and disposal of residuals (i.e., ashes), repair, spare parts, and utilities. These are found to be €36.56/ton and €19.10/ton for the fixed and running costs, respectively, for the German plant (Hogg 2010). Estimated prices for the fixed and variable O&M costs together are €53/ton for the Danish WtE plant (Energinet. dk 2012).

The cost for pretreatment required to prepare the unsorted MSW must also be considered. This is especially important for the fluidized bed boilers, where typically the costs are €10/ton to €30/ton (European Commission 2006). Handling of ash is expensive but necessary. While some ash is used as construction materials, the most contaminated ash must be landfilled. The price varies between the different countries, depending on policies (taxes) and availability. In Sweden, the current price for land-filling is 500 SEK/ton, which equals €53.6/ton (SFS 2015). Other taxes for the waste material intended for incineration (either as direct cost of the material or as a function of the emitted gases, such as greenhouse gases) are used differently in different countries and add to the total cost and also to the diversity of the overall economy.

7.6.2   REVENUES FROM WtE PLANTS

There is a fee associated with the handling of waste. In 2010, the average gate fee in Europe was €87.5/ton, but the fee varies from almost €0/ton up to more than

€100/ton, depending on the country and the specification of the waste (European Commission 2006). The hope is that by increasing the number of WtE plants and with better sorting of the MSW, the gate fee will decrease.

Apart from the income from the fuel itself, the plants could earn income from other products that could also be produced there. These are in the form of electricity, steam, or heat for district heating. The price of electricity varies on an hourly basis. NASDAQ OMX Commodities (2014) gives the selling price on the spot market (taken here as a monthly average for August and September 2014) as €34/MWh for the Nordic countries, €35/MWh for Germany, €45/MWh for the Netherlands, and €45/MWh for the United Kingdom. The average price in the United States for the same period (August–September 2014) was found to be $47.4/MWh (EIA 2014), which equals €36/MWh.

A few of the WtE plants are situated close to other industries that need steam. This could be an additional source of income for the plant but at the cost of lower electricity production. The same is applicable for district heating, with the exception that the loss of electricity production is minor and that the price for the district heating itself is favorable (it can sometimes even be higher than the price for electricity). However, this requires that there is a need for district heating and a network is in place. A rough estimate is that about half the price for district heating is used for maintenance and support of the network but with a price of about 0.8 SEK/kWh for the end user, this represents a revenue of €43/MWh for the heat producer.

7.7   MODERN WtE INSTALLATIONS

It is possible to optimize utilization of the energetic content of fuel by making several improvements to the traditional WtE boiler. One such improvement is to produce both heat and electricity in a unit known as a "combined heat and power plant" or CHP. This increases the energy efficiency of the plant, and as described in Section 7.6, the revenues may be almost the same as for electricity; although the spot price of electricity fluctuates significantly, the price of district heating is more stable. If district heating is available, it is possible to increase the efficiency even further by utilizing a flue gas condenser. This unit condenses the water vapor in the flue gases and produces warm water, but at the cost of a slightly higher electricity demand (since additional fans are required); it also requires a cleaning system for the condensed water. In addition to the production of warm water, the condensation of flue gases also reduces emissions of particulates further and removes all acid gases. Moreover it opens up the possibility of using sulfur recirculation (Andersson et al. 2014) in order to reduce corrosion in the heat exchangers. Another opportunity for improving the boiler is to reduce the exit temperature of the flue gases, which means that more heat is transferred in both the economizer and the air preheater. It is important that the temperature is kept above the dew point of HCl and sulfuric acids to prevent corrosion, but the margin in many plants is currently higher than necessary. Other alternatives are reducing the temperature (and thus the pressure) in the condenser after the steam turbine, increasing the pressure and temperature of the steam, introducing more advanced combustion control, and including a reheating cycle. The reduced temperature in the condenser gives a higher production of electricity and a lower loss of energy but is often determined by the natural conditions and availability of cooling water. Increasing the temperature and pressure of the steam can also increase the production of electricity but there are limits, since the inevitable corrosion of the superheater tubes becomes more severe at higher temperatures. Corrosion can be reduced by using more expensive and corrosion-resistant material (such as Inconel) or a shielded placement (e.g., using the sand that is returned in a circulating fluidized bed boiler or having protective tiles in a grate-fired boiler). Advanced combustion control (i.e., process control with recirculation of the flue gases) provides better utilization of the fuel because it not only ensures complete burnout but also reduces the amount of excess oxygen produced, which, in turn, prevents the formation of NO_x. Finally, the reheating cycle is a way of increasing the production of electricity by reheating the steam after the high pressure stage in the turbine and prior to low pressure expansion. This method is, however, costly and is therefore only applied in large plants.

Typically, the electrical efficiency of these power plants is around 22%–25%, using a steam temperature and pressure of 400°C and 40 bar, respectively. New installations have nevertheless shown that it is possible to reach efficiency values exceeding 30% using more modern technology. One of the largest plants is the AEB Amsterdam plant in the Netherlands, comprising a total of six lines with a capacity of almost 1,400,000 t.p.a. Four of these lines are operating since 1994, with a steam temperature of 415°C and a pressure of 43 bar and two lines since 2007 and operate at 420°C and 130 bar (Martin GmbH 2013). An electrical efficiency of 30.6% was achieved using a number of measures, namely reheating the steam after the high-pressure stage using saturated boiler steam, reducing the oxygen level from 8%–11% to 6% by recirculating the flue gases, lowering the outlet temperature of the flue gases to 180°C from the 200°C–220°C previously used, preheating the condensate in the scrubbers (including the flue gas condenser), and using more corrosive-resistant materials (WSP 2013). These measures aside, the plant is situated close to the harbor and uses cooling water at 25°C, which lowers the condenser pressure to 0.03 bar. The gas is cleaned in several steps: first, SNCR is employed to reduce the formation of NO_x by injecting ammonia into the boiler. An ESP is placed after the boiler to remove particulates, and is followed by an absorption section that uses fine-powdered active carbon and lime, which are removed in a fabric filter. It is here that dioxins/furans and heavy metals are removed. The subsequent scrubber section, with an HCl scrubber followed by an SO_2 scrubber that uses lime, ends with a flue gas condenser (polishing scrubber) in which the heat produced is used to preheat the condensate; the condensed water is then used in the HCl scrubber.

The Zabalgarbi plant in Spain is another example of a plant with high electrical efficiency. The steam pressure used here is 100 bar but the temperature is only 330°C. These are not ideal conditions for a steam turbine, as heavy condensation would be generated even at relatively high pressures. The distinguishing feature of this plant is that it utilizes a nearby gas turbine combined cycle plant to provide the extra heat necessary for the steam (i.e., by using the flue gases after the gas turbine). The final temperature reaches 540°C with the pressure being kept constant at 100 bar; it is claimed (WSP 2013) that an efficiency of 42% is reached (this includes a reheating of the steam after the high-pressure stage). The gas-cleaning system employed is a semidry system. It starts with an injection of ammonia in the high-temperature region to reduce NO_x (flue gases are recirculated, helping to keep the thermal NO_x low). Thereafter the flue gases enter a wet scrubber into which lime and active carbon are injected. In the final stage, the gases pass through a baghouse filter before being released into the air through the stack.

REFERENCES

Andersson, S., E. W. Blomqvist et al. (2014). Sulfur recirculation for increased electricity production in Waste-to-Energy plants. *Waste Management* Volume 34, Issue 1, Pages 67–78. doi: 10.1016/j.wasman.2013.09.002.

Avfall Sverige (2013). Svensk avfallshantering 2013. Retrieved June 15, 2014, from http://www .avfallsverige.se/statistik-index/avfallsstatistik/.

Brunner, C. R. (2002). Waste-to-energy combustion, incineration technologies. *Handbook of Solid Waste Management*. Eds. G. Tchobanoglous and F. Kreith. New York, McGraw-Hill: 13.13–13.84.

de Nevers, N. (1995). *Air Pollution Control Engineering*. Singapore: McGraw-Hill.

EIA (2014). File: Electricity historical data, Independent Statistics & Analysis, U.S. Department of Energy, Washington DC. Retrieved December 3, 2014, from http://www.eia.gov/ electricity/wholesale/.

Energinet.dk (2012). *Technology Data for Energy Plants*. Copenhagen, Denmark: Energistyrelsen.

European Commission (2000). Directive 2000/76/EC on the incineration of waste. *Official Journal of the European Communities* L 332/91.

European Commission (2006). *Integrated Pollution Prevention and Control: Reference Document on the Best Available Techniques for Waste Incineration, August 2006* Retrieved November 15, 2014, from eippcb.jrc.ec.europa.eu/reference/BREF/wi_bref_0806.pdf.

Hasselriis, F. (2002). Waste-to-energy combustion, emission control. *Handbook of Solid Waste Management*. Eds. G. Tchobanoglous and F. Kreith. New York, McGraw-Hill: 13.121–113.176.

Haukohl, J., Ed. (2012). *Waste to Energy State of the Art Report*, 6th edition. Copenhagen, Denmark: ISWA (the International Solid Waste Association).

Hitachi Zosen Inova AG (2014). SCR. Retrieved May 30, 2014, from http://www.hz-inova .com/cms/en/technology-solutions/flue-gas-treatment/denox-systems/scr.

Hogg, D. (2001). Costs for municipal waste management in the EU, Final Report to Directorate General Environment, European Commission, Brussels, Belgium.

IEA (2010). *Projected Costs of Generating Electricity*. Paris, France: International Energy Agency.

Kara, M. (2012). Environmental and economic advantages associated with the use of RDF in cement kilns. *Resources, Conservation and Recycling*, Volume 68, Pages 21–28.

Karlfeldt Fedje, K., S. Andersson, O. Modin, P. Frändegård, and A. Pettersson (2014). Opportunities for Zn recovery from Swedish MSWI fly ashes. *Proceedings of SUM 2014*, May 19–21, Bergamo, Italy.

Kitto, J. B. and S. C. Stultz, Eds. (2005). *Steam*, 41st edition. Barberton, OH: Babcock & Wilcox.

Krüger, B., A. Mrotzek, and S. Wirtz, (2014). Separation of harmful impurities from refuse derived fuels (RDF) by a fluidized bed. *Waste Management*, Volume 34, Issue 2, Pages 390–401.

Martin GmbH (2013). Waste-to-energy plant AEB Amsterdam, the Netherlands. Retrieved December 3, 2014, from http://www.martingmbh.de/media/files/anlagen/update_1/ Amsterdam_08_13.pdf.

Michaels, T. (2010). *The 2010 ERC Directory of Waste-to-Energy Plants*. Washington, DC: Energy Recovery Council.

NASDAQ OMX Commodities (2014). Retrieved September 17, 2014, from http://www .nasdaqomx.com/transactions/markets/commodities.

NOAH AS (2012). Deponiene på Langøya, Holmestrand, Norway. Retrieved October 18, 2012, from http://www.noah.no/OmNOAH/OmLangøya/tabid/554/Default.aspx.

Porteous, A. (2001). Energy from waste incineration—A state of the art emissions review with an emphasis on public acceptability. *Applied Energy*, Volume 70, Issue 2, Pages 157–167. doi:10.1016/S0306-2619(01)00021-6.

Rajendran, K., H. R. Kankanala, R. Martinsson, and M. J. Taherzadeh. (2014). Uncertainty over techno-economic potentials of biogas from municipal solid waste (MSW): A case study on an industrial process. *Applied Energy*, Volume 125, Pages 84–92.

Reza, B., A. Soltani, R. Ruparathna, R. Sadiq, and K. Hewage (2013). Environmental and economic aspects of production and utilization of RDF as alternative fuel in cement plants: A case study of Metro Vancouver Waste Management. *Resources, Conservation and Recycling*, Volume 81, Pages 105–114.

Schlumberger, S. and Bühler, J. (2012). Metal recovery in fly and filter ash in waste to energy plants. *Ash 2012*, January 25–27, Stockholm, Sweden.

Schlumberger, S., Schuster, M., Ringmann, S., and Koralewska, R. (2007). Recovery of high purity zinc from filter ash produced during the thermal treatment of waste and inerting of residual materials. *Waste Management and Research*, Volume 25, Pages 547–555.

SFS 2014:1499 (2015). Lag om ändring i lagen (1999:673) om skatt på avfall, Swedish code of statues, Swedish Government. Retrieved March 30, 2015, from www.notisum.se/rnp/sls/sfs/20141499.pdf.

Stantec (2010). *Waste to Energy—A Technical Review of Municipal Solid Waste Thermal Treatment*. Retrieved May 10, 2014, from http://www.env.gov.bc.ca/epd/epdpa/mpp/pdfs/BCMOE-WTE-Emissions-final.pdf.

Swedish Waste Management (2013). *Energiåtervinning*. Retrieved May 10, 2014 from http://www.avfallsverige.se/avfallshantering/energiaatervinning/.

Tabasová, A., J. Kropáč et al. (2012). Waste-to-energy technologies: Impact on environment. *Energy*, Volume 44, Issue 1, Pages 146–155.

Themelis, N. J. (2012). Current Status of Global WtE. *Proceedings of the 20th Annual North American Waste-to-Energy Conference*, April 23–25, Portland, ME.

WSP (2013). *Review of State-of-the-Art Waste-to-Energy Technologies. Stage Two—Case Studies*. Retrieved November 17, 2014, from http://www.wasteauthority.wa.gov.au/media/files/documents/W2E_Technical_Report_Stage_Two_2013.pdf.

Zheng, L., J. Song, C. L. Yunguang Gao, P. Geng, B. Qu, L. Lin (2014). Preferential policies promote municipal solid waste (MSW) to energy in China: Current status and prospects. *Renewable and Sustainable Energy Reviews*, Volume 36, Pages 135–148.

8 Recent Developments in the Gasification and Pyrolysis of Waste

Tobias Richards

CONTENTS

8.1 BACKGROUND/INTRODUCTION

Thermal treatment of waste material was originally used partly as a pure treatment method and partly to reduce the sheer amount of waste material as much as possible. Arena (2012) lists a number of potential advantages of applying thermal treatment to waste, which, apart from a reduction in both mass and volume, include the possibility of recovering metals (from ash fractions), the destruction of organic contaminants, and a reduced environmental burden. This is also discussed by Consonni and Viganò (2012), who note that gasification has a number of advantages when compared to direct combustion/incineration of waste, namely: the gas formed can be handled more efficiently when combusted in a second stage because this is a homogeneous reaction with a low presence of contaminants (and thereby suitable for the Otto engine, Rankine cycle, gas turbines, or chemical synthesis); the reducing environment in the gasifier improves the quality of the solid residual (e.g., there is no oxidation of the metals) and also prevents the formation of dioxins and furans; and it opens up for possible pressurized gasification, which could improve the degree of efficiency even further and thereby reduce costs.

It has been clear however that waste, especially sorted waste material, contains a significant amount of energy that is possible to utilize. A key hurdle associated with traditional incineration is the high risk of the waste material causing corrosion; this lowers the temperature in the Rankine cycle and thus decreases the electrical efficiency. Higher steam temperatures could be used if gasification is carried out before combustion with an intermediate gas-cleaning stage, which would increase the net production of electricity for a given amount of energy input. Gasification also provides opportunities for producing a syngas that could be used in a gas turbine with an even higher electrical efficiency, or for other purposes, such as manufacturing chemicals. Two further options are to use pyrolysis to make pyrolysis oil (for use as a fuel or, after further processing, as automotive fuel) and as a pretreatment prior to gasification/incineration. Furthermore, the requirements for low levels of toxicity in the remaining material or high costs of posttreatment (i.e., landfill of ashes) in some countries have led to the usage of alternative processes where focus has been placed on high temperatures so that the ashes form a slag. It is in such cases that the gasification of waste has emerged as a competitive alternative.

Malkow (2004) presents a number of options for thermal treatment that are based mainly on pilot or demonstration plants together with some full-scale processes: the processes are within the areas of both pyrolysis and gasification. It was concluded that although these alternative methods are yet to be proven commercially, they possess the potential of increasing the energy efficiency and decreasing both the emissions and amount of hazardous residues.

MSW (municipal solid waste) consists of different materials, all of which behave differently when subjected to heat. Sørum et al. (2001) report the different behavior of the main constituents of MSW: paper, plastic, and wood. It is clear that the onset temperature differs, as does the product's gas/liquid composition. However, even though the difference is great, not many combination effects were found in the laboratory studies undertaken by Dwi Aries et al. (2012) using a mixture of Indonesian waste. This means that a prediction can be made of the amount of gases and liquids that will result from the fraction of each individual material.

Studies made by, for example, Velghe et al. (2011) for MSW and Buah et al. (2007) for RDF (refuse-derived fuel) both concluded that a high temperature was beneficial for the production of pyrolysis oil, the liquid yield increased with increasing temperature, and the chemical composition was suitable for further processing. For MSW, the highest liquid yield was obtained with fast pyrolysis at 510°C; a further increase in temperature lowers the yield.

Several aspects of thermal treatment methods are covered in this chapter. Initially, there is a schematic description of the main steps in thermal treatment; the differences between the processes, from operating conditions to expected outcomes, are examined. This is followed by the current status and then some selected cases that represent different technologies.

8.2 WHAT ARE PYROLYSIS AND GASIFICATION?

8.2.1 PYROLYSIS

Pyrolysis is basically the thermal breakdown of a product (such as waste material) in the absence of oxygen or an oxidizing medium. This is not, however, completely true in the case of waste because the material itself contains oxygen, so a certain degree of oxidation is expected. Furthermore, the fuel often contains some amount of water, which will act as an oxidizing medium if the temperature is sufficiently high.

The overall pyrolysis reaction can be written schematically as

$$\text{Fuel} + \text{heat} \rightarrow \text{gas} + \text{liquid} + \text{solid}$$

In general terms, it could be stated as follows for an organic material:

$$CH_mO_p \rightarrow b_1 CO + b_2 CO_2 + b_3 CH_4 + b_4 C_xH_y + b_5 H_2O + b_6 H_2 + b_7 C$$

The amount of each phase in the product depends on the operating conditions and the composition and morphology of the incoming fuel. Generally, a low temperature (300°C–500°C) and a long residence time (hours) give a high amount of solids (the long residence time ensures high conversion); a moderate temperature (around 600°C) and a short residence time (preferable only a few seconds) tend to give higher amounts of condensable tars (liquid); and a moderate-to-high temperature (>700°C) with a longer residence time (several seconds) ensure a subsequent reaction in the gas phase, with the larger hydrocarbons cracking to form gas products with lower molecular weights. Typically, it is not possible to direct the whole product into one phase; there will always be a distribution. In McKendry (2002) it is stated that processes using biomass as the feedstock for producing solids (charcoal) could reach a yield of 35%, a flash pyrolysis for producing liquids could yield 80%, and a process for producing gas could yield 80%.

Pyrolysis, that is, the degradation of organic material, can be divided into several steps (Thome-Kozmiensky et al. 2012). The first is drying, which occurs mainly between 100°C and 200°C. This is followed by deoxidation where, for example, –OH or –CH$_2$OH substituents are removed by cleavage, followed by a release of CO$_2$, CO, and H$_2$O. It is at this temperature (around 250°C) that desulfurization starts and hydrogen sulfide begins to form. These are typical conditions used when performing mild treatment where the main emphasis is to improve the quality of the solid residual without losing too much energy (i.e., torrefaction). At slightly higher temperatures, around 340°C, aliphatic structures begin to react and depolymerize. Increasing the temperature further to 370°C gives the onset of carbonization, which enriches the carbon in the solid phase. Further cleavage occurs at around 400°C and the C–O and C–N bonds now become involved; pitch also starts to form. This can be described as thermal rearrangement on a molecular level, whereby larger molecules are formed from many parallel reactions (Lewis 1982). Increasing the temperature further to 600°C and above initiates what can be denoted as the "gas evolution phase." Here, the hydrocarbons formed are degraded to more stable gases, such as H_2, CO, CO_2, and CH_4. However, temperature alone is not sufficient to degrade all of the tars formed; larger, condensable hydrocarbons and other measures must be used if these are to be removed or destroyed. The temperature is high enough for some gas–solid phase reactions, including the water gas reaction and the Boudouard reaction.

$$\text{Water gas reaction: } H_2O + C \Leftrightarrow H_2 + CO$$

$$\text{Boudouard reaction: } CO_2 + C \Leftrightarrow 2\,CO$$

In this region there could be a reformation of aromatics from unsaturated aliphatic compounds (olefins), such as the formation of benzene from ethylene; a possible route is showed in Berg et al. 1994. The solid phase undergoes dehydrogenation reactions that further increase the relative content of carbon in the char; at high temperatures virtually all oxygen, sulfur, and hydrogen are removed, so carbon and any inorganic materials remain.

8.2.2 Gasification

Gasification is the thermal degradation of a material performed with the assistance of an oxidizing medium. Its purpose is twofold: to produce an intermediate gas and to increase the degree of conversion of carbon. The gas that is produced has many usages. The oxidizing media used most often are air, oxygen, and steam. It is important that the equivalence ratio, which is the amount of oxygen supplied in relation to what is needed for a stoichiometric combustion (i.e., to produce CO_2 and H_2O from the incoming fuel without there being any surplus oxygen in the outgoing gases), is kept at the right level. Typical values for waste gasification are 0.25–0.35 but, in the case of "moving grate" gasification, it may approach 0.5 (Arena 2012). The heat needed for the reaction (to increase the temperature of the incoming fuel and to overcome eventual endothermic reactions) is either supplied externally or generated by exothermic reactions.

Regardless of the presence of an oxidizing medium, pyrolysis occurs first when the material is heated. These gases can then easily react further (homogeneous gas phase reactions are fast) according to

$$CO + H_2O \Leftrightarrow CO_2 + H_2, \Delta H_{STP} = -41.2 \text{ kJ/mol}$$

$$CH_4 + H_2O \Leftrightarrow CO + 3H_2, \Delta H_{STP} = 206.2 \text{ kJ/mol}$$

$$CO + \tfrac{1}{2} O_2 \rightarrow CO_2, \Delta H_{STP} = -283.0 \text{ kJ/mol}$$

$$H_2 + \tfrac{1}{2} O_2 \rightarrow H_2O, \Delta H_{STP} = -241.8 \text{ kJ/mol}$$

These reactions also represent all other reactions with hydrocarbons that yield CO_2, H_2O, CO, or H_2 as the final products.

The remaining solid material (represented here by carbon) reacts according to

$$C + \tfrac{1}{2} O_2 \rightarrow CO, \Delta H_{STP} = -110.5 \text{ kJ/mol}$$

$$C + O_2 \rightarrow CO_2, \Delta H_{STP} = -393.5 \text{ kJ/mol}$$

$$C + H_2O \Leftrightarrow CO + H_2, \Delta H_{STP} = 131.3 \text{ kJ/mol}$$

The two top reactions are exothermic while the bottom one is endothermic. This means that a gasifier in which air or oxygen is the oxidizing medium provides its own heat, whereas a gasifier using steam needs external heating, for example, the high temperature of the incoming steam or the heated particles in a fluidized bed.

In addition to the reactions listed above, two other gas–solid reactions are important since they occur in a secondary manner. These are

$$C + CO_2 \Leftrightarrow 2 \, CO, \Delta H_{STP} = 172.5 \text{ kJ/mol}$$

$$C + 2 \, H_2 \Leftrightarrow CH_4, \Delta H_{STP} = -74.9 \text{ kJ/mol}$$

Most of the reactions above are written as equilibrium reactions, as may be noted, which means that they can go in either direction, depending on the conditions (i.e., temperature, pressure, and composition). More comprehensive discussions of the gasification concept are to be found in many books on the subject, such as Higman and van der Burgt (2008).

8.3   WHY THERMAL TREATMENT?

Pyrolysis offers an opportunity for producing pyrolysis oil and pyrolysis gas, both of which can be used in efficient conversion cycles. Aside from the energy perspective, the solid materials (i.e., char and metals) can be recycled after separation. The metals will not have been oxidized as a result of the process, which is beneficial for its subsequent value. However, most of the metals are already oxidized, and in order to remove the oxygen, the temperature must be raised substantially; this comes in the form of an energy penalty (and cost) to the process. The benefit of having a relatively low temperature is that different inorganic compounds can be removed (e.g., by leaching) without the application of strong acids, as is the case when there is substantial agglomeration.

Gasification is claimed to have the potential of producing a clean synthesis gas that can be used in gas turbines or gas engines, for example, with high degrees of efficiency. It should also be possible to reach lower emission levels. Moreover, it allows very high temperatures to be used for the ash and thus provides a high degree of vitrification (slagging), which, in turn, further reduces the formation of possible toxic substances (such as dioxins and furans). It also makes the material practically inert and thereby useful as a new material for, for example, construction or roads. Using vitrification, however, incurs a significant penalty to the overall energy efficiency; it is currently employed only in countries with a high penalty for landfill material, such as Japan. In the case of staged gasification, that is, gasification followed by combustion, the main claim is for a higher potential electrical efficiency; other claims are smaller installations and lower NO_x emissions.

8.4 TECHNOLOGY OPTIONS

Based upon the fundamental behavior of thermal treatment, several options are open for both pyrolysis and gasification. This section presents various options that target different goals, with the emphasis being placed on waste treatment/handling. Not all technology that may be employed is described; focus is instead placed on existing technologies that are commercially viable, or close thereto, on a large scale.

8.4.1 PYROLYSIS TECHNOLOGY

As mentioned earlier in the description of the fundamentals of pyrolysis, the technology that should be used varies, depending on the desired output (products). The kinds of pyrolysis can be divided roughly into two groups: slow and fast. Where waste is concerned, however, there are currently no large-scale plants in operation that prioritize the production of pyrolysis oil, which means that they all are using slow pyrolysis.

8.4.1.1 Slow Pyrolysis

In slow pyrolysis the heating rate of the solid material is low and the residence time of the solids is in the order of hours. This ensures a mild treatment and low entrainment of material into the gas phase. Slow pyrolysis is often performed in a rotary kiln (e.g., Mitsui Engineering & Shipbuilding) but can also be achieved in a channel where an external force is applied for transportation purposes (e.g., JFE Thermoselect). A low temperature (around 500°C) requires a longer residence time and gives a solid char with a higher amount of oxygen and hydrogen, but it has a lower energy demand and a less violent reaction during gas devolatilization. Higher temperatures (above 700°C) result in a more carbon-rich char and also reduce the amount of tars in the gas phase. Benefits of using pyrolysis prior to combustion include the reduced necessity of pretreatment for the fuel and the option of separating the solid material before the high temperature combustion, thereby improving utilization of the metals (ferrous and aluminum). Pyrolysis is a first stage in the overall process in both the low- and high-temperature processes. In the former, the solids are separated directly after pyrolysis and then treated separately; in the latter, the solids are also separated from the gas but then pass through a section at an even higher temperature to ensure complete melting.

8.4.2   GASIFICATION TECHNOLOGY

Gasification technology has several alternatives on offer. They can be described according to their mode of operation: fixed bed, fluidized bed, or entrained flow. Where waste material (including MSW) is concerned, only the first two technologies are used due to the high level of pretreatment and the large scale of operation that are necessary for the third technology to produce a high-quality syngas.

8.4.2.1   Fixed Bed Gasification

Fixed bed gasification is generally used for lower throughputs resulting from difficulties in having a large diameter without causing the gases to be channeled. The fuel is fed from the top of the gasifier and lands on a bed of unreacted material. The solid material is transported downward as the reaction proceeds further down into the bed and meets new conditions, such as high temperatures and gases of various compositions. Fixed bed gasifiers can, in most cases, be one of two kinds: updraft or downdraft. This refers to the direction of the gas flow: updraft if it is in the opposite direction to the solid material and downdraft if it is in the same direction. Generally speaking, the downdraft system produces a higher quality gas and is easier to clean for syngas purposes (chemical production) and to avoid damage to the gas turbine. Its design is, however, more complex and it requires more control. The updraft gasifier handles a greater variation in the feedstock quality well (e.g., heat content and moisture content) but, on the other hand, produces a gas with rather high amounts of tars. As a stand-alone unit the updraft gasifier has a high energy efficiency; the gases produced do not need to be cooled extensively before they are cleaned.

8.4.2.2   Fluidized Bed

Fluidized bed gasifiers have the advantage of having a rapid and effective heat and mass transfer within them that distributes the fuel and increases mixing. The drawback is that it is necessary to add a bed material (usually sand) and that the fuel must be pretreated, that is, broken down into rather small pieces no larger than 5 cm in the largest dimension. If there is a buildup of agglomeration (operating around the melting temperature of the inorganic material in the fuel), the sand particles will increase in size, fluidization will not be efficient, and, in a worst-case scenario, there will be a total collapse. Moreover, fluidization demands a certain minimum amount of fluidizing medium (be it air, steam, or oxygen) to work properly. This flow entrains part of the bed material together with the fly ashes; there is also attrition in the bed, which may reduce the size of the material in it, thus forming small particles that are easily entrained. Although most of these are removed in a cyclone after the gasifier so as not to disturb the following process, it will increase not only the need to treat the ash but also (in most cases) the amount of landfill produced. This applies to both fly and bottom ash, in the same way as for traditional combustion. Depending on the velocity of the gas, fluidized bed gasifiers are denoted as being either circulating or bubbling. In the former, the velocity is high enough (5–6 m/s) to entrain most of the sand particles as well as to ensure good mixing in the free board area and a uniform reaction; the sand particles are generally smaller (around 0.25 mm) than in a CFB (circulating fluidized bed) so that they are easier to entrain. This will produce a more uniform gas flow and decrease the risk of the gas being channeled. Valmet is one of the companies that supply this technology. In the latter type, a bubbling fluidized bed keeps the gas velocity lower (1–3 m/s) in order to maintain its particles (up to 1 mm in diameter) close to the bed and thereby decrease the amount of sand that needs to be removed in the cyclone. The idea is that mixing will still be efficient where it is most crucial, that is, in the heterogeneous pyrolysis and gasification reactions, but the drop in pressure is less than in a CFB, which will reduce the mechanical impact on both the bed and the equipment. Examples of companies that supply bubbling fluidized bed gasifiers for treating waste are Ebrara, Kabelco, and Hitachi Zosen.

The bigger suppliers use air as the gasification medium. They all have a subsequent combustion step either directly after gasification or following a gas-cleaning process. The reason for this is the low quality of the feedstock and the relatively low amounts involved, which make it unattractive from a financial perspective to produce synthesis gas that could be treated further. However, Enerkem in 2014 started a facility where 100,000 t.p.a. of waste is treated with the purpose of producing methanol (and later ethanol). A mixture of oxygen and steam is employed as the fluidizing medium in its bubbling bed.

8.4.2.3   Slagging Gasification

Slagging gasifiers operate at high temperatures for inorganic material. The material is melted and forms a molten solution, which is often quenched before further treatment. Any metal that is present can be separated from the molten slag since the smelts have different densities; this could be performed before quenching. Otherwise magnetic separation is undertaken after granulation of the solid material. One of the benefits of slagging gasifiers (or systems with a gasification and subsequent slagging operation) is that the material they form meets the regulations for several usages, such as asphalt mixtures, paving slabs, and roof tiles. It has been reported that the leaching of heavy metals is well below the limits stipulated in, for example, JIS-Japanese Industrial Standard K0058 and the corresponding regulatory test JLT-46 (Arena 2012; WSP 2013). In Japan, the high penalties associated with landfill material mean that ashes are also subjected to posttreatment after conventional combustion in a slagging unit (e.g., utilizing plasma technology to reach the necessary temperatures). Although this lowers the energy efficiency of the plant, it reduces the amount of material sent to landfill sites. A disadvantage of these technologies is that they require lime and coke to be added to the gasifier not only to control the viscosity of the smelt but also to ensure full reduction conditions in the lower section while keeping the temperature sufficiently high. The two main suppliers of this technology are Nippon Steel and JFE.

8.4.2.4   Staged Gasification

Staged gasification is a two step gasification and combustion process, where the flue gases are subjected to minimal treatment in between the two stages. The main advantage here is that combustion can be enhanced; the solid material does not become fully oxidized and may be easier to separate. It is also (as in the case of Entech WtGas) possible to use several units as a pretreatment stage (i.e., low temperature gasification) before combustion. Typically, these units are small and benefit from the fact that the resulting NO_x requires very little cleaning. Examples of companies currently using this technology are Energo, KIV, and Entech WtGas. In some cases, it is not easy to see the difference between staged gasification and staged combustion. The latter occurs in all modern waste-to-energy (WtE) incinerators and has been shown to decrease the formation of NO_x while simultaneously keeping CO emissions low. It is generally considered as being staged gasification if combustion occurs in a separate area (i.e., equipment) and staged combustion if all of the reactions occur in the same unit.

8.4.3 PLASMA GASIFICATION

Plasma is often denoted as the fourth matter of state. It is characterized by the removal of electrons from gas molecules (or atoms), which creates a number of free electrons together with positively charged molecules. Plasma is achieved when the amount of free electrons is high enough to significantly change the electrical characteristics of the gas in question. In the case of gasification, it is generated at temperatures exceeding 2000°C and is generally created by an electric arc. The gas molecules start to dissociate at about 2000°C, and at temperatures above 3000°C, they begin to lose electrons and thereby become ionized (Gomez et al. 2009). This changes the viscosity of the gas, making it more like that of a solid; the electric conductivity will then be of the same order as for metals.

Together with the free electrons this temperature degenerates all organic molecules into their elements efficiently, which eliminates all tars. The generation of a plasma arc is extremely energy-intensive, so several companies only use plasma as an additional feature for other gasifiers. It can be seen as being either a first step in cleaning the gas or a means of providing sufficient heat to ensure that the ash melts (with the same benefits as described for slagging gasifiers). When used on the syngas, the conditions ensure an almost complete breakdown of larger molecules (tars) and make the gas-cleaning process much easier. In many cases, the outcome after plasma treatment meets the stipulations for gas engines and gas turbines without requiring further cleaning.

A number of companies are currently working with plasma in small-scale waste gasifiers but they have not yet been used commercially on a large scale. However, Alter NRG is in the throes of commissioning its huge plasma gasifier in the United Kingdom; it will be the largest waste gasifier in the world, with a target of 350,000 t.p.a. Work is already proceeding on a second unit at the same place, which will make it possible to treat up to 2000 ton of waste per day. Examples of other processes employing plasma gasification are Gasplasma® (Advanced Plasma Power, Swindon, UK), Plasco (Plasco Energy Group, Kanata, ON, Canada), and CHO Power (Europlasma, Oudenaarde, Belgium).

Seen from an operational perspective, the use of plasma gasification (and plasma-assisted gasification in particular) allows the system to be controlled at very short notice. The supply of energy through the plasma is virtually instantaneous, so irregularities in the incoming feedstock will be registered. It should also be noted that, since the energy supply comes from an external energy source (electricity), the amount of syngas generated can be minimized (implying that it is only the unwanted fraction, such as CO_2, that is reduced), which simplifies the gas-cleaning step.

A plasma torch may be constructed in several ways. One is a so-called DC nontransferred arc plasma torch that uses a constricted arc with cold electrodes (e.g., as used by Alter NRG). This basically means that there are direct current electrical discharges that use up to 1×10^5 A and a water-cooled electrode. The electrode is composed of a highly conductive material, such as copper, and the plasma is generated by a strong vortex created by either a magnetic field or a strong swirl in the gas flow (Gomez et al. 2009). Although the water-cooling system allows oxidizing media to be used, it reduces efficiency; the temperature is thus limited to below 8000 K at atmospheric pressure. Another option is to use a "DC-transferred arc plasma torch," as described in, for example, Lemmens et al. (2007). There is only one electrode in this plasma torch, but, as the distance to the other electrode could be significant (in some cases up to 1 m), the plasma arc is consequently transferred over to the reactor. The carrier gas is supplied concentric to the torch electrode. However, it is possible to use graphite electrodes, which enable diatomic gases to be used as the carrier gas (anode torches can only accept monoatomic gases, and are used when no contamination is accepted). This is used by InEnTec in its PEM® technology, where all material is introduced into the plasma chamber after being pregasified by oxygen and steam. In order to decrease the power input from the plasma, additional heat is supplied via joule heating. Although the overall efficiency is low when the plasma and joule heating (also by electricity) use more than 85% of the incoming energy content (Ducharme and Themelis 2010), the chamber does ensure a clean syngas and an inorganic smelt.

8.5 EXAMPLES

There are a number of thermal treatment systems in operation today besides waste incineration. Most installations are in Japan, where gasification has been used together with ash melting in order to achieve very low emissions and increase the use of solid waste (i.e., ash/slag).

The largest facility today is nevertheless to be found in Finland, and is followed by a number of plants in Japan (Table 8.1). However, it should be mentioned that Germany has been active in this field and has had two large facilities in operation. The one in Karlsruhe, using the Thermoselect technology, closed in 2004; the other, which was run by SVZ and used its own concept based on the Lurgi slagging gasifier, closed in 2006. There are a number of plants currently under construction and development, which should be in operation in the period 2015–2017; some of them will, according to signed contracts, be larger than today's plants and exceed 300,000 ton of waste p.a. In the case of pyrolysis, the Mitsui Babcock plant (Mitsui R21 in Toyohashi, Japan) is probably the largest facility currently in operation, with a yearly intake of up to 120,000 ton of waste p.a. This unit combines pyrolysis with separate combustion and ash melting.

8.6 A DESCRIPTION OF THE VARIOUS TECHNOLOGIES

8.6.1 VALMET

Valmet's technology (previously Metso, before the divergence of some of its activities to Valmet, Espoo, Finland, in 2013) is used in one facility for waste gasification, which is currently the largest of its kind in the world. This gasifying unit handles

TABLE 8.1
World's Largest Waste Gasification Facilities until 2013

Place	Size (t.p.a.)	Manufacturer	Year of Installation	Process
Lahti, Finland	250,000	Valmet (Metso)	2012	Fluidized bed gasification followed by gas combustion
Kitakyushu, Japan	215,000	Nippon Steel	2007	Direct melting (fixed bed) and gas combustion
Kurashiki, Japan	170,000	JFE (Thermoselect)	2005	Gasification followed by separate ash melting and gas combustion
Narumi, Japan	159,000	Nippon Steel	2009	Direct melting (fixed bed) and gas combustion
Sagamihara, Japan	158,000	Kabelco	2010	Fluidized bed gasification followed by slagging combustion
Shizouka, Japan	150,000	Nippon Steel	2010	Direct melting (fixed bed) and gas combustion

Source: WSP, *Review of state-of-the-art waste-to-energy technologies. Stage Two—Case studies,* http://www.wasteauthority.wa.gov.au/media/files/documents/W2E_Technical_Report_Stage_Two_2013.pdf, 2013; Gasification Technologies Council, *World database,* http://www.gasification.org/what-is-gasification/world-database/, 2014.

250,000 ton of sorted MSW and industrial waste per annum. The gasifier (Kymijärvi II) is located in Lahti, Finland, and is designed for a fuel input of 160 MW (two lines that each handle 80 MW). It uses gasification before the gas-cleaning step, followed by combustion in a gas boiler (Figure 8.1). The gasifier is a CFB that uses air as the oxidizing medium and operates at around 850°C–900°C (Lahti Energia 2014); the bed materials are sand and lime. After gasification, the gas produced is cooled down in a heat-recovery unit to approximately 400°C before hot filtration takes place. The filtration units are of the candle filter type, and remove metals (both alkali and heavy metals) and particles of matter. The subsequent combustion stage operates at a steam temperature of 540°C and a pressure of 121 bar. The final gas-cleaning operation comprises catalytic NO_x reduction (selective catalytic reduction or SCR) followed by a dry-cleaning step using activated carbon and a bag filter. The plant's annual production of electricity is 50 MW and of district heating is 90 MW, giving a net electrical efficiency of 31%. The fuel used by the company, which is sorted waste from industry, shops, construction sites, and households, consists mainly of unclean plastic, cardboard, wood, and paper. The waste is shredded into strips 2 cm–4 cm long before being fed into the gasifier.

8.6.2 NIPPON STEEL

Nippon Steel owns several of the largest waste gasifiers in the world and has more than 30 units in operation (Nippon Steel 2013). They all use the same technology, so the case chosen here is its largest facility, the Shinmoji plant in Kitakyushu City

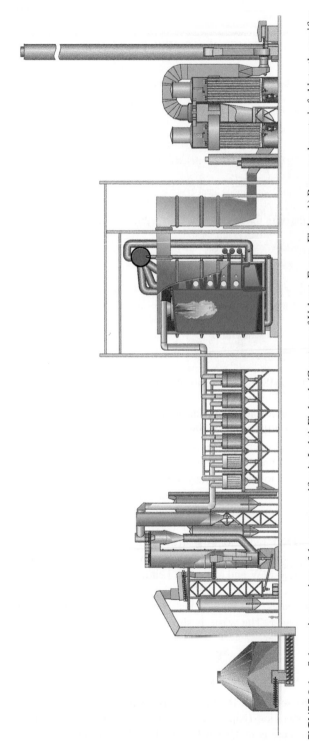

FIGURE 8.1 Schematic overview of the waste gasifier in Lahti, Finland. (Courtesy of Valmet, Espoo, Finland.) Processed waste is fed into the gasifier and the gases produced are cleaned before being combusted in the gas boiler.

(Fukuoka Prefecture, Japan), which treats around 215,000 t.p.a. Its technology is based on an updraft fixed-bed gasifier with the addition of an ash-melting stage. The fixed bed means that there will be a temperature profile in the reactor. Initially, the material is dried and heated up to 300°C. Further heating to about 1000°C decomposes the material as the solids proceed downward. The final stage for transforming the solids into gas is to add air and oxygen-enriched air (O_2 content of around 36%) to ensure complete combustion of the remaining char (mainly carbon). The technology is designed to reach temperatures of 1700°C–1800°C in the bottom section to ensure that the solid material is melted and can be removed as molten slag. The gases formed during the process are transferred to the combustion chamber. The steam conditions are 400°C and 40 bar. A bag house filter, together with an SCR NO_x control system, is used to treat the flue gas.

The slow heating process and high final temperatures have the consequence that the fuel requires very little pretreatment; it only needs to meet the requirements set by the feeding system. Lime is fed in together with the fuel to regulate the viscosity of the final melt (Tanigaki et al. 2012). Coke is also introduced at the top and acts as a reducing agent.

The molten material is quenched with water before passing through a magnet separator to divide the material into metals and slag. Ash particles, which may contain some unreacted carbon, are collected via a cyclone before the product gases enter the combustion boiler. These particles are fed back into the gasifier in the lower section (high temperature and oxygen-rich environment) for complete conversion; this reduces both the amount of coke needed and the amount of fly ash in the boiler.

8.6.3   THERMOSELECT

The Thermoselect process combines pyrolysis and gasification in the first unit and then utilizes the gases formed in different ways, depending on the situation. The largest plant (170,000 t.p.a.) is in Kurashiki, Japan, and the gases that are produced are transferred to the gas engines in a neighboring steel works. The fuel is first compacted by a hydraulic press and then pushed into a degassing (pyrolysis) channel, which is heated indirectly up to 800°C (Malkow 2004). When the degassing process is complete, the gases and solids enter the high-temperature gasifier. True to the rule of gravity, the solids fall and the gases rise. The solid material then passes through an oxygen inlet, which combusts the remaining carbon completely; it also raises the temperature to 2000°C, which vitrifies the solids and forms a molten slag. The slag is treated further with external burners so that it remains at 1600°C and can be homogenized, allowing the metal and slag to be separated before being stored in a bin. The gases are held at 1200°C for at least 2 seconds before being rapidly quenched to around 70°C in an oxygen-free environment (Sumio et al. 2004). After cooling, the gases pass through a scrubber and a sulfur-removal system before they can be utilized. The last unit was installed in 2006.

8.6.4   KABELCO

Kabelco has its largest waste gasifier (158,000 ton/year) in Sagamihara, near Tokyo, Japan. Its technology utilizes a bubbling fluidized bed followed by a swirling melting system. The waste is first crushed before it enters the gasifier via an airtight lock hopper. The gasifier is fed with air as the oxidizing medium, which may be preheated if the fuel being used has a low heating value. In the gasifier metals, such as aluminum and iron, and other incombustible material are removed at the bottom of the sand bed; the remaining gases, along with some entrained particles, proceed on to the melting furnace where they are combusted at 1200°C using air. The flow is introduced tangentially, which creates a swirling flow. This, in turn, forces the molten slag (from the entrained inorganics) outward to the wall, which flows down, and ends up in a rapid water cooler followed by a granulator. Both the gasifier and the furnace walls are lined with refractory material, with the latter also being equipped with extra water cooling for extended durability. Following combustion, the gases pass through a heat-recovery system and then a dry flue gas cleaning system; in the first step, active coal is added before the baghouse filter. The filtered gases are then led to a second baghouse filter after the addition of slaked lime. They are reheated with steam and aqueous ammonia is added for a catalytic removal of NO_x before being released via the chimney stack.

8.6.5 ENERKEM

Enerkem recently (June 4, 2014) inaugurated its first large-scale commercial facility, located in Edmonton, Canada. The feedstock is approximately 100,000 t.p.a. of sorted MSW. It produces not only energy carriers, such as steam and heat, but also chemicals (ethanol in this particular case). A bubbling fluidized bed, supplied with oxygen and steam as the fluidizing media, is used for thermal conversion. The temperature in the bed is controlled at 700°C–750°C and the pressure is slightly increased, to about 2 bar. The low temperature means that relatively cheap construction material can be used, thereby keeping costs down. While little information is available regarding its process, the company uses, according to an earlier patent (issued in 2009), a cyclone after the gasifier, after which the gases are sent to a gas quench before being subjected to scrubbing and posttreatment processes (in a catalytic converter).

8.6.6 MITSUI ENGINEERING & SHIPBUILDING

Mitsui Engineering & Shipbuilding has licensed a technology in its facilities that was first invented by Siemens. One of the company's largest plants is situated in Toyohashi, Japan, and handles about 400 ton of waste material every day (equals about 120,000 t.p.a.). Inaugurated in 2002, it is still in operation. The waste is first shredded and then injected into a rotating pyrolysis drum. The retention time is about 1 hour; the drum is heated by air from the subsequent boiler, which increases the temperature to about 450°C (Mitsui Recycling 2003). After pyrolysis, the solids are removed and cooled down to 80°C before being sorted into steel (ferrous metals), aluminum, and carbon (including noncombustible inorganics) fractions. The material containing carbon is mixed with recycled boiler ash and fly ash from the boiler and gas-cleaning system before being transported to the combustion chamber, into which the pyrolysis gases are also injected. Combustion takes place at 1300°C, which is well above the melting point of the material; this ensures that a molten slag is formed, which can be tapped off in a water bath before granulation. Flue gases are recirculated (via the pyrolysis reactor) to assure a high burnout (good mixing) and a low formation of NO_x. The flue gases pass through a heat-recovery boiler where steam at 400°C and 40 bar is produced for the steam turbine. Finally, flue gases are cleaned using two bag filters to remove dust and chlorine.

8.6.7 WESTINGHOUSE PLASMA GASIFICATION

The Westinghouse plasma gasification technology is promoted by Alter NRG and employs a plasma-assisted gasifier. The plasma torches are situated in the lower part of a fixed bed gasifier that provides energy to the process; it ensures complete melting of the inorganics (temperatures around 2000°C), which exit the bottom at about 1650°C. The plasma torches heat up the incoming air to over 5500°C (Willis et al. 2010). In order to withstand such high temperatures, the gasifier must have three layers of refractory lining in the top sections and a special silicon carbide lining in the high-temperature melting zone.

The gasifier is also supplied with air (from the sides of the bed), which provides a final temperature somewhere between 900°C and 1100°C for the outgoing gases; the tendency is to use lower temperatures to avoid the carryover of molten particles. The company has a plant at Mihama-Mikata, Japan (commissioned in 2003), that treats about 6000 ton of MSW per annum and more units are under way. The plant in Eco-Valley, on the island of Hokkaido in Japan, could treat 220 ton/day but it was closed in 2013 due to a shortage of fuel. It is interesting to note that this technology will be used in a large project in the Tees Valley, United Kingdom, that is currently under construction and is planned to be commissioned during 2015. Its estimated volume is 350,000 t.p.a., making it the world's largest waste gasifier (Air Products 2014).

8.7 DISCUSSION

There is a higher awareness in society today regarding recycling material and the waste hierarchy. This has led to a serious questioning of traditional waste incineration from the perspective of not only emissions but also the utilization of material. Incinerated matter has lost most of its material value; it becomes a medium in energy processes, such as the production of steam or supply of hot water. Each of these processes is important and necessary, but it has to originate from a selected resource without tampering significantly with its inherent potential; the potential here embraces all possible measures of utilization including, but limited to, reuse, reassembly, construction material, raw material feedstock, and also its use as an energy source. Although gasification generally has a higher public acceptance due to the potential of its technology compared to incineration, there is no proof yet that it is an economic viable process on a large scale.

Development is still necessary if there is to be an option for thermal treatment that is superior to the present waste combustion system. A study by the International Solid Waste Association (ISWA 2013) showed that the energy efficiency and emissions of the alternative methods are either comparable to, or less favorable than, traditional

WtE facilities, even though it was difficult to obtain reliable data. For the sake of comparison, the best available technology was used, and not average technology. It should be mentioned that the ISWA study was based upon existing units and was not a comparison between theoretical potentials. Young (2010), however, concluded that a combination of pyrolysis and gasification processes would be the most viable method from economical, material, and efficiency perspectives. In a paper by Leckner (2015), it is concluded that only fixed or fluidized bed gasifiers are viable options for WtE facilities when MSW is used as the fuel. Plasma gasification is excluded due to its high electricity demand and the suspension gasifier (entrained gasification) needs too much pretreatment, which is costly. It is further noted that MSW gasifiers are relatively small, which prevents the gases produced from being used as synthesis gas but they can instead be used for producing electricity in gas engines either directly or in downstream Brayton or Rankine cycles.

It should be observed though that the total system must be analyzed in order to determine which method is the most appropriate. It embraces energy efficiency, emissions, production of ash, production of slag, possibility of utilizing the remaining solids (e.g., as building material, leaching ashes to recover important and valuable metals, landfill), cost of the technology, reliability of the process, footprint of the process, acceptance of society, and so on. In many cases, comparisons are based on only one, or maybe a few, of the criteria listed here, which will not give the complete picture. Each of these criteria is important and cannot be compared directly. The goals that are being evaluated should be focused upon: Are they more single-targeted goals (e.g., reduced greenhouse gases, limited landfill usage, and reduced costs) or more multifaceted, such as increased sustainability (e.g., environmental, social, and economic goals)?

Take, for example, the goal of reducing landfills. Incineration has long been promoted as a measure for reducing the need for landfill by up to 90% (by volume) but it is still necessary for the remaining material (mainly ashes). Processing this material further, however, would improve overall performance. It could involve separating out even more valuable material and returning it to society's material cycle (e.g., leaching ash to recover copper and zinc) or turning the material into a new and different material (e.g., by melting the ashes). Thermal treatment in a reducing environment provides opportunities to improve both of these alternatives. From an energy perspective, a separation of material, especially with the intent of regaining specific metals, is enhanced in a low-temperature process where the metals have not been oxidized and are not sintered together. This is the case in pyrolysis or low-temperature gasification. Another option is to proceed all the way to the molten stage (possibly as a posttreatment method) and perform the separation there. This will lead to a good separation of the major metals, but comes with the cost of an energy penalty and the remaining material after the separation will be a mixture. Although this new material is downgraded from the original material in the sense that it is a mixture of different materials that are fixated into an inert matrix, it nevertheless fulfills a purpose because it meets a functional demand (such as construction material). This downgrading (loss of potential) should then be evaluated vis-à-vis alternative ways of meeting this demand together with an analysis of what would happen to the material if it were not used in this way (i.e., going to landfill instead).

There are two main technologies for producing a molten slag product after gasification, namely energy provided externally via plasma torches or heat generated internally by combustion in an oxygen-rich environment. In the former case, energy, in the form of electricity, must be supplied to the system, which will increase its parasitic load of electricity (the amount of electricity needed for the actual process reduces the amount of electricity eventually supplied to the grid). The latter case, however, has the drawback of providing more CO_2 in the gas phase (by the combustion of carbon in the lower section to ensure a high temperature). One way of controlling this is by introducing coke to the system, which converts the CO_2 to CO higher up in the gasifier. This is an endothermic reaction that lowers the temperature, so careful control must be provided to ensure that the conditions are correct. Melting ash is an energy-demanding process, as shown, but it generates a new type of construction material that further reduces the need for landfill, and this has to be considered when evaluating the different alternatives. Ash is melted in many of today's grate-fired waste boilers, but the temperature is not, however, high enough to totally sinter the material into an almost inert matrix. In some places (in Japan), there are ash smelters (often plasma based) that treat incineration ash, creating a product with properties similar to those of the molten slag from gasifiers.

Gasification is a way of either providing useful products from combustible waste other than energy carriers or increasing the efficiency of the conversion process. An example of the former is the production of a synthesis gas that may be processed into chemicals (CO and H_2 are excellent building blocks for a variety of substances). The latter can be exemplified by the usage of a gas turbine with the potential of increasing the electric output significantly. There are currently quite a few projects under construction and a number of projects that are in the planning phase (with signed contracts). Many of them use technologies that have been tested either in other applications or on a smaller scale and will thereby act as pioneers. A number of these small companies have faced difficulties in fulfilling their plans, often associated with financing bodies or permits, which has slowed down the growth and spread of the technology. Others are on their way to increasing the scale of production and, within a couple of years, will prove how efficient the processes are on a larger scale.

Another company worth mentioning as well as those described in this chapter is Chinook Sciences (United Kingdom). It has yet to provide a large MSW gasifier but has successfully handled automobile shredder residues and other waste; it is to commission a facility in the Midlands, United Kingdom, later this year. Chinook Sciences has signed a contract with the United Arab Emirates to deliver a gasifier, which would eventually handle 1 million t.p.a. of waste. Its process is a pressurized, combined pyrolysis and gasification unit in which the temperature, amount of oxygen, and steam are all controlled to achieve the composition of the syngas desired. The key concept of the process is that it operates at a low temperature, with the gases being transferred to a second gasifier operating at a high temperature to crack all the tars and produce a high-quality syngas. After heat recovery (by gas cooling) and cleaning, the gases are sent to a set of gas engines to produce power.

REFERENCES

Air Products. 2014. Tees Valley. Retrieved November 20, 2014. http://www.airproducts.co.uk/microsite/uk/teesvalley/facilities.htm.

Arena, U. 2012. Process and technological aspects of municipal solid waste gasification. A review. *Waste Management*, 32: 625–639. doi:10.1016/j.wasman.2011.09.025.

Berg, C., S. Kaiser, T. Schindler, C. Kronseder, G. Niedner-Schatteburg, and V. E. Bondybey. 1994. Evidence for catalytic formation of benzene from ethylene on tungsten ions. *Chemical Physics Letters*, 231:139–143.

Buah, W. K., A. M. Cunliffe, and P. T. Williams. 2007. Characterization of products from the pyrolysis of municipal solid waste. *Process Safety and Environmental Protection*, 85:450–457. doi:10.1205/psep07024.

Consonni, S. and F. Viganò. 2012. Waste gasification vs. conventional waste-to-energy: A comparative evaluation of two commercial technologies. *Waste Management*, 32: 653–666. doi:10.1016/j.wasman.2011.12.019.

Dwi Aries, H., I. Indarto, S. Harwin, and R. Tri Agung. 2012. Thermogravimetric analysis and global kinetics of segregated MSW pyrolysis. *Modern Applied Science*, 6:120–130. doi:10.5539/mas.v6n1p120.

Ducharme, C. and N. Themelis. 2010. Analysis of thermal plasma assisted waste to energy processes. *Proceedings of the 18th Annual North American Waste-to-Energy Conference*, May 11–13, Orlando, FL. NAWTEC18-3582 pp. 101–106. doi:10.1115/NAWTEC18-3582.

Gasification Technologies Council. 2014. *World database*. Retrieved November 18, 2014. http://www.gasification.org/what-is-gasification/world-database/.

Gomeza, E., D. Amutha Rania, C. R. Cheesemanb, D. Deeganc, M. Wisec, and A. R. Boccaccinia. 2009. Thermal plasma technology for the treatment of wastes: A critical review. *Journal of Hazardous Materials,* 161:614–626. doi:10.1016/j.jhazmat.2008.04.017.

Higman, C. and M. van der Burgt. 2008. *Gasification* (2nd edition). Boston, MA: Elsevier.

ISWA. 2013. Alternative Waste Conversion Technologies. White paper, retrieved November 10, 2014, from http://www.iswa.org/index.php?eID=tx_iswaknowledgebase_download&documentUid=3155.

Lahti Energia. 2014. Technology description. Retrieved November 18, 2014. http://www.lahti-gasification.com/power-plant/power-plant-technology.

Leckner, B. 2015. Process aspects in combustion and gasification Waste-to-Energy (WtE) units. *Waste Management* 37:13–25. doi:10.1016/j.wasman.2014.04.019.

Lemmens, B., H. Elslandera, I. Vanderreydta, K. Peysa, L. Dielsa, M. Oosterlinckb, and M. Joosb. 2007. Assessment of plasma gasification of high caloric waste streams. *Waste Management*, 27(11):1562–1569. doi:10.1016/j.wasman.2006.07.027.

Lewis, I. C. 1982. Chemistry of carbonization. *Carbon*, 20(6):519–529.

Malkow, T. 2004. Novel and innovative pyrolysis and gasification technologies for energy efficient and environmentally sound MSW disposal. *Waste Management*, 24:53–79. doi:10.1016/S0956–053X(03)00038-2.

McKendry, P. 2002. Energy production from biomass (part 2): Conversion technologies. *Bioresource Technology*, 83:47–54. doi:10.1016/S0960-8524(01)00119-5.

Mitsui Recycling. 2003. Pyrolysis, gasification and melting process. Retrieved November 19, 2014. http://www.ieabcc.nl/workshops/Tokyo_Joint_Meeting/02_Mitsui.pdf.

Nippon Steel. 2013. Reference List of Direct Melting System. *Nippon Steel & Sumikin Engineering*. Retrieved November 18, 2014, from http://www.eng.nssmc.com/english/business/environment/pdf/system_e.pdf.

Patent, WO/2009/132449. Retrieved November 20, 2014. http://patentscope.wipo.int/search/en/detail.jsf?docId=WO2009132449&recNum=1&docAn=CA2009000575&queryString=ALLNUM:(WO2009/132449%2520)&maxRec=1.

Sørum, L., M. G. Grønli, and J. E. Hustad. 2001. Pyrolysis characteristics and kinetics of municipal solid wastes. *Fuel*, 80:1217–1227. doi:10.1016/S0016-2361(00)00218-0.

Sumio, Y., M. Shimizu, and F. Miyoshi. 2004. Thermoselect waste gasification and reforming process. *JFE Technical Report No. 3*, July, pp. 21–26.

Tanigaki, N., M. Kazutaka, and O. Morihiro. 2012. Co-gasification of municipal solid waste and material recovery in a large-scale gasification and melting system. *Waste Management*, 32:667–675. doi:10.1016/j.wasman.2011.10.019.

Thome-Kozmiensky, K. J., N. Amsoneit, M. Baerns, and F. Majunke. 2012. Waste, 6. Treatment. *Ullmann's Encyclopedia of Industrial Chemistry*. doi:10.1002/14356007. o28_o06.

Velghe, I., R. Carleer, J. Yperman, and S. Schreurs. 2011. Study of the pyrolysis of municipal solid waste for the production of valuable products. *Journal of Analytical and Applied Pyrolysis*, 92:366–375. doi:10.1016/j.jaap.2011.07.011.

Willis, K. P., S. Osada, and K. L. Willerton. 2010. Plasma gasification: Lessons learned at Ecovalley waste facility. *Proceedings of the 18th Annual North American Waste-to-Energy Conference* May 11–13, Orlando, FL, 2010, NAWTEC18-3515, pp. 133–140, doi:10.1115/NAWTEC18-3515.

WSP. 2013. *Review of state-of-the-art waste-to-energy technologies. Stage Two—Case studies*. Retrieved November 17, 2014. http://www.wasteauthority.wa.gov.au/media/files/documents/W2E_Technical_Report_Stage_Two_2013.pdf.

Young, G. C. 2010. *Municipal Solid Waste to Energy Conversion Processes: Economic, Technical and Renewable Comparisons. Economic, Technical, and Renewable Comparisons*. Hoboken, NJ: Wiley.

9 Metal Recycling

Christer Forsgren

CONTENTS

9.1 BACKGROUND

Metals have an essential role in our modern society. In everyday life, we depend on metals in our infrastructure, buildings, vehicles, household utilities, and electronic devices. Scarcity of metals would have a large impact on our way of life and secure access to metals is crucial for politicians all over the world.

9.1.1 HISTORY

More than 5000 years ago, humans started to understand that all material consists of elements. The knowledge that heating specific minerals together with charcoal generates elemental metals is a very important factor in the evolution of the society that we have today. Almost all the metals present in different minerals are in an oxidized form, often as sulfides or oxides. Very few metals have value as oxidized elements and have to be reduced to elemental form in order to be useful. For example, elemental copper can be produced by reducing copper sulfide and elemental aluminum by reducing aluminum oxide from the naturally occurring mineral bauxite.

$$CuS_2 + 2O_2 \rightarrow Cu + 2SO_2 \qquad (9.1)$$

$$2Al_2O_3 + 3C \rightarrow 2Al + 3CO_2 \qquad (9.2)$$

In example (9.1), the reaction generates heat, that is, it is exothermic, whereas in example (9.2), the reaction requires external energy, that is, it is endothermic.

In today's society, some metals do have value in an oxidized form as well. One example is ferrous, where the elemental form is used to produce steel, and the oxidized form, as $FeCl_3$ and $FeSO_4$, is used to purify water.

Metals with a density higher than 5 kg/L are called heavy metals. In their oxidized form, many of these are environmentally toxic. The use of metals, particularly technology metals in industry and consumer products, has grown rapidly in recent decades. For example, more than 80% of the global mining production of precious group metals (PGMs), rare earth metals (REM), indium, and gallium, since 1900, has been undertaken in the last three decades.

9.1.2 SOURCES FOR METALS

As a rule of thumb, 10 times more steel than copper and aluminum is recycled each year.

9.1.2.1 Minerals

Historically all metal products use minerals as the source of raw material. Mineral ore is found in the earth's crust and is extracted by various types of mining. Like all natural resources, the amount of mineral ore is limited. As the high metal content minerals in the earth's crust are extracted, first the cost of mining minerals increases and the price of metals produced from virgin minerals rises.

9.1.2.2 Recycling

Metals are found all around us and the highest concentrations of metals in the world are today found in cities. The concentration of copper in electronic waste is normally more than 10 times higher than in the ore currently produced by mining. Quality requirements, turnover rate, and recycling costs determine whether mineral-based or recycled metals are used. Current use of REM is almost exclusively based on mineral mining, rather than recycling, while more than 50% of ferrous-based products come from recycled material.

As can be seen in example (9.2), reducing oxidized metals generates emissions of carbon dioxide, one of the major greenhouse gases. It is almost always more energy-efficient to produce metals from waste instead of mineral sources. For aluminum, recycling consumes about 5% of the energy used to produce it from mineral ore. For each kilogram of copper produced from ore, more than 20 kg more carbon dioxide is emitted, compared to recycling.

Total recycling efficiency depends on the efficiency of collection, preprocessing, and metallurgical recovery. An example with a collection rate of 30%, a preprocessing efficiency of 60%, and efficiency in the metallurgical process of 95% creates a total recycling rate of only 17% ($0.3 \times 0.6 \times 0.95 = 0.17$; Gunn 2014).

Metals have an "eternal" life cycle compared with cellulose- and plastic-based products that normally can be recycled no more than 5–10 times without significant downcycling.

9.1.3 LEGISLATION

In the main, there are two types of legislation that impact on the recycling of metals: extended producer responsibility (EPR) and environmental protection.

In many parts of the world, EPR legislation requires producers and/or importers to design products with reduced environmental impact and to ensure that the last owner can discard products in a convenient way, without cost. There are also requirements for collection and recycling rates for metal packaging (tin and aluminum containers), end-of-life vehicles (ELVs), and others that require recycling of a certain percentage of the weight. Often >90% of the metal content is recycled.

Legislation aims to protect the environment from metals that, mostly in their ionic form, can leach into water. There are often limitations on the concentration and mobility of metals put on landfill sites, since many of them are toxic. Other legislation limits the emission of metals into the environment in the form of small particles and dust, in both elemental and oxidized forms. The reason for landfilling metals is normally that recycling costs are higher than the cost of metal from ore with the same amount of contamination. Society has learnt over the years that some metals have a very negative impact on society and the environment. This is why metals like arsenic, mercury, and cadmium have been banned in most applications and why they should be landfilled separately rather than recycled.

9.1.4 VALUE

9.1.4.1 Financial

The value of an individual metal depends on many parameters such as the cost of exploration, scarcity, use, and speculations. All larger base metals, apart from ferrous, are traded every day on commodity exchanges of which the London Metal Exchange is the most important.

9.1.4.2 Environmental

Some metals are essential for the biocycle, such as selenium and potassium in soil. Other metals should be kept at the lowest possible levels, due to their toxic characteristics. Recycling requires, in general, less use of energy and emits fewer greenhouse gases than mining. Avoiding the landfilling of metals, that could contaminate groundwater, is also a good reason for recycling.

9.1.4.3 Strategic

In some countries some of the metals are found in high concentrations in the ore. Normally it is only possible to mine certain metals if there are high concentration deposits with low concentrations of difficult-to-handle contaminants and the possibility of low extraction costs. Examples include cobalt in Congo (Africa), rare earth elements (REEs) in China, and palladium in Russia. Recycling is an effective tool in reducing dependence on imports from countries where there could be political implications. Recently, China restricted exports of REE to Japan. As REE is used in modern electronics, its availability is crucial to all producers.

Some metals are extracted from minerals in conditions that are very bad for the health of workers and make a large impact on nature. Minerals containing metals that are used to finance civil war or criminals are called conflict minerals (one classic example is mining of cobolt in the Democratic Republic of Congo).

9.2 COLLECTION

The first step in the recycling chain is to collect waste fractions from both households and industry. Losses of material due to insufficient collection are often larger than in the separation, sorting, and refining processes.

9.2.1 HOUSEHOLDS

Metals collected from households are mainly ferrous alloys and aluminum and to some extent copper.

9.2.1.1 Curbside

Curbside collection is often based on collecting one fraction at a time, or using different colored plastic bags (depending on the type of waste) that are automatically separated at a later stage. A truck picks up the material that is left on the curb once, or several times, a week. Metal-containing fractions mainly consist of packaging and, in some cities, waste from electrical and electronic equipment (WEEE).

9.2.1.2 Door to Door

This is similar to curbside collection but the waste is stored in different types of containers, often plastic, that are emptied and, in some cities, weighed, before being emptied into a truck.

9.2.1.3 Recycling Centers/Container Parks

Many different fractions are collected for recycling, often by municipality-owned companies or EPR organizations. Larger units like bicycles and stoves can be collected separately and sent for recycling.

9.2.1.4   Material Recycling Facilities Collection

Material Recycling Facility (MRF) fitted collection consists of often only two fractions, "Recyclables" and "Waste". The "Recyclables" fraction is sorted in a dedicated plant by hand and/or by machines. The "waste" fraction is incinerated or landfilled.

9.2.2   INDUSTRY

Metal scrap is generated in almost all types of industry. The engineering, construction, and demolition industries normally generate large volumes.

9.2.2.1   Mixed Material

In some companies often due to lack of space, different material can be collected in the same container and then sent to a recycling company that separates the different materials using machines.

9.2.2.2   Source Separation

Depending on the type of metals used in the industry/company, containers for metal waste can be separated at source. The value of the different alloys is higher if they are separated in this way. A handheld X-ray fluorescence can be used to monitor the content of metal alloys. This type of instrument can monitor the content of different metals, normally with a detection limit of 0.01%, in less than 30 seconds.

9.2.3   HANDLING OF SCRAP

To reduce the size of scrap, stationary and mobile shears are used. Cast iron and closed containers, like metal flasks, should be avoided in these units for safety reasons. Cast iron can be very stiff and hard, damaging both shears and shredders. Closed containers can explode. Crushers and balers are used to increase the density of metal scrap.

Small particles, like turnings or grinding swarf, are compacted in briquetting equipment, which makes handling much simpler and reduces losses due to dusting in the next stage of recycling.

Granulators are used to separate plastics from metals in items such as cables. Plastics and metals can be sorted by density.

9.3   LOGISTICS

9.3.1   TRANSPORTATION

How, and how far, different metals are transported for recycling depends on the density, both of the metal itself and how much the pieces are compacted, market prices, and taxes among other factors. Generally, the higher the value and density of the material, the further it can be transported while still being commercially competitive. For low-value waste, transportation could be a large cost, thus limiting alternative treatment.

9.3.1.1 Truck

Containers carrying metals are normally transported by truck, especially for short distances. Although regulations in different countries vary, a trailer truck normally carries about 25 tons and is always limited by weight, rather than volume, which could be the case for other waste fractions of lower density, like plastics. Some metals are baled for easier handling.

9.3.1.2 Train

Electric trains have the advantage of low environmental impact and low costs for long-distance transportation. Since few scrap generating plants, recyclers, and metal smelters have rail connections, trucks are often used for part of the transportation. One railcar can normally take about 60 tons.

9.3.1.3 Ship/Barge

Shipping is a good alternative, not only between continents, but also where large volumes need to be transported. Sea containers and loading metal scrap directly onto the ship are most common. Shipping shows low CO_2 emissions per ton of scrap.

9.3.2 CONTAINER SELECTION

There are many types of metal containers. Design is often optimized for different types of trucks as well as the distance of shipping. Steel containers are used for handling many different metals. Transporting metal scrap in steel containers generates high noise levels, especially for aluminum. The amount of disturbance caused to neighbors is a factor that must be addressed.

9.4 STAGES OF RECYCLING

The stages of recycling are very similar irrespective of the type of material being recycled. The first stage is often manual and involves the dismantling of often small components. The second stage often utilizes a hammer mill that turns large pieces into fragments no larger than fist size. These are often identified either by the human eye or by a sensing machine. The last stage is to sort one metal from another, metal from plastics, and metal from glass (Figure 9.1).

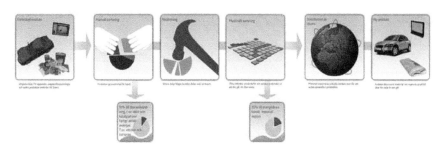

FIGURE 9.1   SIS, separation-identification-sorting.

9.4.1 SEPARATION

9.4.1.1 Manual Dismantling

Manual dismantling removes items that can be reused as well as material classified as hazardous waste, such as mercury switches, and material that is difficult to remove by machine. Manual dismantling is also a good option for the extraction of very valuable material. For complex mixtures, machines often cannot compete with sorting by experienced staff. Manual dismantling is mainly used in countries that have low labor costs.

9.4.1.2 Shredders, Mills, Crushers

Many complex products, like cars and electronic waste, need to be mechanically treated in order to separate metals from plastics and different metals from each other. A drawback of using shredders is that the hammers hit randomly, which means that pieces of plastic can become mixed into the metal fraction and vice versa. Plastics in metal fraction burn when metals are melted. It is critical when metal becomes mixed into the plastic fraction, which makes sorting more difficult as it requires melt filtration to remove the metals before a clean plastic fraction can be generated. How well different complex products separate into different materials is called liberation. Manual dismantling gives the possibility of very good liberation but is often, depending on labor costs, a more expensive alternative.

Shredders are big hammer mills with up to 10,000 hp of power. After the hammer mill, different sorting steps generate a magnetic fraction, a heavy fraction, and a light fraction (Figure 9.2). The magnetic fraction is sent to a steel mill after further removal, manually or mechanically, of copper-containing items like small

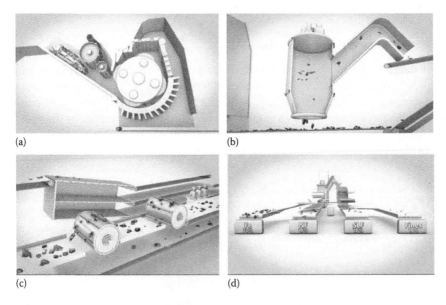

(a)   (b)

(c)   (d)

FIGURE 9.2 Hammer mill shredding (a), separation of light fraction (b), sieving of light fraction in the rear. Magnetic separation and manual copper removal of heavy fraction in the front (c), average outcome after sorting (d).

electrical motors. The light fraction is removed by pressure and separated in a cyclone. It mainly contains plastics, wood, foam, small pieces of aluminum, and copper. After screening, metals are sorted by Eddy current magnets. The residue is called shredder light fraction or automotive shredder residue, also called fluff, which is either landfilled or energy recovered. The heavy fraction, also called nonferrous fraction, is treated by a combination of sink–float and sensor-based sorting in order to recycle/recover different metals.

To avoid explosions and fires at shredding facilities, closed containers and residual fuels in petrol tanks should be presorted (Figure 9.3).

9.4.2 IDENTIFICATION

9.4.2.1 Manual

Eyes and fingers are used to manually identify different materials. When a material is identified, the next step is sorting, which can be both manual and automatic.

9.4.2.2 Automatic Sensors

The development of sensors is fast and often more than one is used in the same equipment. Sensors are able to process material very quickly. Different sensors are used, depending on the type material being identified:

1. *X-ray.* Identifies different material by transmission and/or reflection of X-rays.
2. *Metal detectors.* Detect metals in a waste mixture on a conveyor belt and removes them by jet pulse at the end of the process.
3. *Near infrared/ultraviolet light.* Light is absorbed/reflected and mainly used to identify different polymers.
4. *Color.* Can be used to sort copper from steel as they do not have the same color.

FIGURE 9.3 Shredder plant.

9.4.3 Sorting

It is normally best to remove the material that is present in the lowest concentration, as sorting requires energy and equipment capacity.

9.4.3.1 Magnets: Permanent and Electromagnetic

Some elements are paramagnetic, which means that they are influenced by a magnetic field. Ferromagnetism is used to sort paramagnetic material from other material. Magnets can be used to sort most ferrous material from other metals like copper, aluminum, and stainless steel. Magnets can have different strengths; a "super magnet" is very strong and is mainly alloyed with neodymium (Nd). This type of magnet can even be used to separate some stainless steel, which normally is regarded as a non-paramagnetic material. An electromagnet is a type of magnet in which the magnetic field is generated by electric current. Electromagnets are often used on cranes to lift paramagnetic material, where the magnetic strength can be changed by turning on and off the electricity. Using magnets is the most common method to sort metal scrap.

9.4.3.2 Eddy Current Separators

Aluminum and copper can be separated from plastics, glass, rubber, and others using an Eddy current separator (ECS). It is possible to induce a magnetic field, for a short period of time, in Al and Cu, normally by rotating strong magnets at the end of a conveyor belt, making the two metals jump up and then fall into a container located some distance away from the belt. In the same magnetic field, Al absorbs twice as much energy as copper and this can be used to sort pieces of Al from Cu. This process works even better with pieces between 5 mm and 100 mm in size and best if the size distribution is small (Nijkerk and Dalmijn 2001).

9.4.3.3 Sink–Float

By adding small high-density particles in water, the total density of the mixture can be increased. The highest density can be achieved by adding about 35 vol% of ferrosilicon, with a density of 3 kg/dm^3. Magnetite and barite (barium sulfate) are alternative additives. Baths of different density may be used to separate different kinds of metals in a sink–float process. In the first step, light metals such as magnesium (Mg; density 1.7 kg/dm^3) float while all other metals sink. In the next step, the density is increased further by adding more small particles, so that aluminum (density 2.7 kg/dm^3) will float and other metals will sink. By changing the density incrementally, different metals can be sorted. This technology works with pieces as small as 1 mm in size.

The same technology is also used to sort different plastics. By adding only about 5 vol% of particles in water, a density of 1.2 kg/dm^3 can be reached, making it possible to separate ABS plastics that float, from rubber and thermosets that sink (Figure 9.4)

9.4.3.4 Shaking Tables

Shaking tables are used for particles smaller than 10 mm. By shaking the slightly tilted table, heavier particles move upward and lighter particles downward and are collected at different positions around the table. The amplitude, frequency, and degree of tilt are adjusted to reach the desired sorting.

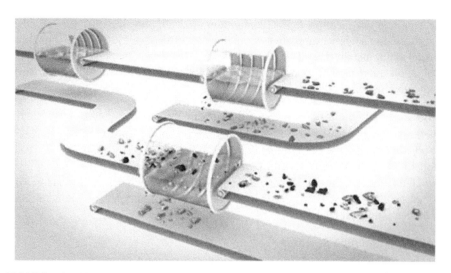

FIGURE 9.4   Sink–float plant.

- *Dry shaking tables.* Light fractions, such as polyurethane (PUR) foam, plastics, and textiles can be separated from heavier, often metal-containing fractions, using air flow from below, which separates these particles in a cyclone or baghouse filter. Dust emissions require the closed design of this equipment.
- *Wet shaking tables.* These tables are often used in the mining industry. In recycling, these are mainly used to separate light plastic fractions, like printed circuit boards (with little or no copper and plastics) from heavier, metal-containing fractions.

9.4.3.5   Material-Recycling Facilities

These are mainly used for different fractions generated by households. A MRF plant uses a combination of manual and automatic processes to separate different metals, plastics, paper, cardboard, and glass. Most automated plants need to process at least 100,000 tons of waste per year in order to cover costs.

9.4.4   REFINING

In order to produce "pure" metals, a refining step/stage is required. Often a combination of different technologies is used to produce metal of the right quality and content of alloying elements.

9.4.4.1   Pyrometallurgical Processes

The melting points of metals differ greatly. Low melting point metals are lead, which melts at 163°C, aluminum at 659°C, copper at 1083°C, and steel at 1371°C. Heating scrap to different temperatures is one method of separation. Boiling point is used for the separation of mercury (357°C), cadmium (765°C), and zinc (907°C). All other metals used in society evaporate at temperatures well above 1500°C. For some high melting point metals, like silver (Ag) and gold (Au), the melting point is also used for separation. In general, all metals that are in oxidized form end up in the slag while elemental metals end up in the melt.

9.4.4.2 Lead Smelters

The low melting point of 163°C and high boiling point of 1740°C of lead makes it a very simple material to form and therefore it is used in many applications. These properties also make lead easy to recycle; however, its toxic character sets environmental restrictions. Lead acid batteries constitute the largest volume of waste treated at lead smelters today. There are smelters that also take lead-containing minerals and some that only treat lead scrap.

9.4.4.3 Aluminum Smelters

Secondary aluminum is used mainly for metal-casting applications like motor blocks and car rims. Alloyed secondary aluminum cannot be used to produce thin-walled products like foils. These smelters produce two main types of products: Cu alloyed and Si alloyed. Most plants use two types of ovens: melting furnaces and converters. In a melting furnace, metal scrap is melted by using an oxy-fuel burner. Salt is added to remove contaminants from the aluminum into the floating salt slag. Salt slag is also a protection to prevent the oxidation of aluminum by oxygen in the air. Melted aluminum is then poured into a converter, where some contaminants can be removed by adding chlorine gas. Necessary alloying metals such as Cu and Si can be also added in the converter. The aluminum leaves the plant in the form of aluminum bars (at room temperature) or as molten aluminum in large insulated flasks, which are sent directly to the customer.

The presence of organic material, like oil and plastics, in aluminum scrap could cause oxidation of aluminum, which is why amounts must be controlled.

The gas produced by furnaces must be cleaned of dust and acidic gases such as hydrochloric acid and sulfur dioxide.

As aluminum is easily oxidized, the process must be conducted in a reducing atmosphere. The produced salt slag is reactive; it contains small droplets of elemental aluminum and carbides/nitrides that can react with many different things. In contact with water, the salt removes the protective aluminum oxide layer from the surface of small elemental aluminum particles generating hydrogen gas. The salt content also makes the slag unsuitable for landfilling. Special plants have been developed to treat this (Figure 9.5).

9.4.4.4 Copper Smelter

About one-third of copper production comes from recycled copper. Different types of furnaces are used to melt copper-containing scrap. Melted Cu together with alloying metals are, in a second stage, cast into electrodes. These are put into electrolysis cells containing an acid where oxidized Cu ions move in a solution from one electrode to the other where they are reduced to elemental copper. PGMs do not

FIGURE 9.5   Aluminum recycling.

dissolve and end up in sludge at the bottom of the electrolytic cell and are sent for further refining, mainly hydrometallurgical. Copper from transformers, electrical motors, and cables is very clean and can be used directly in some applications, which is why they are not mixed with the large volume of mixed electronic scrap that is the main fraction used in the first melting furnace.

Brass is an alloy of 65% copper and 35% zinc. This is mainly kept separate and sent to special brass metal producers. Scrap with bronze, which is made from 80% copper and 20% tin, is handled in a similar way. If scrap containing these copper alloys is fed into a copper smelter, the metals can still be recycled, though the recycling cost is higher than if they are kept as a separate stream.

Some copper smelters can handle quite high concentrations of plastic, and also release energy from the combustion, while others have large limitations in content of plastics. Gas cleaning with wet scrubbers and textile filters is an effective method of removing acidic gases, like hydrogen chloride, that are generated from PVC (polyvinyl chloride; used as insulation on copper cable); hydrogen bromide, which is generated from brominated flame retardants (often used in printed circuit boards); and ABS plastics, used in electric equipment. Another metal used for fire protection is antimony, which ends up in the slag after melting. The use of antimony is growing, though recycling is very limited and most is landfilled with slag. Slag from copper smelters is very stable and can be used for construction purposes.

All type of copper products can be manufactured from recycled/recovered copper as there is no downcycling, unlike aluminum and steel; not all products can be produced from these recycled metals. The main difference is the electrochemical process used for copper refining where >99.995% copper can be produced.

9.4.4.5   Steel Smelter

In developed parts of the world, magnetic scrap accounts for half of the raw material used in steel production. Blast furnaces that use iron ore as raw material can also use clean scrap, normally to a level of 10 w%–20 w%. At scrap yards, different steel alloys are kept separate and sent to different steel mills, depending on what kind of alloy they produce. Stainless steel scrap is the main raw material for stainless steel mills.

Electric arc furnaces (EAFs) are the most commonly used type of oven for steel scrap recycling. The possibility to reduce oxidized steel, like rust, is very limited at a steel smelter even if some elemental carbon is added. Copper is an element that should not be present in steel scrap as concentrations of copper above 0.3 w% significantly decrease the strength of the steel produced. Tin is another element that should be avoided in steel scrap as it causes similar problems. Galvanization is a problem for most steel mills, which is why zinc-containing material should be sorted separately.

EAFs normally use gas cleaning, which separates out dust, but these are sensitive to plastics like PVC, which contain mercury and chlorine that can form dioxins during the thermal process.

Steel foundries can use recycled scrap directly in their process if it is cut to a size suitable for their furnaces. Some induction furnaces require scrap of high density in order to function properly. The material also must be clean from contaminants that can evaporate into the air as most foundries do not have advanced air pollution control equipment.

9.4.5   Hydrometallurgical Processes

This technology is mainly used for high-value materials in relatively small volumes. Copper is one metal that is extracted by the leaching of minerals. For commercial recycling, this method is mainly used for PGMs and to some extent for REEs.

9.4.5.1   Leaching

In the leaching process, oxidation potential, temperature, and the pH of the solution are important parameters. These are often manipulated to optimize dissolution of the desired metal component into the aqueous phase. Acid or sometimes alkali is often used. The smaller the particles are, the more efficient and rapid the leaching process is.

9.4.5.2   Solution Concentration and Purification

A leaching liquid, based on water, is mixed in a "mixer-settler" with an organic solvent, such as kerosene, that does not dissolve in water. In the organic solvent, a metal-chelating additive can selectively form a complex with one type of metal ion. The solvent is then gravimetrically separated from the leachant. The leachate, solvent, and metal chelate are reused in the process.

9.4.5.3   Metal Recovery

From the solvent fraction, metals can be recovered by another stage of extraction, electrolysis (electrowinning), and/or precipitation. This technology can generate very pure metals.

The raw material needs to be free from organic fractions that normally consume used chemicals. Ammonia is used to selectively recover copper from mixtures of metals.

9.5 RECYCLING OF SPECIFIC METALS AND SECONDARY USE

9.5.1 BASE METALS

9.5.1.1 Ferrous-Based

The annual global consumption of steel is about 1500 million tons. In emerging economies like China and Brazil, the use of recycled steel is very low as most of it is used for infrastructure such as houses, bridges, and railroads, which have long life cycles. In large parts of the European Union, the use of recycled steel exceeds 50%.

Scrap ferrous metals are classified into different classes, primarily depending on the size of the particles and the content of alloying metals. Recycling often incurs some downcycling, due mainly to increased copper content.

Ferrous waste in small particles requires dry storage or very limited storage time to prevent corrosion. In enclosed storage, such as ships, this corrosion can consume oxygen and cause suffocation.

9.5.1.2 Aluminum

The annual global consumption of aluminum is about 50 million tons. Sorting is based on alloy content and size. Primary aluminum, produced from bauxite, is used to manufacture thin-walled products such as foils. Almost all cast thick walled products are based on recycled material.

9.5.1.3 Copper

The annual global consumption of copper is about 22 million tons. One-third is recycled material. Copper is mainly used with a >99.9% quality. Alloys of copper and tin, called bronze, as well as alloys of copper and zinc, called brass, are also commonly produced from recycled material. No downcycling is incurred by standard recycling processes, including electrolysis.

9.5.2 PRECIOUS GROUP METALS

PGMs are consumed at an increasing rate and mining often takes place in unsustainable conditions. Despite this, less than 50% of PGMs are recycled. The main challenge, in terms of recycling, is that small amounts of material are used in a vast number of products. The recycling rate of Au and Pd from WEEE in Europe is currently less than 20%.

9.5.2.1 Gold and Silver

If separately collected, these metals can be recycled by melting. In products, these elements are often found together with copper that has a slightly higher melting point of 1083°C. Since Au and Ag are nobler, extraction of Cu generates a concentration of Au/Ag that, with a difference of 100° in melting point, can be separated either by melting or by hydrometallurgical processes (see Section 9.4.5).

9.5.2.2 Platinum Group Metals

Ruthenium (Ru), Rhodium (Rh), Platinum (Pt), Palladium (Pd), Iridium (Ir), and Osmium (Os) are mainly used in catalysts. High melting temperatures for Pd (1555°C), Pt (1768°C), and Os (3027°C) make them possible to separate by melting. Their noble characteristics make these elements very difficult to oxidize.

9.5.3 Rare Earth Elements

Only 20 years ago, almost no products contained REEs. Today all electronic products contain greater or lesser quantities of REEs. More than 90% of all REEs are produced in China. In European Union, there is only one plant, which is in Latvia. REE minerals almost always contain radioactive elements that make sustainable mining very costly. Today less than 1% of the REEs used in society are recycled.

REEs are present, in small amounts, in many different minerals. The reason they are called "rare" is because it is rare to find only one, or a few of them, in one mineral. The cost to separate different REEs from each other is high since the extraction characteristics are similar. Many different separation steps are needed.

In the periodical table of elements, most of the REE elements can be found in the upper row of the two rows below the main table. These elements are divided into light and heavy REEs.

Light REEs: La, Ce, Pr, Nd, Pm, and Sm
Heavy REEs: Sc, Y, Eu, Gd, Tb, Dy, Ho, Er, Tm, Yb, and Lu

9.5.3.1 Magnets: Neodymium-Based

The largest amount (more than 50%) of the REE value used today is Nd-containing permanent magnets. In total about 6000 tons of Nd magnets are produced every year. The recycling rate is less than 1%.

There are two main production technologies for Nd magnets: sintering and using an organic binder. These magnets are very sensitive to corrosion, which is why many are covered by a nickel surface. This, together with the organic binder, makes recycling difficult as small amounts of these contaminants affect the suitability of the material for the production of new magnets. The fact that the magnet is very brittle and acts as a magnet results in a thin layer of magnetic particles on a large part of the ferrous material. When this material is sent to a steel mill, the Nd ends up in the slag.

For recycling purposes, one option is to heat the magnet to a temperature where it no longer functions as a magnet, the Curie temperature, which is about 400°C in this case. Samarium (Sm) alloyed magnets used at higher temperatures have a Curie temperature over 700°C but are not as strong as Nd-based magnets.

Nd magnets are used in hard disks, DVDs, CDs, electrical motors, loudspeakers, air-conditioning equipment, and offshore windmills, among other applications.

9.5.3.2 Fluorescent Powder

Fluorescent powder is used in the backlight of LCDs, the backlight of plasma TVs, and CRT, and for fluorescent lighting.

Europium (Eu) and Yttrium (Y) are the main REEs in fluorescent lighting and can be recycled by hydrometallurgical methods. The content of mercury needs to be addressed separately as this element is very toxic.

The cost of recycling different REEs from these sources is higher than the raw material value, which is why almost no recycling takes place today.

9.5.3.3 Batteries

Ni metal hydride batteries and some Li ion batteries contain significant amounts of REEs. Their value, compared to the cost of recycling, is still not great enough to make this recycling viable. From NiM hydride batteries, Ni is recovered and from Li ion batteries, Co and Cu. The amount of Li in a Li ion battery is normally less than 2 w%.

Today, REE oxides can be recycled by pyrometallurgical processes.

9.5.4 SPECIAL METALS

The recycling of REEs and special metals is currently very limited. Rhodia, Umicore, Dowa, Aurubis, Boliden, and Xstrata (Gunn 2014) are some companies that are active in this area. Alkali metals (group 1 in the periodical table of elements) and alkaline earth metals (group 2) are not generally used as elemental metals. When recycled, they are usually used as construction materials as their value is low.

9.5.4.1 Lithium

Out of the total lithium production, 95% comes from Chile, Argentina, China, and Australia, which amounted to about 30,000 tons in 2009. Large reserves have been found in Bolivia. A brine of lithium (Li) carbonate is pumped up from deposits below ground and dried by the sun. Use of Li in batteries is expected to increase dramatically, which is why recycling is required in the long term. No recycling of Li takes place today. In a Li ion battery, Li can be found in different forms of salts. The anode is often made out of elemental carbon and fluorine-containing polymers (PVdF). Salts in the electrolyte makes recycling of the small amounts of Li present in a battery expensive.

9.5.4.2 Magnesium

Magnesium is used in light-weight, cast products like chain saws, often in alloys with aluminum. The low density makes it easy to sort by sink–float technologies. As magnesium is not a noble element, it requires storing in dry conditions.

9.5.4.3 Titanium

About 90% of the titanium (Ti) currently used is in the form of titanium oxide (TiO_2), a white pigment used in paint and plastics. Small amounts of elemental Ti, used in aviation, chloride corrosion resistant products, and other lightweight products, are recycled. There is presently no recycling of the pigment.

9.5.4.4 Tantalum

Tantalum is used in capacitors. If collected separately, it is possible to recycle it by pyrometallurgical processes.

9.5.4.5   Tungsten

Tungsten is used in drilling tools and carbide inserts. This is currently recycled by smelting and is used in similar applications.

9.6   RECYCLING EXAMPLES

The metal wheel is a simplified description of how carrier metals like Fe, Al, Pb, and Cu enable recycling of other metals. Slag from one type of metal smelter can also be used for recycling of metals in another type of smelter. Primary smelters using different types of ore can often use recycled material if the quality is good enough.

This is exemplified in Figure 9.6. The blue inner wheel circle contains the carrier metal in the smelter. Next wheel circle outward is gray and it shows metals that can be dissolved in the carrier metal provided they are in elemental form. The white

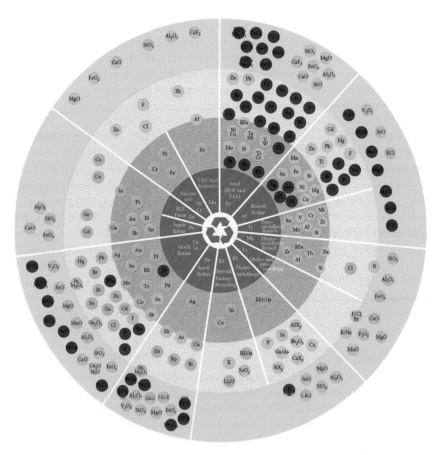

FIGURE 9.6   The metal recycling wheel. (Modified from Reuter, M. A. et al., *Metal Recycling: Opportunities, Limits, Infrastructure*, A Report of the Working Group on the Global Metal Flows to the International Resource Panel UNEP, 2013.)

wheel circle represents elements that are mainly found in dust, slag, and slime. They should, if possible, be further treated by a hydrometallurgical process. Furthest out is the green wheel circle where mainly benign low-value products are present. The elements are represented with different colors, depending on whether they can be recovered or not. Green circles show elements that mainly can be recovered, yellow circles are mainly elements in an alloy or a compound in an oxidized product that probably will be lost, and the red circles are elements that mainly will be lost and in some cases even negative for the carrier metal.

9.6.1 RECYCLING OF A FOOD CAN

Metal packaging is covered by an extended producer liability system in many parts of the world. Since metals are magnetic, they can easily be separated from other fractions by an overband magnet at low cost. The most important element to separate from steel in the quality control system is copper that could ruin the quality of the steel produced from recycled cans. In most parts of the world, tin is replaced by an epoxy thermoset layer to prevent corrosion and has very little negative impact on the recycling of steel.

9.6.2 RECYCLING OF A CAR (ELV)

At a car dismantler, parts that can be reused are removed. Large plastic components and/or glass are, in some countries, removed for material recycling. Catalytic converters and tires are removed and recycled. In most countries, all liquids (battery, fuel, oil, etc.) are classified as hazardous waste and treated accordingly. Pyrotechnical units, such as airbags and pretensioners, are either removed or initiated. The increased use of electronics in modern cars has also seen the removal of computers, electrical motors, generators, printed circuit boards, and cables, which can be recycled as electronic waste. The number of small electrical motors in a modern car often exceeds 50. Since small motors contain magnetic material, they are often removed from the shredded material as a ferrous fraction. Some type of sorting is required to remove copper-containing material from the ferrous fraction in the motor as this reduces the quality of the steel. Some parts of the safety cage in modern cars contain high-strength steel (often manganese alloy) that could be separated, although this is not currently carried out.

After the removal process, the car body is shredded in a hammer mill. After the hammer mill, the material is often separated into: (1) magnetic fraction, (2) heavy fraction (mainly nonmagnetic metals like stainless steel, copper, and aluminum, as well as rubber, stones, and fiber composites), and (3) light fraction (plastics, foam, textiles, and small pieces of metal). The heavy fraction is further identified and sorted, primarily to recover nonmagnetic metals. The light fraction is often sieved to improve the separation of Cu and Al with an ECS. Fines fraction (<10 mm) generated by sieving, is often landfilled, a plastic fraction mainly containing polypropylene is material recycled, and the residual polymer fraction, energy recovered.

From 2015, the recycling requirements for ELVs, in the European Union, is 95 w% with a maximum 10 w% of energy recovery (Figure 9.7).

FIGURE 9.7   Scrap car dismantling and compaction.

9.6.3   RECYCLING OF ELECTRICAL CABLES

Cable-recycling plants can recycle both production waste from cable producers, scrap from cable installations, and end-of-life cables from demolition. In the first two fractions, the type of insulation is known, which makes it possible to recycle plastics for use in similar applications. Old PVC cables from demolition contain softeners that are currently banned, which sets limits on recycling. The electrical conductor in cables is either copper or aluminum, which is why these cables are source-separated and treated separately. Feral line, a combination of aluminum and steel, is also treated separately. There are two main cable recycling technologies: (1) cable stripping, in which plastics are separated from the metal by slicing it off with equipment that uses knives, similar to peeling a banana; and (2) cable granulation, which can handle larger volumes. The mechanical recycling starts with a coarse granulator. The next step is a fine granulator, which turns the cable into pieces of <5 mm. Plastics are separated from metals at a dry table due to their differences in density. The plastic fraction can be further separated by gravimetrical separation in water, as polyethylene floats, and PVC, rubber, and Halogen Free Flame Retardant, polyolefin sink in water. The recycled copper is of very high quality.

9.7   CONCLUSIONS AND THE FUTURE

The recycling of metals is a very important part of the circular economy that we should all be striving for. Compared to production technologies, the technology of recycling complex products is undeveloped. Globally, recycling research is still at a very low level and very little has happened in comparison with the development of new products. In the European Union, new research funds have been initiated to increase recycling research.

Even though the same products are sold across all markets, recycling levels and technologies differ vastly between countries.

Today, the raw material market based on waste is global, but, due mainly to increased market prices and the recession, a trend of protectionism is apparent in many countries.

With lower costs and higher capacity, different forms of robotics will be used both for dismantling and for sorting over time.

REFERENCES

Gunn, G. 2014. *Critical Metals Handbook.* New York: John Wiley & Sons.

Nijkerk, A. A. and W. L. Dalmijn. 2001. *Handbook of Recycling Techniques.* The Hague, the Netherlands: Nijkerk Consultancy.

Reuter, M. A., C. Hudson, A. van Schaik, K. Heiskanen, C. Meskers, and C. Hagelüken. 2013. *Metal Recycling: Opportunities, Limits, Infrastructure.* A Report of the Working Group on the Global Metal Flows to the International Resource Panel UNEP.

10 Material and Energy Recovery from Waste of Electrical and Electronic Equipment
Status, Challenges, and Opportunities

Efthymios Kantarelis, Panagiotis Evangelopoulos, and Weihong Yang

CONTENTS

10.1 INTRODUCTION

As technology innovation progresses and the innovation cycles become shorter, the production of electronic equipment increases and so its replacement, which makes electrical and electronic equipment (EEE) a rapidly growing stream. The United Nations Environment Program (UNEP) has estimated that 20–50 million tons of e-waste are generated worldwide annually, which has become a serious risk for living organisms and environment (UNEP 2005). The annual growth rate of waste from electrical and electronic equipment (WEEE) has been estimated at about 3% to 5% every year, which is approximately three times faster than the conventional solid waste streams (Eurostat 2012). This situation arises from the need of newer, faster, and more efficient technological equipment and more demanding applications. Therefore, older equipment becomes obsolete and is discarded, leading to the continuously growing e-waste stream. The severity of WEEE growth has been noted by U.S. Environmental Protection Agency (USEPA), which reports that 438 million new electronic products were sold while around 2 million tons of electronic products were ready for end-of-life management (USEPA 2011; Figure 10.1).

A total of 43 million TVs, computers, and printers; 15.4 million white goods; 1 million air-conditioners; and 40 million mobile phones were discarded in China in 2009 (Ongondo et al. 2011) while almost 3.5 million tons of WEEE were collected in

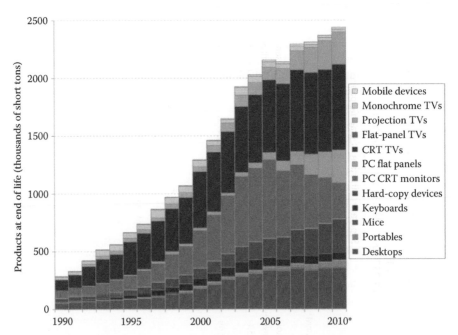

FIGURE 10.1 Quantity of electronic products ready for end-of-life management in the United States. *Results for 2010 are projected based on estimates from previous years. (Data from USEPA, *Electronics Waste Management in the United States through 2009*, United States Environmental Protection Agency, Washington, DC, 2011.)

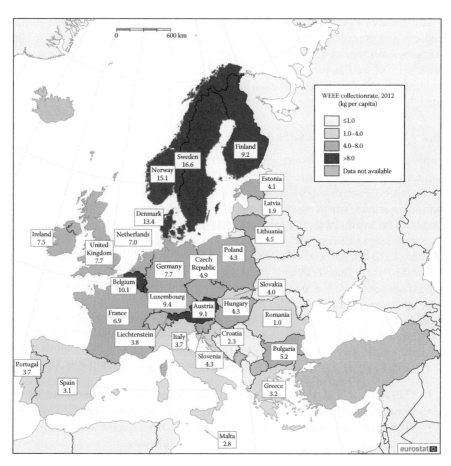

FIGURE 10.2   WEEE collection rate in Europe in 2012. (Adapted from Eurostat, http://ec
.europa.eu/eurostat/documents/342366/351758/weee-collection.pdf, 2012.)

the European Union in 2012 as each European citizen discards 14–24 kg of WEEE
every year (Eurostat 2012; Figure 10.2).

   WEEE consists of several valuable materials such as plastics, metals, ceramic
materials, and rare earth metals as well as toxic components such as heavy metals
and persistent organic pollutants. Therefore, the benefits from proper waste treatment
of WEEE are straightforward by providing the industry with materials, minimizing
the cost of extraction of virgin materials, and at the same time minimizing the environmental risks from hazardous components. In order to understand the problem of
WEEE and its lack of efficient handling, it is necessary to trace its life cycle. Initially,
each electrical and electronic device takes a while before it is considered as waste.
In the United States, more than 4 million tons were stored in 2009 before ending up
for end-of-life management (USEPA 2011). Some obsolete devices are traded nation-
ally while others are shipped to developing countries where older technology can be
used for a few more years.

The reuse of electronics in developing countries can extend the lifespan of the products and contribute to regional development in terms of following up to the technological innovations; however, usually these countries lack in environmental restrictions as well as proper waste treatment facilities. Therefore, exporting large quantities of old electronics will contribute to improper and uncontrolled treatment, which leads to major environmental hazards (Swedish EPA 2011).

10.2   STATUS AND LEGISLATION

European Union tries to reduce growing WEEE stream by enacting several directives for its management, use, and recycling (Directives 2002/96/EC and 2012/19/EU among others). Each member state must adopt national legislations in order to meet the specific objectives, which allow them to tailor legislations according to their different conditions on a national level. Specifically, the restriction of the use of certain hazardous substances in EEE restriction of hazardous substances (RoHS) I, introduced on February 13, 2003, sets maximum limits for Pb, Hg, Cd, Cr^{+6}, polybrominated biphenyls, and polybrominated diphenyl ethers used in specified types of EEE. This has prevented high amount of banned substances from being disposed and potentially released to the environment. RoHS directive implementation by the producers requires changes in product design worldwide. RoHS revision in 2011 set new goals and restrictions to chemical compounds for the forthcoming years (The European Parliament and the Council of the European Union 2003, 2011; European Union 2014a).

The more important elements of new directive are summarized as follows:

- The rules will gradually be applied to all EEE, cables, and spare parts until 2019
- Review of RoHS list until July 2014 and periodically thereafter
- More transparent rules for allowing exemptions from the substance ban
- Improving the correlation with the REACH regulations
- Photovoltaic panels are excluded from new RoHs, in order EU to meet its objectives for increased share of energy produced from renewable sources and increased energy efficiency (European Union 2014a)

The aim is to extend producers' responsibility by requesting to finance the collection, treatment, recovery, and proper disposal of WEEE according to product characteristics, while at the same time to reduce the cost for waste handling on municipality level. This implies that producers must develop more environmental friendly products with more recyclable materials and simpler to dismantle, which will eventually reduce the environmental impacts (Swedish EPA 2011). Similar legislations about extended producer responsibility (EPR) have been introduced in Japan, South Korea, Taiwan, and China and in several states of the United States (Atasu, Özdemir, and van Wassenhove 2012; Swedish EPA 2011).

The Netherlands and Sweden introduced extended producers' responsibility from the early stages of the EU directive. In the Netherlands, all the e-waste categories, except from the information and communications technology, are managed by an EPR organization while in Sweden the management is performed by nonprofit organizations (Gottberg et al. 2006).

According to the European Union, EEE is categorized as listed in Table 10.1.

TABLE 10.1

**Categories of EEE and Indicative Equipment According to Directive 2012/19/
EU and the Targets for Recovery, Recycling, and Reuse**

	Category	Indicative Equipment	Targets for Recovery	Targets for Recycling and Reuse
1.	Large household appliances	• Appliances used for refrigeration, conservation, and storage of foods (refrigerators, freezers, etc.) • Appliances used for cooking and other processing of food (cookers, electric stoves, electric hot plates, microwaves, etc.) • Appliances used for heating rooms, beds, seating furniture (electric radiators, electric heating appliances, etc.) • Appliances used for exhaust ventilation and conditioning equipment (air-conditioners, electric fans, etc.) • Washing machines, cloth dryers, dish washing machines	>80%	>75%
2.	Small household appliances	Appliances used for sewing, knitting, weaving, and other processing for textiles (vacuum cleaners, carpets sweepers, etc.)	>70%	>50%
3.	IT and telecommunications equipment	• Centralized data processing (mainframes, minicomputers, printer units, etc.) • Personal and laptop computers • Printing and copying equipment • Pocket and desk calculators • Other equipment for the collection, storage, processing, presentation, or communication of information by electronic means • Equipment for transmitting sound images or other information by telecommunications (cellular phones, Telex, telephones, etc.)	>75%	>65%
4.	Consumer equipment and photovoltaic panels	Products and equipment for the purpose of recording or reproducing sound or images including signal or other technologies for the distribution of sound and image than by tele-communications (radio and television sets, video cameras, video recorders, audio amplifiers, musical instruments, etc.)	>75%	>65%

(Continued)

TABLE 10.1 (*Continued*)

Categories of EEE and Indicative Equipment According to Directive 2012/19/ EU and the Targets for Recovery, Recycling, and Reuse

	Category	Indicative Equipment	Targets for Recovery	Targets for Recycling and Reuse
5.	Lighting equipment		>70%	>50%

6.	Electrical and electronic tools[a]	• Equipment for turning, sanding, grinding, sawing, cutting, shearing, drilling, making holes, punching, folding, bending, or similar processing of wood, metal, and other materials
		• Equipment for riveting, nailing, screwing, removing rivets, nails, screws
		  – Tools for welding, soldering
		  – Tools for mowing and gardening activities, etc.
			>70%	>50%

7.	Toys, leisure, and sports equipment	Electric trains and cars, video games, and consoles Computers for biking, diving, running, etc.	>70%	>50%
8.	Medical devices[b]	Appliances for detecting, preventing, monitoring, treating, alleviating illness, injury, or disability (nuclear medicine equipment, dialysis equipment, radiotherapy, cardiology equipment, etc.)	Not defined	
9.	Monitoring and control instruments	Monitoring and control instruments used in household and industrial installations (smoke detectors, heating regulators, measuring weight, adjusting appliances, etc.)	>70%	>50%
10.	Automatic dispensers	All appliances that deliver automatically all kinds of products	>80%	>75%

[a] With the exception of large-scale stationary industrial tools.
[b] With the exception of implanted and infected products.

Figure 10.3, illustrates the percentages of the total e-waste mass generated in 2005 in each category. It should be noted that the percentages may have changed due to the technological developments, which can affect WEEE composition. As expected, the first category represents almost half of the total waste fraction.

Studies indicate that it is not practical to have a dumpster for each of the 10 different categories of WEEE, and some simplification has to be made by categorizing

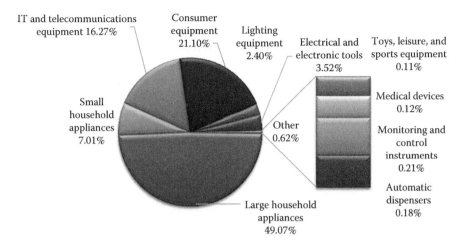

FIGURE 10.3 Average compositional breakdown of e-waste generated by EU-27 member states in 2005 of the 10 different WEEE categories. (Data from Huisman, J. et al., *2008 Review of Directive 2002/96 on Waste Electrical and Electronic Equipment, Final Report,* Study No. 07010401/2006/442493/ETU/G4, United Nations University, Bonn, Germany, 2007.)

WEEE into the following groups (Bridgwater and Anderson 2003; Dalrymple et al. 2007):

1. Refrigeration equipment. Requires specialist treatment under the ODS regulations
2. Large household appliances (excluding refrigeration equipment), which have high metal content and can be easily reprocessed together
3. Equipment containing CRTs (cathode ray tubes) Must be handled separately due to broken monitor glass
4. Linear and compact fluorescent tubes. To prevent contamination and enable recycling
5. All other WEEE

10.3 COMPOSITION

The challenges associated with WEEE treatment are related not only to the rapidly growing quantity but mainly to the complexity of WEEE composition, and in-depth characterization of the streams is required for development of environment-friendly and economically viable recycling processes (Menad, Guignot, and van Houwelingen 2013). Lack of publicly available information of WEEE composition poses uncertainties to companies on how to make the necessary choices in terms of recycling technologies and equipment. There are plenty of studies focusing on the chemical composition and the characterization of WEEE fractions. Widmer et al. (2005) summarize data from e-waste stream recycled by the SWICO/S.EN.S recycling system in Switzerland, which is shown in Table 10.2.

TABLE 10.2

**E-Waste Material Composition Recycled by the SWICO/
S.EN.S Recycling System**

Material	Percentage
Metals	60.20%
Plastics	15.21%
Metal–plastic mixture	4.97%
Cables	1.97%
Screens (CRT and LCD)	11.87%
Printed circuit boards (PCB)	1.71%
Others	1.38%
Pollutants	2.70%

Source: Widmer, R. et al., *Environ. Soc. Impacts Electron. Waste Recycling,* 25(5), 436–458, 2005; Ongondo, F.O. et al., *Waste Manag.,* 31(4), 714–730, 2011.

Characterization of the composition of small WEEE fraction that can be discarded in the municipal electronic waste bins (by sampling of 5 tons of WEEE) from different regions of Germany is shown in Table 10.3 (Dimitrakakis et al. 2009). The sampling, based on land development structure index, results in a more objective chemical composition of WEEE.

The composition of WEEE varies depending on the age and type of the discarded item, and several methods of characterization of the different fractions have been reported. These include investigation of material composition by means of: FTIR (Fourier transform infrared spectroscopy) for identification of the polymer type, EDXRF (energy-dispersive X-ray spectroscopy) for identification of the heavy metals and the halogens, HPLC-UV/MS (high-performance liquid chromatography–ultraviolet/ mass spectrometry) for flame retardants, and GC-HRMS (gas chromatography/high-resolution mass spectrometry) for PDBB/F (polybrominated dibenzo-*p*-dioxins and dibenzofurans) detection, Raman spectroscopy, and so on (Schlummer et al. 2007; Taurino et al. 2010). The application of combined methods provides a faster analysis of WEEE plastics. Faster and more efficient characterization of the chemical composition of WEEE can improve the recycling techniques in terms of separation and blending of different flows for material reuse (Taurino et al. 2010).

Most WEEE types contain varying quantities and types of plastics (Table 10.4). The need for different polymers in the different electronic components makes their recycling even more complicated. The plastics that are commonly encountered in EEE are acrylonitrile butadiene styrene (ABS), polycarbonate (PC), PC/ABS blends, high-impact polystyrene (HIPS), and polyphenylene oxide blends (Dalrymple et al. 2007).

There are two different polymer fractions that are available on the market: a well-defined polymer product that comes from plastic fractions during dismantling (e.g., during the course of CRT glass recovery) and a second kind of the polymer-containing

TABLE 10.3

Average Material Composition on WEEE (% w/w)

Material	Composition (Dimitrakakis et al. 2009)	Composition (Association of Plastics Manufacturers in Europe [APME] 2004)
Plastics	33.64	19
Ferrous metals	15.64	38
Nonferrous metals	3.81	28
Electronic components	23.47	
"Bonded" materials	8.02	
Cables	5.43	
PWBs	2.89	
Others	3.89	10
Batteries	1.53	
Rubber	0.54	
LCDs	0.14	
Glass		4
Wood		1

Source: Dimitrakakis, E. et al., *J. Hazard. Mater.,* 161(2/3), 913–919, 2009; APME, *Plastics—A Material of Choice for the Electrical and Electronic Industry-Plastics Consumption and Recovery in Western Europe 1995*, APME, Brussels, Belgium, 2004.

TABLE 10.4

Main Polymers Used in the Manufacture of the Most Common WEEE Items Collected

WEEE Item	Polymer Composition
Printers/faxes	PS (80%), HIPS (10%), SAN (5%), ABS, PP
Telecoms	ABS (80%), PC/ABS (13%), HIPS, POM
TVs	PPE/PS (63%), PC/ABS (32%), PET (5%)
Toys	ABS (70%), HIPS (10%), PP (10%), PA(5%), PVC (5%)
Monitors	PC/ABS (90%), ABS (5%), HIPS (5%)
Computer	ABS (50%), PC/ABS (35%), HIPS (15%)
Small household appliances	PP (43%), PA (19%), ABS-SAN (17%), PC (10%), PBT, POM
Refrigeration	PSandEPS (31%), ABS (26%), PU (22%), UP (9%), PVC (6%)
Dishwashers	PP (69%), PS (8%), ABS (7%), PVC (5%)

Source: Buekens, A. and Yang, J., *Chemical Feedstock Recycling,* 10(16), 415–434, 2014.

fraction from WEEE processing, which is a by-product of the metal recovery processes, and is isolated from the bulk WEEE by means of shredders, magnetic separators, and cyclones (Cui and Forssberg 2003).

Epoxy resins are widely used materials in electronics manufacture and they are used in conductive adhesives, flip chip encapsulation, bonding of leads, die coatings, surface-mounting adhesives, encapsulation, and conformal coatings. The largest single flame retardant, cost wise, is tetrabromobisphenol-A (TBBPA) used predominantly in polychlorinated biphenyls (PCBs; Weil and Levchik 2004). The amount of TBBPA in a finished resin used in electronic equipment may contain about 18%–21% Br (Weil and Levchik 2004). Other flame retardants commonly used in WEEE are shown in Table 10.5, while their toxicity and recyclability are shown in Table 10.6.

The different classes of PCB according to National Electrical Manufacturers Association (NEMA) are listed in Table 10.7.

TABLE 10.5
Characteristics of Commercial Flame Retardants

Commercial Flame Retardants		Br Chemical%	Br%	Sb_2O_3
Decarbromophenyl oxide	DBDPO	11–13	9–11	2–5
Tetrabromophthaldiphenyl ethane	TBPME	14–16	9–11	2–5
Brominated oligomer epoxy	BOE	17–21	9–11	2–5
Octabrophenyl oxide	OCBDPO	18–20	14–16	2–5
Tribromophenoxy ethane	TBPE	21–24	15–17	2–5
Tetrabromobisphenol-A	TBBPA	20–23	12–14	2–5

Source: Menad, N. et al., *Resources Conser Recycling,* 24(1), 65–85, 1998.

TABLE 10.6
Environmental Toxicity and Recyclability of Commercial Flame Retardants

Commercial Flame Retardants	Melting Range[°C]	Environmental Toxicity	Recyclability
DBDPO	300–315	Dioxin/Furan	Excellent
TBPME	445–458	No Issue	Excellent
BOE	120–140	No Issue	Good
OCBDPO	70–150	Dioxin/Furan	Excellent
TBPE	223–225	Potential	Fair
TBBPA	197–181	Dioxin/Furan	Poor
ATO(Sb_2O_5)	656	Carcinogenic	Excellent

Source: Menad, N. et al., *Resources Conser Recycling,* 24(1), 65–85, 1998.

TABLE 10.7

NEMA Classes of Copper Clad Laminate PCB

NEMA Grade	Resin, Reinforcement	Description	Typical Uses
XXP, XXXPC	Phenolic, paper	Hot or cool punching	Inexpensive consumer items like calculators
FR-1 and FR-2	Phenolic, paper	Flame retardant XXP/XXXPC	Where flame retardancy is required
FR-3	Epoxy resin, paper	Flame retardant	High insulation resistance
CEM-1,-3	Epoxy, glass cloth, or glass core	Punchable epoxy, properties between XXXPC and FR-4	Radios, smoke alarms, lower cost than FR-4
G-10	Epoxy, paper, glass	Excellent electricals, water-resistant, not f. r.	Computers, telecom, cost above FR-2
FR-4	Epoxy, glass cloth	Like G10 but flame retardant, Tg ~130°C	Where flame retardancy is required
FR-5	Epoxy, glass cloth	Like FR-4, more heat resistant, higher Tg than FR-4	Military, aerospace, where specified
FR-6	Polyester resin, glass mat	Flame resistant	Low capacitance or high-impact apps.

Source: Weil, E. and Levchik, S., *J. Fire Sci.*, 22(1), 25–40, 2004.

WEEE contains several compounds that can be associated with environmental and health hazards The most toxic metals contained in this fraction of waste are antimony (Sb), barium (Ba), beryllium (Be), cadmium (Cd), chromium (Cr), and lead (Pb; Tsydenova and Bengtsson 2011). All these metals are used in electronics manufacturing process as additives for different polymers (e.g., Pb is used as stabilizer on PVC cables and Sb for flame-retardant formulation), some for improving metals' properties (e.g., Cr improves hardness of steel and Be improves Cu strength) while others such as Cd and Ba for their electrical properties (Swedish EPA 2011). The heavy metals as well as their oxides are highly toxic and they are associated with carcinogenesis, skin diseases, and respiratory infections. Therefore, health and safety measures and extensive air and flue gas–cleaning system are necessary in the whole recycling process. Chemical analysis of metals and other compounds include techniques such as leaching, loss of ignition, and inductively coupled plasma atomic emission spectroscopy (ICP-AES; Yamane et al. 2011). Typical analyses of WEEE components are listed in Table 10.8.

The annual demand to production ratio of metals used in EEE is listed in Table 10.9. As shown, some metals such as Ru and In consume a vast majority of the produced metals and this indicates the importance of their recycling from WEEE.

TABLE 10.8

Reported Metals Composition of Personal Computers and PCB (ppm)

Element	Composition personal computers (ewasteguide.info 2014a)	Composition PCB (Kim et al. 2004)	Composition PCB (Evangelopoulos 2014)
Pb	62,988	13500	49611
Al	141,723		
Ge	16		
Ga	13		
Fe	204,712	14,000	10,300
Sn	10,078	32,400	1,530
Cu	69,287	156,000	338,690
Ba	315		1,645
Ni	8,503	2,800	1,340
Zn	22,046	1,600	9,410
Ta	157		
In	16		
V	2		14.8
Be	157		
Au	16	420	6.61
Eu	2		
Ti	157		1,372
Ru	16		
Co	157		3.23
Pd	3	100	11.6
Mn	315		78
Ag	189	1240	398
Sb	94		40.8
Bi	63		
Cr	63		237
Cd	94		0.23
Se	16		
Nb	2		
Y	2		
Hg	22		3.65
As	13		0.264
Pt			0.01
Mo			0.187

TABLE 10.9

Annual Production and Consumption of Metals by EEE Production and Their Ratio

Metal	Production (tons/year)	Demand for EEE (tons/year)	Demand/Production (%)
Ag	20,000	6,000	30
Au	2500	300	12
Pd	230	33	14
Pt	210	13	6
Ru	32	27	84
Cu	15,000,000	4,500,000	30
Tin	275000	90,000	33
Sb	130000	65,000	50
Co	58000	11,000	19
Bi	5600	900	16
Se	1400	240	17
In	480	380	79

Source: Buekens, A. and Yang, J., *Chemical Feedstock Recycling,* 10(16), 415–434, 2014.

10.4 ENERGY, MATERIALS, AND FEEDSTOCK RECYCLING OPTIONS FOR WEEE

Recycling of WEEE has been proven to be a valuable option not only for metal recovery but also for the energy savings compared to the extraction of virgin ore. Table 10.10 summarizes the recycled materials energy savings to virgin materials.

The precious metals, whose extraction consumes significant amounts of energy due to low concentration in mining ores, in some cases can be easier recovered. For example, the amount of Au that can be recovered from 1 ton of e-waste from personal computers can exceed the amount of Au that can be recovered from 17 tons of gold ore (Rankin 2011). A range of techniques is currently applied for retrieving components and materials from WEEE. The implementation of a technology or a combination of more technologies depends on the material characteristics and overall process economics.

The overall recycling process from final user includes a collection and transportation step to a suitable treatment facility. The treatment essentially consists of a scheme of sorting/disassembly (step 1), size reduction (step 2), and separation/refining (step 3; Figure 10.4).

TABLE 10.10
Recycled Materials Energy Savings Over Virgin Materials

Material	Energy Savings (%)
Al	95
Cu	85
Fe and Steel	74
Pb	65
Zn	60
Paper	64
Plastics	>80

Source: Cui, J., *Mechanical Recycling of Consumer Electronic Scrap*, Luleå University of Technology, Luleå, Sweden, 2005.

10.4.1 DISMANTLING AND SORTING

The first step of WEEE treatment is achieved almost exclusively by manual intervention (primary recycling) (Figure 10.5).

The primary recycling includes manual dismantling where whole components of the e-waste fractions are separated into metals, plastics, glass, and hazardous materials (Figure 10.6). The weight of the total fraction is reduced significantly since the metallic and plastic covers can be dismantled, sorted, and transferred directly for reuse or recycling. Materials like CRT-glass and LCD screens contain Hg and are treated in special mercury-recovery facilities or incinerated in authorized hazardous waste treatment units with extensive flue gas–cleaning system. Batteries and conductors are sent to other processes for recovery of Cd, Ni, Hg and Pb (Swedish EPA 2011). This step remains the most popular because simplicity and low cost are ensured (Achilias et al. 2012).

Manual sorting slows down the next steps of the recovery and recycling process and can become a bottleneck for the whole process. Therefore, several new technologies have been introduced for making the dismantling and sorting processes automated. One solution for automatic sorting is based on infrared spectroscopy (Figure 10.7). Lamps with selected range of infrared wavelength (instead of ordinary visible light) are used and the materials reflect the light differently according to their chemical structure (Grietens 2009).

The reflection of different materials like those encountered in plastic-coated cardboard, ordinary cardboard, and different kinds of plastics, even if their color looks similar, can be distinguished. Those differences in wavelengths activate a jet of pressurized air, sending the selected material to a different conveyor from the rest e-waste fraction (Xenics NV 2014).

Two main wavelength regions are used, the near-infrared (NIR) spectrum and the mid-infrared (MIR); the NIR distinguishes different materials based on the reflection of the light.

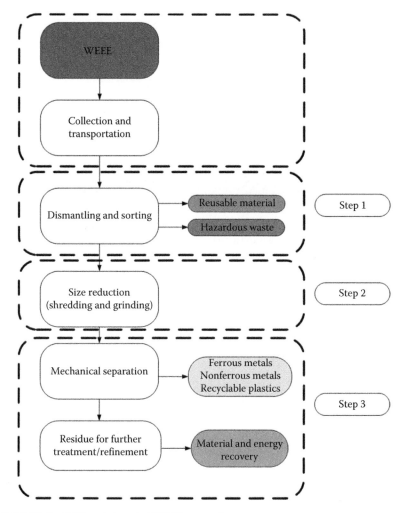

FIGURE 10.4 Different steps in WEEE processing.

MIR works on a similar principle to NIR, but projects light in the MIR range. French company Pellenc ST piloted this technology in 2008 as a more efficient way to separate paper and cardboard. The new MIR method brings efficiency levels up to 90%, which is an improvement of around 30% (Capel 2008). Other automated techniques include VIS (visual spectrometry), which recognizes all colors that are visible and works for both transparent and opaque objects; EM (electromagnetic), which sorts metals with electromagnetic properties, as well as metals from nonmetals, and recovers stainless steel or metallic compounds; RGB (red, green, blue), which sorts specifically in the color spectrums of red, green, and blue for specialized applications; CMYK (cyan, magenta, yellow, key), which sorts paper or carton that has been printed using CMYK; and X-ray, which sorts by recognizing the atomic density of materials (Capel 2008).

FIGURE 10.5   Manual dismantling of personal computers. (Data from E-waste Recyclers. com, *E-waste Recyclers*, http://www.e-wasterecyclers.com/E_Waste_Dismantling.html, 2014.)

The hazardous components and materials that must be removed from any collected WEEE streams can be categorized as follows (The European Parliament and the Council of the European Union 2003):

- Capacitors containing PCBs
- Mercury-containing components such as switches or backlighting lamps
- Batteries
- PC boards of mobile phones and of other devices if the surface area of the circuit board is greater than 10 cm^2
- Toner cartridges
- Plastics containing brominated flame retardants
- Asbestos waste and components that contain asbestos
- CRTs
- Freons and hydrocarbons
- Gas discharge lamps
- LCDs, together with their casing where appropriate, of a surface area greater than 100 cm^2 and all those backlighted with gas discharge lamps
- External electrical cables
- Components containing refractory ceramic fibers
- Components containing radioactive substances above exemption thresholds
- Electrolyte capacitors containing substances of concern

Sorting and segregation of different fractions of WEEE and hazardous components in the early processing chain pays dividends in the later stages of treatment (DTI Global Watch Mission 2006).

10.4.2 SHREDDING AND GRINDING

The second processing step (Figure 10.4) includes a variety of impaction and shredding methods that are well advanced (Dalrymple et al. 2007). Size reduction can be

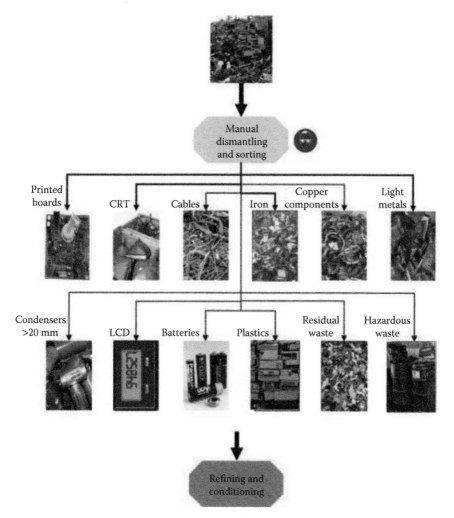

FIGURE 10.6 General scheme of different fractions obtained by manual dismantling and sorting of WEEE. (Data from ewasteguide.info, http://ewasteguide.info/manual_dismantling_and_sorting, 2014b.)

achieved by mechanical shredding, crushing, and grinding processes, providing size homogeneity of the e-waste mixture. The size homogeneity is essential for all the solid waste–sorting processes since the total decreases and the efficiency of sorting increases. Furthermore, the proper particle size of the shredded fraction usually is less than 5 mm, which can be achieved through several different shredders in row (Cui and Forssberg 2003; Figure 10.8).

Maximum separation of materials can be achieved by shredding the waste to small (or fine) particles, generally below 5 mm or 10 mm (Cui and Forssberg 2003). Lab and industrial-scale studies on personal computers and PCBs have shown that after secondary shredding, the main metals present are in the −5 mm fraction

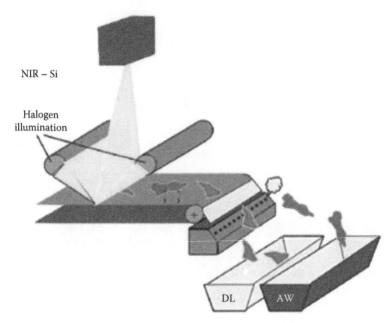

FIGURE 10.7   Schematic sorting of different waste according to their light reflection. DL, dump Load; AW, accepted ware. (Data from Grietens, B. Get it sorted: Efficiency in separation systems, *Waste Management World*, 10, 21–24, 2009.)

FIGURE 10.8   Different industrial shredders. (Data from Taskmaster®, http://www.franklin-miller.com/solid-waste-shredders.html, 2014.)

and show liberation of about 96.5%–99.5% (Cui and Forssberg 2003; Zhang and Forssberg 1997).

10.4.3 MECHANICAL SEPARATION AND SORTING

The next step for WEEE processing is the sorting/separation, based on the differences of physical characteristics of the materials such as weight, size, shape, density, electric conductivity, and magnetic characteristics (Cui and Forssberg 2003). The magnetic separation is performed with a magnetic drum, which attaches the ferrous materials to its surface and deposits all the nonferrous metals and the other materials to a different fraction. Recent developments include sophisticated magnetic separators with rare earth alloy permanent magnets that have the ability to provide very high-field strengths and gradients (Cui and Forssberg 2003). Separation based on the electric conductivity sorts the material into their different conductivity characteristics. There are three main techniques that can perform this separation, namely, the corona electrostatic separation, the triboelectric separation, and the most commonly used Eddy current separator (Figure 10.9). The Eddy current separator applies a powerful magnetic field to the materials in order to repel the nonferrous metals like Cu and Al after which the rest material continues to the density separation (Cui and Forssberg 2003).

Density separation separates the low-density recyclable plastics on the top of a bath from the flame-retardant-containing high-density plastics, which sink (Dalrymple et al. 2007). The techniques in step 2, coupled with the different available separation methods in step 3, can achieve significant recovery of materials. However, the heterogeneity and high complexity of WEEE make it difficult to obtain high-recycling rates for some materials and further processing is required (Figure 10.10).

There are several recycling routes for the treatment of WEEE, using pyrometallurgy, hydrometallurgy, pyro-hydrometallurgy, electrometallurgy, and bio-(hydro)-metallurgy (or combinations). The two major routes for WEEE processing are the hydrometallurgical and the pyrometallurgical technologies, which typically are followed by electrorefining or electrowinning for selected metal separation (Khaliq et al. 2014).

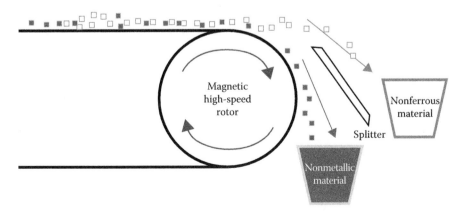

FIGURE 10.9   Eddy current separator.

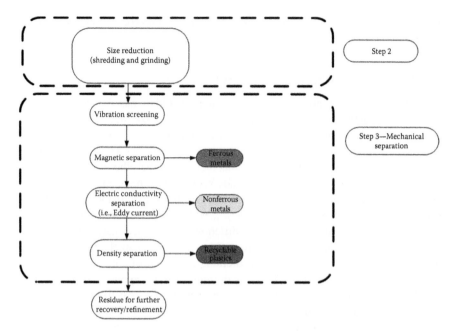

FIGURE 10.10   Schematic of the first steps in a typical e-waste recycling process. (Adapted from Swedish EPA, *Recycling and Disposal of Electronic Waste Health Hazards and Environmental Impacts*, Swedish EPA, Stockholm, Sweden, 2011.)

10.4.4   PYROMETALLURGICAL TREATMENT

The pyrometallurgical process is mainly used for ore metal extraction and refining, but some companies have extended their operations by implementing "integrated" smelters, which means the combination of several extra chemical units for processing WEEE (Cui and Forssberg 2003). The mineral processing could provide an alternative route for metal recovery from WEEE; this method especially increases the recovery rates of precious metals in order to improve the mechanical separation.

The pyrometallurgical processing includes a furnace or a molten bath where the shredded scrap of the WEEE fraction is burned in order to remove the plastic materials while the refractory oxides and some metal oxides form a slag phase. After this procedure, the recovered materials are refined by chemical processing. The plastic fraction is not recovered as material but it is used, along with other flammable materials, as energy provider in order to reduce the total cost of the process. It should be noted that, since the mineral processing is designed to treat and purify metal ore, application of this method to WEEE faces some limitations because of the differences in particle size involved and the material content in those systems.

Industries that process e-waste together with metal scrap by pyrometallurgical treatment include the Noranda (Noranda process) at Quebec, the Boliden Ltd. (Rönnskär smelter) in Sweden, and Umicore at Hoboken, Belgium. Pyrometallurgical processes are efficient since they maximize the recovery rates of precious metals, but plastics in the WEEE cannot be recovered in terms of material. Moreover, some metals such as Fe and Al cannot be easily recovered since they end up in the slag as oxides. The environmental hazards are high since dioxins are generated during the smelting process. Furthermore, the ceramic materials present in the feed increase the volume of the slag and cannot be handled easily. Therefore, extra steps of treatment are necessary and recovery and purity of precious metals with this method can only partially be achieved (Khaliq et al. 2014).

10.4.5 HYDROMETALLURGICAL TREATMENT

The techniques used for metal recovery from WEEE and primary ores are very similar and the same acid and caustic leaching is applied (Khaliq et al. 2014). Leaching is the process of extracting a soluble constituent from a solid by means of a solvent. For WEEE, the leaching primarily removes the base metals and other impurities, and second, different acid solutions (such as HNO_3, H_2SO_4, and HCl) are used for precious metals recovery. Further purification can be achieved by means of precipitation of impurities, solvent extraction, adsorption, and ion exchange. The final solutions are treated by electrorefining, chemical reduction, or crystallization for metal recovery (Khaliq et al. 2014; Luda 2011). The hydrometallurgical treatment can be divided into four main steps (Cui and Zhang 2008):

1. *Mechanical treatment.* Before the chemical treatment, the mechanical treatment is necessary for granulating the total fraction.
2. *Leaching.* The WEEE fraction passes through a series of acid or caustic leaches, where the soluble component is extracted from a solid by means of a solvent. Strong acids are mainly used because of their high efficiency and their ability to leach not only base metals but also precious metals. The most common leaching agent for base metals is HNO_3; for Cu, H_2SO_4 or aqua regia; for precious metals (such as Au and Ag), the most common chemicals include thiourea or cyanide; and for palladium, HCl and $NaClO_3$ are mostly used.
3. *Separation/purification.* The leached metals solution pass through separation and purification processes to discard impurities.
4. *Precious metals recovery.* The final step of the hydrometallurgical process is electrorefining, chemical reduction, or crystallization for the separation of the precious metals.

Oxidative H_2SO_4 leaching of PCBs in mobile phones dissolves Cu and Ag partially, while oxidative HCl leaching dissolves Pd and Cu. Cyanidation is reported to recover Au, Ag, Pd, and Cu (Quinet et al. 2005). The overall optimized flow sheet permitted the recovery of 93% of Ag, 95% of Au, and 99% of Pd (Quinet et al. 2005). Moreover, Cu, Pb, and Sn recovery from PCB can be achieved by dissolution in acids and electrochemical treatment in order to recover the metals separately (Veit et al. 2006). Another approach of hydrometallurgical treatment involves leaching with NH_3/NH_5CO_3 solution to dissolve Cu. The remaining solid residue is then leached with HCl to recover Sn and Pb. Heating of the produced $CuCO_3$ converts it to CuO (Liu et al. 2009).

Extraction efficiencies of around 90% for Y, Cu, Au, and Ag; 93% for Li; and 97% for Co, with respective purities of 95%, 99.5%, 80% (Au and Ag), 18%, and 43% have been reported for different hydrothermal treatment of WEEE; however, the relatively low impurities are not suitable for direct commercialization (minimum purity of 99% would be needed), but these products are still marketable to companies for final refinement. Moreover, a final wastewater treatment step should be considered (Rocchetti et al. 2013). Life cycle assessment of hydrometallurgical treatment of PCBs identified equivalent emissions during treatment of WEEE, which are listed in Table 10.11.

A summary of different studies involving hydrometallurgical treatment of WEEE is shown in Table 10.12.

TABLE 10.11

Emissions for the Treatment of 100 tons WEEE Residues: PCB Granulate from PCBs

Impact Categories	PCB Granulate from PCBs
Global warming (kg CO_2-eq.)	7.8×10^5
Abiotic depletion (kg Sb-eq.)	2.2
Acidification (kg SO_2-eq.)	2044
Eutrophication (kg phosphate-eq.)	137
Ozone layer depletion (kg R11-eq.)	0.05
Photochemical ozone creation (kg ethane-eq.)	154

Source: Rocchetti, L. et al., *Environ. Sci. Technol.,* 47(3), 1581–1588, 2013.

TABLE 10.12

Studies Related to Hydrothermal Treatment of WEEE

Reference	E-Waste Component	Leaching Agent(s)	Recovered Metals
Quinet et al. (2005)	PCB (mobile phones)	H_2SO_4, HCl, HCN	Cu, Ag, Pd, Au
Liu et al. (2009)	PCB	NH_3/NH_5CO_3, HCl	Sn, Pb, Cu
Park and Fray (2009)	PCB	Aqua regia	Au, Ag, Pd
Sheng and Etsell (2007)	PCB	HNO_3, epoxy resin, aqua regia	Au
Kogan (2006)	E-waste scrap	HCl, H_2SO_4, $MgCl_2$, H_2O_2	Al, Sn, Pb, Cu, Ni, Zn, Ag, Au, Pd, Pt
Mecucci and Scott (2002)	PCB	HNO_3	Pb, Cu, Sn after electrodissolution in HCl

Even though the reported recovery rates of different metals are quite high, limitations of hydrometallurgical treatment have been identified, which slow down its industrial implementation. These are summarized as follows (Khaliq et al. 2014):

1. The time scale of hydrometallurgical processing is long, which can implicate the overall recycling scheme.
2. The fine particle size that is needed for efficient dissolution results in 20% loss of precious metals during the liberation process.
3. Several leachants that are used, such as CN, are extremely dangerous and high safety standards are required. Moreover, extremely corrosive halide leaching requires specialized equipment to avoid corrosion.
4. There is loss of precious metals during dissolution and subsequent treatment, which affects the overall recovery efficiency and process economics.

The industrial processes for metal recovery from WEEE are based on combined technology from pyrometallurgy, hydrometallurgy, and electrometallurgy. In most cases, the WEEE is blended with other metallic fraction and is processed on either Cu or Pb smelters. In the final stage, the pure Cu is produced by electrorefining while the precious metals are separated and recovered by hydrometallurgical techniques. Industrially, the implementation of such treatment can be found at the Umicore, Noranda, Boliden, Kosaka, Kayser, and Metallo Chimique (Khaliq et al. 2014).

10.4.5.1 Umicore

Umicore's integrated smelters constitute the world's largest recycling facility for the extraction of precious metals (Hagelüken 2006). The refinery is located at Hoboken in Belgium and can process a high range of different metals such as precious metals (Au, Ag, Pt, Pd, Rh, Ir, Ru), special metals (Se, Te, In), secondary metals (Sb, Sn, Ar, Bi), and of course base metals (Cu, Pb, Ni). The average annual capacity of the unit is 250,000 tons of feed material with an annual production of over 50 tons of platinum group metals, 100 tons of Au, and 2400 tons of Ag. The fraction of WEEE that ends up in this unit is either nonferrous fraction or mostly PCB. The plastic and organic content partially replaces the coke as an energy source and the resultant Cu from the smelter is subjected to leaching and electrowinning, where it is purified. The slag from the Cu smelter is fed to the Pb blast furnace that produces impure Pb, which is purified in a different step. Finally, the precious metals from all the other processes are fed to the precious metals refinery where cupellation occurs (Khaliq et al. 2014). The flow sheet of the whole process is presented in Figure 10.11.

10.4.5.2 Boliden

Boliden's Rönnskär smelter is located in Skelleftehamn in northern Sweden and it has an annual production of more than 200,000 tons of Cu, 13 tons of Au, and 400 tons of Ag. During the last years the Kaldo furnace was modified in order to process more low-grade WEEE fraction such as PCB and nonferrous metals. The total capacity of the unit was increased from 45,000 tons/y to 120,000 tons/y (Boliden Group 2012).

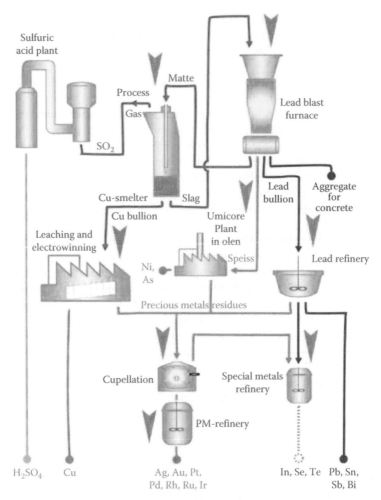

FIGURE 10.11 Flow sheet of the Umicore's integrated metal smelters and metal refinery. (Data from Umicore SA, *Exploring Umicore Precious Metals Refining*, http://ambientair.rec. org/documents/umicore.pdf, 2005.)

The crushed WEEE fraction of high Cu content is fed directly into the electric smelting furnace and while PCB and nonferrous fraction of WEEE are fed into the Kaldo furnace where they are combusted. The plastics contained in the fraction contribute to the process's energy requirements. The Kaldo furnace produces a mixture of Cu, Ag, Au, Pt, Pd, Ni, Se, and Zn, which are further processed to the anode casting plant, the electrorefinery, and the precious metals plant. The flow sheet of the whole process is shown in Figure 10.12.

10.4.5.3   Noranda

Noranda's smelter (Horne) is located in Rouyn-Noranda in Quebec, Canada. This commercial pyrometallurgical process is the largest and most advanced recycling

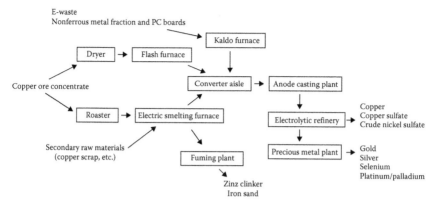

FIGURE 10.12 Schematic of the processes used to recover copper and precious metals from ore concentrate, copper scrap, and e-waste at Boliden's Rönnskär smelter in Sweden. (Data from Swedish EPA, *Recycling and Disposal of Electronic Waste Health Hazards and Environmental Impacts*, Swedish EPA, Stockholm, Sweden, 2011.)

plant of its kind in North America and has the capacity to process complex feeds such as the WEEE. More specifically, the feed material consists of certain industrial Cu scrap and WEEE, such as PCB either low or high grade, cell phones, pins/punchings, lead frames/trims/bare boards, sweeps, insulated consumer wire/degaussing wire, and copper yokes. The total annual capacity is 840,000 tons of Cu and precious metal–bearing materials (Glencore Recycling 2014).

The process requires pretreatment of the WEEE together with the Cu concentrate. The molten bath in the reactor reaches 1250°C. Plastics and other combustible materials contained in WEEE are burned to produce energy. All the impurities including the Fe, Pb, and Zn are converted into oxides and end up to a silica-based slag. The melt liquid Cu is further purified in the converter, while the precious metals–containing residue is processed by electrorefining of anodes (Khaliq et al. 2014; see Figure 10.13).

10.4.6 Thermochemical Treatment

As mentioned above, WEEE is a mixture of various materials whose effective separation and recovery is the key for development of a sustainable system. Thermochemical treatment is another alternative for treating mixed plastic waste streams, enabling the separation of the organic and inorganic matter. Thermochemical treatment includes combustion, gasification, and pyrolysis processes.

10.4.6.1 Combustion

Unlike other fossil-fueled incineration plants, waste to energy (WtE) plants have significantly lower energy efficiencies (13%–24%) due to lower steam temperatures, fouling, and slagging. Moreover, acidic gases such as HCl, SO_x, NO_x, HF, and volatile organic compounds (VOCs) (such as polyaromatic hydrocarbons, PCBs, and polychlorinated dibenzodioxins and dibenzofurans [PCDD/Fs]) are emitted. The solid residues of the final process cannot be recovered and constitute a serious problem due to their heavy metal content (Malkow 2003).

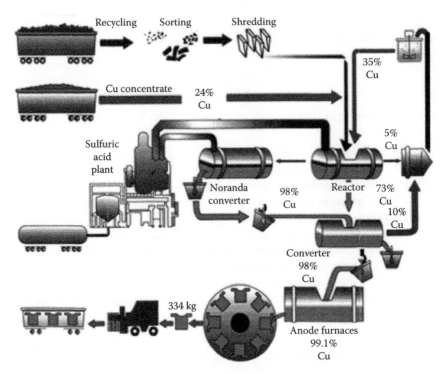

FIGURE 10.13 Schematic diagram of the Noranda smelting process. (Data from Khaliq, A. et al., *Resources,* 3, 152–179, 2014.)

Among the potential pollutants, dioxins attract the most attention. The formation of PCDD/Fs is catalyzed by Cu present in the fly ash (where Cl_2 is more active than HCl; Ni et al. 2012) in a process known as de-novo synthesis of PCDD/Fs, and maximum formation occurs at about 300°C (Kikuchi et al. 2005). Those emissions during combustion pose an environmental threat and a well-advanced air pollution control system should be coupled to WtE facility. Emissions during combustion of WEEE components are shown in Table 10.13.

The destruction of PCDD/Fs and PCBs can be achieved by application of high-temperature processing (>1300°C); however, the costs (both capital and operating) associated with such treatment have incentivized researchers for other routes involving catalytic processes at lower temperatures (Van der Avert and Weckhuysen 2002).

The two main catalytic routes reported in the literature include total oxidation of chlorinated hydrocarbons at temperatures between 300°C and 550°C over supported noble metal catalysts (e.g., Pt, Pd, and Au) and catalytic hydrodechlorination, in which chlorinated hydrocarbons are transformed in the presence of hydrogen into alkanes and HCl. Commonly used catalysts are Ni, Pd, and Pt. Although hydrodechlorination offers economic and environmental advantages, it is not used often (Bonarowska et al. 2001; Coute et al. 1998; Feijen-Jeurissen et al. 1999; Nutt et al. 2006).

TABLE 10.13

Yields of Gases and Volatiles (mg/kg) during Combustion of Different WEEE Components

	WCB (600°C) (Ortuño et al. 2014)	NmfWCB (600°C) (Ortuño et al. 2014)	WCB (850°C) (Ortuño et al. 2014)	NmfWCB (850°C) (Ortuño et al. 2014)	EW (500°C) (Moltó et al. 2009)	EC (500°C) (Moltó et al. 2009)	PCB (1400°C) (Ni et al. 2012)
HBr	12,700	45,900	11,800	59,600			~1800
Br$_2$	1,000	3,700	6,700	5,800			~400
HCl	400	5,300	500	6,500			
Cl$_2$	100	500	100	300			
CO	89,000	288,000	94,800	295,600	39,000	24,000	
CO$_2$	441,600	595,000	677,400	506,200	210,000	90,000	
Methane	2,910	8,310	30	30	1687	576	
Ethylene	930	1,540	–	–	512	127	
Propylene	310	340	–	–	337	191	
Acetylene	330	830	–	150	20	24	
Propyne	120	250	–	–	–	–	
Benzene	650	1,290	–	–	18	68	
Toluene	–	1,070	–	–	–	268	
Bromomethane	200	3,740	–	–			
Acetone	–	5,490	–	2,220			
Cyanogen Bromide	–	–	160	750			
Total PCDD/F and PCBs (pg who2005 TEQ/g)					6.806	3.057	

Source: Ni, M. et al., *Waste Manag.*, 32(3), 568–574, 2012; Ortuño, N. et al., *Sci. Total Environ.*, 499: 27–35, 2014; Moltó, J. et al., *J. Anal. Appl. Pyrolysis*, 84(1), 68–78, 2009. WCB: Waste circuit boards; Nmf: no metal fraction; EW: electronic waste; EC: electronic circuits; PCB: printed circuit boards; (–) indicates value not detected or <10 mg/kg.

WEEE combustion could include combustion in cement kilns or power plants. However, this application faces several limitations because of the Cl and certain heavy metals' content and particle size of mixed plastic waste from WEEE (University of United Nations 2007).

Combustion of WEEE is not a sustainable option anymore, since it contributes only to energy recovery, which is the last means of recycling and so has to be regarded as last resort option.

10.4.6.2   Gasification

Gasification refers to partial oxidation (reaction with a controlled amount of oxygen and/or steam) of carbonaceous material at elevated temperatures to produce a mixture of syngas (CO and H_2), CO_2, and other light hydrocarbons. It is way to convert waste into energy carriers with simultaneous transformation into a less voluminous substance, resulting in a more sustainable and effective waste management. The produced gas can be used in many applications such as lime and brick kilns, metallurgical furnaces, as raw material syngas, synthetic natural gas (SNG), Fisher–Tropsch fuels, and other chemical syntheses or even combusted in gas turbines, providing higher efficiencies and market opportunities (Malkow 2003; Yamawaki 2003). The remaining solid material in the form of slag (high-temperature gasification >1000°C) could be used as additive in building materials. However, due to stricter rules and requirements for higher recycling rates both for plastics and metals, this does not represent a viable option for WEEE anymore.

There have been many processes through the years that have used plastic containing waste as feedstock (mainly MSW). Most of the processes employed high temperatures for destruction of unwanted halogenated hydrocarbons. Some of the processes (among others) are: Texaco process, Thermoselect, SkyGas, and Sustec Schwarze Pumpe—Sekunddärohstoff Verwertungs Zentrum process (Kantarelis et al. 2008; Goodship and Stevels 2012; University of United Nations 2007). However, the requirements for low metal content and low halogen content (should not exceed 10% in order to avoid corrosion) in the feedstock make them inappropriate for WEEE treatment for energy and material recovery. Moreover, the costs associated to preparation and treatment limit the interest of this recovery option of mixed plastic waste (University of United Nations 2007). Nevertheless, gasification (especially steam gasification) at lower temperatures (<900°C) coupled with advanced gas treatment could serve for both material and energy recovery.

10.4.6.3   Pyrolysis

Pyrolysis is a method for enabling separation of organic and inorganic matter of WEEE. It refers to the thermal decomposition of matter in the absence of air and aims to provide value-added materials (reusable monomers, hydrocarbons having carbon number distribution in the range of C_1–C_{50}, and valuable aromatic solvents like benzene, toluene, etc.; Ray and Thorpe 2007) and separate the plastic and inorganic fractions, which is in accordance with the principles of sustainable development (Achilias et al. 2007). Life cycle analysis studies have shown that pyrolysis provides significant resource savings without high impact on climate change or landfill space (Alston and Arnold 2011) and therefore it is regarded as the most promising processing route for WEEE plastics (Brebu and Sakata 2006; Luda et al. 2005).

During pyrolysis, the long polymeric chains break down to smaller fragments (molecules) whose quantity and nature depend on the process conditions. The degradation of the plastic fraction enables the separation of the organic, metallic, and glass fiber fractions, which makes recycling of each fraction more viable. For most plastics, pyrolysis starts at around 300°C; however the actual onset depends on additives and characteristics of the different materials. The decomposition is a series of complex reactions and the required heat can be provided by combustion of gases produced after removal of some valuable and some hazardous compounds. Higher temperatures (>600°C) favor the production of smaller and simpler gaseous molecules, while low temperatures (~400°C) favor the formation of liquid products (oil; Buekens 2006). Other authors distinguish those pyrolysis regimes in *thermal cracking* (>700°C) where gaseous products are aimed and *thermolysis* (400°C–500°C) where liquid products are desired (Achilias et al. 2007). Those limits are somehow arbitrary and depend on the actual feedstock and process conditions. Buekens has categorized the factors that affect the products distribution as shown in Table 10.14.

Pyrolysis of plastics is well studied and these studies provide essential information about different degradation mechanisms. However, the plastics present in WEEE differ greatly from model compounds of flame-retarded plastics, that have been used in several studies. The most common WEEE component that has been studied as a basis for pyrolytic recycling of WEEE is PCB or epoxy resins that represent their main polymeric fraction (Evangelopoulos 2014; Lin and Chiang 2014; Luda et al. 2007, 2010; Mankhand et al. 2012; Quan et al. 2009; Sousa and Riedewald 2014).

Phenols and substituted phenols are the most abundant compounds found in the liquid products as well as brominated compounds (mono- and di-brominated

---

TABLE 10.14
Factors Affecting the Product Distribution of Pyrolysis

Factor	Effect
Chemical composition of material	The primary pyrolysis of a product relates directly to the composition of the material and also the mechanism of the decomposition (thermal or catalytic)
Pyrolysis temperature and heating rate	Higher heating rates and temperatures favor the formation of smaller molecules
Residence time	Longer residence time favors secondary reactions that lead to increased gas yields, coke, and tar formation
Reactor type	Determines the heat and mass transfer rates and the residence time of the products
Pressure	Lower pressures minimize the condensation reactions between the reactive vapors and this results in less coke and heavy ends formation

Source: Buekens, A., Introduction to feedstock recycling of plastics, In: J. Scheirs and W. Kaminsky, eds., *Feedstock Recycling and Pyrolysis of Waste Plastics:Converting Waste Plastics into Diesel and Other Fuels,* Wiley and Sons, New York, 3–42, 2006.

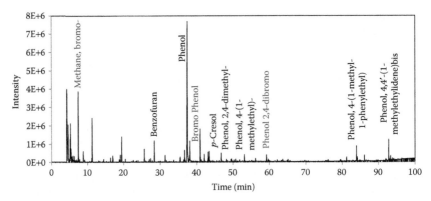

FIGURE 10.14 Pyrogram obtained from analytical pyrolysis of PCB. (Data from Evangelopoulos, P., *Pyrolysis of Waste Electrical and Electric Equipment [WEEE] for Energy Production and Material Recovery*, KTH, Stockholm, Sweden, 2014.)

phenols; bisphenol A; mono-, di-, tri-, and tetra-brominated bisphenol A; Lin and Chiang 2014; Luda et al. 2007, 2010; see Figure 10.14).

Generally, higher temperatures and longer times favor debromination at the expense of liquid yield (Evangelopoulos 2014; Lin and Chiang 2014; Luda et al. 2007, 2010). Reduction of brominated phenols in the pyrolysis oil adds value to the whole recycling process. The size of the PCB particles has been identified as a crucial parameter for the pyrolysis behavior and the onset of pyrolysis. Heat transfer limitations occur when the particle size of the material exceeds 1 cm^2 (Quan et al. 2009).

Mankhand et al. (2012) indicated that pyrolysis of PCB improves the Cu recovery by leaching, compared to nontreated boards. Moreover, a combined approach of pyrolysis with simultaneous separation of metals and glass using molten salts is also under development (Sousa and Riedewald 2014). HIPS and ABS (halogenated or not) are other polymers that have been studied as model compounds that represent the behavior of plastics in WEEE (Hall and Williams 2006a, 2006b). Characterization of products indicated that many halogenated compounds are found in the liquid product. Hall and William's study (2007) on mixed WEEE fraction from WEEE recycling site indicates that pyrolysis gases were mainly halogen free, while liquid fraction contained valuable chemicals as phenol, benzene, toluene, and so on.

The resultant pyrolysis oil due to the presence of flame retardants is heavily contaminated by halogenated compounds and must be further processed to achieve a marketable product (Guan et al. 2008). Understanding the formation of brominated compounds and facilitating their removal from the final product are major considerations in developing processes for mixed plastic waste (Goodship and Stevels 2012). The necessity for dehalogenation of the final oil product implies that upgrading of the latter or dehalogenation measures during the pyrolysis process should be applied. Dehalogenation is of major importance for the pyrolysis process and different options are discussed in Section 10.4.6.3.2. The above-mentioned studies indicate the versatility of pyrolysis in the overall chain of WEEE recycling.

10.4.6.3.1   Pyrolysis Technologies

Research on pyrolysis has been conducted using several reactor types like fluidized bed reactors, rotary kilns, tubular or fixed bed reactors, and stirred batch reactors. The product distribution of each reactor depends on the process conditions as summarized in Table 10.14. Most of the studies employed batch or semibatch equipment and therefore it is hard to study the influence of operating parameters on a real continuous process. The main scientific findings have already been presented and in this section pyrolysis technologies that differ from the rest are discussed.

10.4.6.3.1.1 Vacuum Pyrolysis
A method to facilitate thermal decomposition of the plastic matter of WEEE is the application of vacuum pyrolysis. Vacuum pyrolysis studies of PCB scraps aiming for solder recovery as well as separation of glass fibers and metals have been reported (Long et al. 2010; Zhou and Qiou 2010; Zhou et al. 2010). The vacuum during pyrolysis reduces the decomposition temperature and shortens the residence time of the produced vapor, which minimizes secondary cracking and condensation reactions among the vapors maximizing the liquid product. Cu recovery rates of 99.86% were reported with Cu grade of 99.5% (Long et al. 2010), while recovery of solder is possible after heating the solid residue at 400°C and applying centrifugal forces (Zhou et al. 2010).

The obtained oil mainly consisted of phenolic and benzofuranic compounds. However, the pyrolysis oils contained a significant number of brominated compounds such as 2- and 2,6-dibromophenol. Therefore, the oils could be used as fuels or chemical feedstock after proper treatment (Long et al. 2010; Zhou and Qiou 2010; Zhou et al. 2010).

Vacuum pyrolysis has been studied as a part of a treatment steps for In recovery from LCD panels. Results from pyrolysis indicate that liquids could be used as chemical feedstock or fuels (Ma and Xu 2013).

10.4.6.3.1.2 Microwave Pyrolysis
Microwave pyrolysis is another alternative that has been studied for treatment of WEEE. Microwave irradiation is an efficient way to heat up the material and initiate its thermal decomposition. WEEE material subjected to microwave irradiation has shown mass reduction, indicating that it is a favorable method to treat residual wastes and increase the solid metal to organics ratio, which is valuable for further recycling treatment (Andersson et al. 2012). Addition of activated carbon (AC) has been found to enhance the heating of the material, by absorption of the microwave irradiation, improving decomposition of products (Sun et al. 2011). Liquid products contained significant amount of Br; however, significant debromination can be achieved downstream of the pyrolysis reactor by using $CaCO_3$ (Sun et al. 2011).

10.4.6.3.1.3 Commercial-Demonstration Projects
Several projects to commercialize pyrolysis technologies through the years have been attempted. The closest process to commercialization has been reported to be the Haloclean process (with a capacity of 3000 t/y; Figure 10.15), which converts the WEEE into debrominated pyrolysis oil, noble metal rich solid residue, and gaseous bromine (DTI Global Watch Mission 2006).

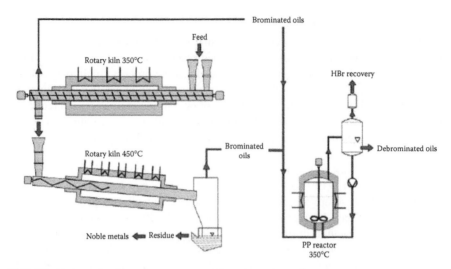

FIGURE 10.15 The Haloclean process. (Data from DTI Global Watch Mission, *WEEE Recovery: The European Story*, 2006; Hornung, A. et al., *Haloclean and PYDRA—A Dual Staged Pyrolysis Plant for the Recycling Waste electronic and Electrical Equipment [WEEE]*, Skelfteå, Sweden, 2003.)

Another process for WEEE treatment that has been reported to be in operation in larger scale is the catalytic depolymerization process (CDP), where e-waste is converted into diesel fuel by pyrolysis, followed by catalytic treatment. A key part of the process is the heating of the material, which is achieved by means of mechanical friction (Figure 10.16). This technique is said to give very accurate control of the temperature in the process, and prevent overheating, which would lead to the production of dioxins. However, little information is available about the reliability and wear of the different process components (DTI Global Watch Mission 2006). During this process high-grade diesel is produced. Part of the produced diesel is directly converted into electricity by means of powerful generators and is then used to supply the entire recycling plant (DTI Global Watch Mission 2006; Koehnlechner 2008).

The Veba Combi Cracking Unit-KAB is another example of a pyrolysis-hydrogenation process able to process WEEE with high halogen content. The process was initially designed for coal liquefaction and it was further modified to treat Vacuum Residuum(VR)VR into synthetic crude. Since 1988, it has been substituted with chlorine-containing waste, especially PCBs (APME and Verband Kunstofferzeougende Industrie 1997). The plant was able to treat 10 t/h of waste but it had to close down in 2003 because of process economics (Goodship and Stevels 2012; University of United Nations 2007).

Another plant that has been closed down because of economical competiveness is the BASF thermolysis plant (Goodship and Stevels 2012; University of United Nations 2007). The overall treatment/recycling loop of WEEE, including pyrolysis step, can be represented as shown in Figure 10.17.

FIGURE 10.16   Detail of CDP plant (Hamos). (Adapted from DTI Global Watch Mission, *WEEE Recovery: The European Story*, 2006.)

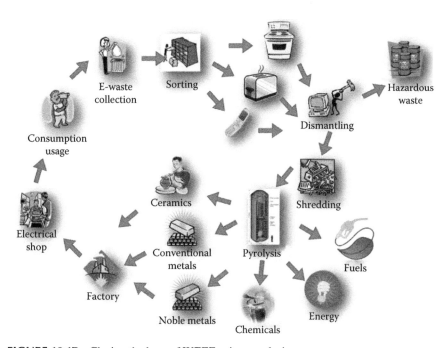

FIGURE 10.17   Closing the loop of WEEE using pyrolysis.

10.4.6.3.2   Dehalogenation of Pyrolysis-Oil-Catalytic Upgrading

As presented above with the help of research studies and demonstration attempts, the maximum output of the pyrolysis process can be achieved only when the liquid product is marketable or can be used for fueling recycling plant. Therefore, dehalogenation of the liquid product is required.

In fact, it is a crucial step, for the recycling of WEEE plastics, to remove organic halogen in an efficient manner (Blazsó et al. 2002). The dehalogenation could be carried out prior, during, or after the decomposition of WEEE plastics (Yang et al. 2013).

10.4.6.3.2.1 Dehalogenation Prior to WEEE Decomposition   This method of dehalogenation requires a two-stage pyrolysis process. At the first stage a low-temperature treatment (~300°C) liberates halogenated compounds, followed by a second-stage pyrolysis at higher temperatures to produce oil with no/reduced halogen content. However, a distinction between Cl- and Br-containing compounds should be made. While Cl-containing compounds exhibit a low temperature release of HCl at the low-temperature region, Br-containing retardants without synergistics (Sb_2O_3) do not decompose earlier than the rest of the WEEE material and their decomposition occurs at the same temperature range and thus different methods of dehalogenation must be used. Nevertheless, behavior similar to Cl-containing flame retardants is observed when Sb_2O_3 is present (Yang et al. 2013). The introduction of a two-stage process implies that the capital and operating costs are higher compared to one-stage pyrolysis process and that the composition of the processed material is well known and no significant fluctuations in the composition/or processed material are expected.

10.4.6.3.2.2 Dehalogenation during WEEE Decomposition   Dehalogenation during the course of pyrolysis can be achieved by controlling the pyrolysis parameters and/or inclusion of additives/or catalysts. The control of the reaction parameters involves the slow heating of the material and the long residence time of the produced vapors. In that way the organic brominated compounds that are formed in the first stages of the decomposition spend longer time in the hot zone, which allows for their destruction (Miskolczi et al. 2008). However, lower yields of oil are to be expected, while the long size of the reactor suggests increased costs compared to typical fixed or fluidized bed reactors.

In a similar approach to prolong the residence time of produced vapors in hot environment, Hlaing et al. (2014) introduced a reflux condenser operating at 200°C at the top of the pyrolysis chamber, and managed to reduce the bromine content in the final liquid product by almost 10 times.

Dehalogenation by introduction of additives in the pyrolysis reactor was achieved by Blazsó et al. (2002) while pyrolyzing PCB scraps in the presence of NaOH or sodium-containing silicates. Introduction of the additives resulted in an enhanced bromomethane evolution and suppression of formation of brominated phenol. Ca-based additives during pyrolysis of brominated ABS have shown good activity in reducing the bromine and antimony content in liquid products at the expense of the final liquid yield (Jung et al. 2012).

10.4.6.3.2.3 Catalytic Pyrolysis of WEEE In this dehalogenating method, a suitable catalyst is used to carry out the cracking reactions and transform the organic halogenated compounds during the pyrolysis (in situ). The presence of the catalyst lowers the reaction temperature and time. Additional benefit of catalytic degradation is that the products are of a much narrower distribution of carbon atom number with a peak at lighter hydrocarbons (Panda et al. 2010). Various combinations of cracking catalysts and adsorbents for halogenated compounds ($CaCO_3$ and red mud) decreased as well the amount of all heteroatoms in pyrolysis oils of PCBs; after pyrolysis at 300°C–540°C, the oils were passed into a secondary catalytic reactor (Vasile et al. 2008).

Y- and ZSM-5 zeolite catalysts have been used for the removal of organobromine compounds during pyrolysis of HIPS and ABS that were flame-retardant, using decabromodiphenyl ether. Generally, the use of zeolite catalysts increases the amount of gaseous hydrocarbons produced during pyrolysis and decreases the amount of pyrolysis oil, while increased amount of coke is formed on the catalyst surface. Valuable products like styrene and cumene are reduced; however, other useful compounds as naphthalene are formed. Both zeolites (and especially the Y-) are found to be less effective at removing antimony bromide from the volatile pyrolysis products (Hall and Williams 2008). HIPS and PC (both model compounds and in commercial from CDs fraction) were tested in presence of ZSM-5 and MgO catalysts. Catalytic treatment seemed to lower the monomeric fraction (compared to thermal pyrolysis) in favor of other desirable chemicals. No information regarding halogens content in the liquid was presented (Antonakou et al. 2014).

Waste FCC catalysts were also used by Hall et al. (2008) in order to see the influence on brominated HIPS and ABS plastics. Unlike the Y-zeolites investigated by the same researchers, the fluid catalytic cracking (FCC) catalyst did manage to reduce the bromine content. However, it did not alter the product distribution significantly.

Blazso and Czegeny (2006) have investigated the introduction of other zeolite catalysts during the analytical pyrolysis of TBBPA flame retardant in a try to identify debrominating effects. The catalysts used included NaY, 13×, 4A, and Al-MCM-41 zeolites. It was found that small pore size zeolites like 4A show limited debromination activity while medium-sized pores of NaY showed enhanced activity toward debrominated products. Surprisingly Al-MCM-41 did not show any debromination activity but all brominated bisphenols have been cracked to bromophenol and dibromophenol. Other types of catalysts studied include FeOOH, $Fe(Fe_3O_4)$-C and $Ca(CaCO_3)$-C with mixed reported activities toward debrominated products, depending on the feedstock tested (Brebu et al. 2004, 2005).

From an economic perspective, finding the optimum catalysts and thus reducing the cost even further will make this process an even more attractive option. This option can be optimized by reuse of catalysts and the use of effective catalysts in lesser quantities. This method seems to be the most promising to be developed into a cost-effective commercial polymer-recycling process to solve the acute environmental problem of plastic waste disposal (Panda et al. 2010).

10.4.6.3.2.4 Hydrodehalogenation–Pyrolysis Oil Upgading The dehalogenation of produced pyrolysis oil has the advantage of decoupling the oil production and upgrading processes and the goal is to remove the halogenated compounds after optimized pyrolysis process (Yang et al. 2013). This can be either by implementation of the catalytic routes mentioned in catalytic pyrolysis section or by the hydrodehalogenating process. During hydrodehalogenation, pyrolysis oil reacts with a hydrogen-donating media in the presence of catalyst, producing a hydrogenated liquid product and valuable hydrogen halides. As mentioned previously, typical catalysts include supported Ni, Pd, and Pt.

Vassile et al. (2007) have investigated the use of commercial hydrogenation DHC-8 and NiMo/AC catalysts. Results showed that brominated compounds were eliminated after processing, with Br being converted to HBr. Wu et al. (2005) demonstrated that hydrodehalogenation of chlorbenzene is feasible in liquid phase in the presence of NaOH and different catalysts (Ni/AC, Ni/γ-Al$_2$O$_3$, Ni/SiO$_2$, Raney Ni) at relatively mild conditions. Several other halogenated hydrocarbons have been investigated and demonstrated that hydrodehalogenation route is feasible, producing a marketable organic product.

10.5 CONCLUSIONS

Sustainable WEEE recycling should aim for maximum output of different recycled items by comprehensive processes that include recovering and recycling of the inorganic and organic fractions. Complexity of WEEE poses a serious challenge for the implementation of one efficient recycling route.

A disassembly stage is required to remove hazardous components with manual dismantling still being core operation, and automated dismantling processes under development. Crushing and separation are the key points for improving successful further treatment.

Physical recycling is important recycling method with low environmental burden and cost. Nevertheless, the separation between different fractions of WEEE has to be enhanced.

Metal recovery can be performed by traditional pyrometallurgical approaches on metal-concentrated WEEE fractions. Even though pyrometallurgical processing is attractive and commercially available, in order to be of economic interest for the Cu smelters, the Cu content should be higher than 5 wt%, which is not generally the case for WEEE (Goodship and Stevels 2012).

Compared to pyrometallurgical processing, hydrometallurgical treatment targets to recycle specific metals with high recovery yields. However, environmental concerns, process requirements, and economics slow down its implementation in industrial scale.

High-temperature gasification technologies have been tested for long time with several limitations due to halogen and metal content of the WEEE. While in short term it does not seem probable that WEEE gasification could be viable, lower gasification temperatures coupled with advanced gas treatment facilities could prove effective in the future.

The pyrolytic approach is attractive because it allows recovery of valuable compounds in all product streams (gases, oils, and residue). The main limitation of pyrolysis is the halogenated oil that is obtained. Therefore, the development in catalytic processing/upgrading of oil (in or ex situ) is the most crucial step for the maximum exploitation of all product streams and economic viability. Recent developments in catalytic processing indicate that WEEE pyrolysis will represent an economical and sustainable route. Moreover, pyrolysis can be benefited by new developments and requirements (RoHS) in electronics industry and the use of more environmental friendly substances (European Union 2014b).

REFERENCES

Achilias, D.S., Andriotis, L., Koutsidis, I.A., Louka, D.C.A., Nianias, N.P., Siafaka, P., Tsagkalias, I., and Tsintzou, G. 2012. Recent advances in the chemical recycling of polymers (PP, PS, LDPE, HDPE, PVC, PC, Nylon, PMMA). In: D. Achilas, ed., *Material Recycling—Trends and Perspectives.* InTech: 1–64. Available from: http://www.intechopen.com.

Achilias, D. et al. 2007. Chemical recycling of plastic wastes made from polyethylene (LDPE and HDPE) and polypropylene (PP). *Journal of Hazardous Materials,* 149(3): 536–542.

Alston, S. and Arnold, J. 2011. Environmental impact of pyrolysis of mixed WEEE plastics part 2: Life cycle assessment. *Environmental Science and Technology,* 45(21): 9386–9392.

Andersson, M., Wedel, M.C.F., and Christéen, J. 2012. Microwave assisted pyrolysis of residual fractions of waste electrical and electronics equipment. *Minerals Engineering,* 29: 105–111.

Antonakou, E. et al. 2014. Pyrolysis and catalytic pyrolysis as a recycling method of waste CDs originating from polycarbonate and HIPS. *Waste Management,* DOI:10.1016/j.wasman.2014.08.014.

Association of Plastics Manufacturers in Europe (APME). 2004. *Plastics—A Material of Choice for the Electrical and Electronic Industry-Plastics Consumption and Recovery in Western Europe 1995.* Brussels, Belgium: APME.

APME and Verband Kunstofferzeougende Industrie. 1997. *Feedstock Recycling of Electrical and Electronic Plastics Waste.* Available from: http://www.plasticseurope.org/Document/feedstock-recycling-of-electrical-and-electronic-plastics-waste.aspx?Page=DOCUMENT&FolID=2.

Atasu, A., Özdemir, Ö., and Van Wassenhove, L. N. 2012. Stakeholder perspectives on E-waste take-back legislation. *Production and Operations Management,* 22(2): 382–396.

Blazsó, M. and Czégény, Z. 2006. Catalytic destruction of brominated aromatic compounds studied in a catalyst micro-bed coupled to gas chromatography/mass spectrometry. *Journal of Chromatography A,* 1130(1): 91–96.

Blazsó, M., Czégény, Z., and Csoma, C. 2002. Pyrolysis and debromination of flame retarded polymers of electronic scrap studied by analytical pyrolysis. *Journal of Analytical and Applied Pyrolysis,* 64(2): 249–261.

Boliden Group. 2012. Recycling of electronic scrap. Cited October 13, 2014. Available from http://www.boliden.com/Operations/Smelters/E-scrap-project/.

Bonarowska, M., Malinowski, A., Juszczyk, W., and Karpinski, Z. 2001. Hydrodechlorination of CCl2F2 (CFC-12) over silica-supported palladium–gold catalysts. *Applied Catalysis B: Environmental,* 30(1/2): 187–193.

Brebu, M. and Sakata, Y. 2006. Novel debromination method for flame-retardant high impact polystyrene (HIPS-Br) by ammonia treatment. *Green Chemistry,* 8: 984–987.

Brebu, M. et al. 2004. Thermal degradation of PE and PS mixed with ABS-Br and debromination of pyrolysis oil by Fe- and Ca-based catalysts. *Polymer Degradation and Stability,* 84(3): 459–467.

Brebu, M. et al. 2005. Removal of nitrogen, bromine, and chlorine from PP/PE/PS/PVC/ABS-Br pyrolysis liquid products using Fe- and Ca-based catalysts. *Polymer Degradation and Stability,* 87(2): 225–230.

Bridgwater, E. and Anderson, C. 2003. *CA Site WEEE Capacity in UK.* Network Recycling. Available from: http://webarchive.nationalarchives.gov.uk/20060715175411/dti.gov.uk/files/file30304.pdf.

Buekens, A. 2006. Introduction to feedstock recycling of plastics. In: J. Scheirs and W. Kaminsky, eds., *Feedstock Recycling and pyrolysis of Waste Plastics: Converting Waste Plastics into Diesel and Other Fuels.* New York: Wiley and Sons: 3–42.

Buekens, A. and Yang, J. 2014. Recycling of WEEE plastics: a review. *Chemical Feedstock Recycling,* 10(16): 415–434.

Capel, C. 2008. Waste sorting—A look at the separation and sorting techniques in today's European market. *Waste Management World,* 9(4). Available from: http://www.waste-management-world.com/articles/print/volume-9/issue-4/features/waste-sorting-a-look-at-the-separation-and-sorting-techniques-in-todayrsquos-european-market.html.

Coute, N., Ortego Jr., J., Richardson, J., and Twigg, M. 1998. Catalytic steam reforming of chlorocarbons: Trichloroethane, trichloroethylene and perchloroethylene. *Applied Catalysis B: Environmental,* 19(3/4): 175–187.

Cui, J. 2005. *Mechanical Recycling of Consumer Electronic Scrap.* Luleå, Sweden: Luleå University of Technology.

Cui, J. and Forssberg, E. 2003. Mechanical recycling of waste electric and electronic equipment: A review. *Journal of Hazardous Materials,* 99(3): 243–263.

Cui, J. and Zhang, L. 2008. Metallurgical recovery of metals from electronic waste: A review. *Journal of Hazardous Materials,* 158(2/3): 228–256.

Dalrymple, I. et al. 2007. Recycling technologies for the treatment of end of life printed circuit boards (PCBs). *Circuit World,* 33(2): 52–58.

Dimitrakakis, E., Janzb, A., Bilitewski, B., and Gidarakos, E. 2009. Small WEEE: Determining recyclables and hazardous substances in plastics. *Journal of Hazardous Materials,* 161(2/3): 913–919.

DTI Global Watch Mission. 2006. *WEEE Recovery: The European Story.* Available from: http://www.welding.cz/pajky/dok06/dti.pdf.

The European Parliament and the Council of the European Union. 2003. Directive 2002/96/EC of the European Parliament and of the Council on Waste Electrical and Electronic Equipment (WEEE). *Official Journal of the European Union* L 037, February 13, pp. 24–39.

The European Parliament and the Council of the European Union. 2011. Directive 2011/65/EU of the European Parliament and of the Council. *Official Journal of the European Union* L 174, July 1, pp. 88–110.

European Union. 2014a. *Environment: Fewer Risks from Hazardous Substances in Electrical and Electronic Equipment.* Cited October 13, 2014. Available from http://europa.eu/rapid/press-release_IP-11-912_en.htm.

European Union. 2014b. *Waste Electrical and Electronic Equipment (WEEE).* Cited October 10, 2014. Available from http://ec.europa.eu/environment/waste/weee/index_en.htm.

Eurostat. 2012. *Waste Electrical and Electronic Equipment (WEEE).* Cited October 21, 2014. Available from http://ec.europa.eu/eurostat/c/portal/layout?p_l_id=664648&p_v_l_s_g_id=0.

Evangelopoulos, P. 2014. *Pyrolysis of Waste Electrical and Electric Equipment (WEEE) for Energy Production and Material Recovery.* Stockholm, Sweden: KTH.

E-waste Recyclers.com. 2014. *E-waste Recyclers.* Cited October 13, 2014. Available from http://www.e-wasterecyclers.com/E_Waste_Dismantling.html.

ewasteguide.info. 2014a. *Valuable Substances in e-Waste.* Cited October 13, 2014. Available from http://ewasteguide.info/valuable-substances.

ewasteguide.info. 2014b. *Manual Dismantling and Sorting.* Cited October 13, 2014. Available from http://ewasteguide.info/manual_dismantling_and_sorting.

Feijen-Jeurissen, M. et al. 1999. Mechanism of catalytic destruction of 1,2-dichloroethane and trichloroethylene over γ-Al_2O_3 and γ-Al_2O_3 supported chromium and palladium catalysts. *Catalysis Today,* 54(1): 65–79.

Glencore Recycling. 2014. *Horne Smelter.* Cited October 13, 2014. Available from http://www.glencorerecycling.com/EN/Facilities/Pages/Horne.aspx.

Goodship, V. and Stevels, A. eds. 2012. *Waste Electrical and Electronic Equipment (WEEE) Handbook.* Cambridge: Woodhead Publishing.

Gottberg, A. et al. 2006. Producer responsibility, waste minimisation and the WEEE Directive: Case studies in eco-design from the European lighting sector. *Science of the Total Environment,* 359(1/2): 38–56.

Grietens, B. 2009. Get it sorted: Efficiency in separation systems. *Waste Management World,* 10 (5): 21–24.

Guan, J., Li, Y.S., and Lu, M.X. 2008. Product characterization of waste printed circuit board. *Journal of Analytical and Applied Pyrolysis,* 83(2): 185–189.

Hagelüken, C. 2006. Recycling of electronic scrap at Unicore precious metals refining. *Acta Metallurgica Slovaca,* 12: 111–120.

Hall, W.J., Norbert Miskolczi, N., Onwudili, J. and Williams, P. 2008. Thermal processing of toxic flame-retarded polymers using a waste fluidized catalytic cracker (FCC) catalyst. *Energy and Fuels,* 22(3): 1691–1697.

Hall, W.J. and Williams, P. 2006a. Fast pyrolysis of halogenated plastics recovered from waste computers. *Energy and Fuels,* 20(4): 1536–1549.

Hall, W.J. and Williams, P. 2006b. Pyrolysis of brominated feedstock plastic in a fluidised bed reactor. *Journal of Analytical and Applied Pyrolysis,* 77(1): 75–82.

Hall, W.J and Williams, P. 2007. Analysis of products from the pyrolysis of plastics recovered from the commercial scale recycling of waste electrical and electronic equipment. *Journal of Analytical and Applied Pyrolysis,* 79 (1/2): 375–386.

Hall, W.J. and Williams, P. 2008. Removal of organobromine compounds from the pyrolysis oils of flame retarded plastics usingpyrolysis oils of flame retarded plastics usingpyrolysis oils of flame retarded plastics using zeolite catalysts. *Journal of Analytical and Applied Pyrolysis,* 81(2): 139–147.

Hlaing, Z.Z., Wajima, T., Uchiyama, S., and Nakagome, H. 2014. Reduction of bromine compounds in oil produced from brominated flame retardant plastics via pyrolysis using a reflux condenser. *International Journal of Environmental Science and Development,* 5(2): 207–211.

Hornung, A., Koch, W., and Seifert, H. 2003. *Haloclean and PYDRA—A Dual Staged Pyrolysis Plant for the Recycling Waste Electronic and Electrical Equipment (WEEE).* Metals and Energy Recovery, International Symposium in Northern Sweden. Skellefteå, Sweden. June 25–26. Available from: http://www.bsef-japan.com/index/files/W.%20Koch%20%20%20%20HALOCLEAN%20%26%20PYDRA.pdf.

Huisman, J. et al. 2007. *2008 Review of Directive 2002/96 on Waste Electrical and Electronic Equipment, Final Report,* Study No. 07010401/2006/442493/ETU/G4. Bonn, Germany: United Nations University.

Jung, S., Kim, S., and Kim, J. 2012. Thermal degradation of acrylonitrile–butadiene–styrene (ABS) containing flame retardants using a fluidized bed reactor: The effects of Ca-based additives on halogen removal. *Fuel Processing Technology,* 96: 265–270.

Kantarelis, E., Zabaniotou, A., Donaj, P., and Yang, W. 2008. *Pyrolysis and Steam Gasification of Electrical Cables Shredder Residue.* Thessaloniki, Greece: Aristotle University of Thessalonik.

Khaliq, A., Rhamdhani, M.A., Brooks, G., and Masood, S. 2014. Metal extraction process for electronic waste and existing industrial routes: A review and Australian perspective. *Resources,* 3(1): 152–179.

Kikuchi, R., Sato, H., Matsukura, Y., and Yanamoto, T. 2005. Semi-pilot scale test for production of hydrogen-rich fuel gas from different wastes by means of a gasification and smelting process with oxygen multi-blowing. *Fuel Processing Technology,* 86(12/13): 1279–1296.

Kim, B. et al. 2004. A process for extracting precious from spent printed circuit boards and automobile catalysts. *JOM,* 56: 55–58.

Koehnlechner, R. 2008. Gold Diggers. *Recycling Magazine,* 18: 6–11.

Kogan, V. 2006. Recovery of precious metals from electronic scrap by hydrometallurgical processing. Europe, Patent No. WO2006013568 A2.

Lin, K. and Chiang, H.L. 2014. Liquid oil and residual characteristics of printed circuit board recycle by pyrolysis. *Journal of Hazardous Materials,* 271: 258–265.

Liu, R., Shieh, R., Yeh, R., and Lin, C. 2009. The general utilization of scrapped PC board. *Waste Managment,* 29(11): 2842–2845.

Long, L. et al. 2010. Using vacuum pyrolysis and mechanical processing for recycling waste printed circuit boards. *Journal of Hazardous Materials,* 177(1–3): 626–632.

Luda, M.P. 2011. Recycling of printed circuit boards. In: S. Kumar, ed., *Integrated Waste Management—II.* InTech: 285–298. Available from: www.intechopen.com.

Luda, M.P., Balabanovich, A., and Zanetti, M. 2010. Pyrolysis of fire retardant anhydride-cured epoxy resins. *Journal of Analytical and Applied Pyrolysis,* 88(1): 39–52.

Luda, M.P., Balabanovich, A., Zanetti, M., and Guaratto, D. 2007. Thermal decomposition of fire retardant brominated epoxy resins cured with different nitrogen containing hardeners. *Polymer Degradation and Stability,* 92(6): 1088–1100.

Luda, M.P., Euringer, N., Moratti, U., and Zanetti, M. 2005. WEEE recycling: Pyrolysis of fire retardant model polymers. *Waste Management,* 25(2): 203–208.

Ma, E. and Xu, Z. 2013. Technological process and optimum design of organic materialsvacuum pyrolysis indium chlorinated separation from waste liquid crystal display panels. *Journal of Hazardous Materials,* 263(Part 2): 610–617.

Malkow, T. 2003. Novel and innovative pyrolysis and gasification technologies for energy efficient and environmentally sound MSW disposal. *Waste Management,* 24(1): 53–74.

Mankhand, T., Singh, K., Gupta, S., and Das, S. 2012. Pyrolysis of printed circuit boards. *International Journal of Metallurgical Engineering,* 1(6): 102–107.

Mecucci, A. and Scott, K. 2002. Leaching and electrochemical recovery of copper, lead and tin from scrap printed circuit boards. *Journal of Chemical Technology and Biotechnology,* 77(4): 449–457.

Menad, N., Björkman, B., and Allain, E.G. 1998. Combustion of plastics contained in electric and electronic scrap. *Resources, Conservation and Recycling,* 24(1): 65–85.

Menad, N., Guignot, S., and van Houwelingen, J. 2013. New characterisation method of electrical and electronic equipment wastes (WEEE). *Waste Management,* 339(3): 706–713.

Miskolczi, N. et al. 2008. Production of oil with low organobromine content from the pyrolysis of flame retarded HIPS and ABS plastics. *Journal of Analytical and Applied Pyrolysis,* 83(1): 115–123.

Moltó, J., Font, R., Gálvez, A., and Conesa, J. 2009. Pyrolysis and combustion of electronic wastes. *Journal of Analytical and Applied Pyrolysis,* 84(1): 68–78.

Ni, M. et al. 2012. Combustion and inorganic bromine emission of waste printed circuit boards in a high temperature furnace. *Waste Management,* 32(3): 568–574.

Nutt, M., Heck, K., Alvarez, P., and Wong, M. 2006. Improved Pd-on-Au bimetallic nanoparticle catalysts for aqueous-phase trichloroethene hydrodechlorination. *Applied Catalysis B: Environmental,* 69(1/2): 115–125.

Ongondo, F.O., Williams, I.D. and Cherrett, T.J. 2011. How are WEEE doing? A global review of the management of electrical and electronic wastes. *Waste Management,* 31(4): 714–730.

Ortuño, N., Conesa, J., Moltó, J., and Font, R. 2014. Pollutant emissions during pyrolysis and combustion of waste printed circuit boards, before and after metal removal. *Science of the Total Environment,* 499: 27–35.

Panda, A., Singh, R., and Mishra, D. 2010. Thermolysis of waste plastics to liquid fuel: A suitable method for plastic waste management and manufacture of value added products—A world prospective. *Renewable and Sustainable Energy Reviews,* 14(1): 233–248.

Park, Y. and Fray, D. 2009. Recovery of high purity precious metals from printed circuit boards. *Journal of Hazardous Materials,* 164: 1152–1158.

Quan, C., Li, A., and Gao, N. 2009. Thermogravimetric analysis and kinetic study on large particles of printed circuit board wastes. *Waste Management,* 29(8): 2353–2360.

Quinet, P., Proost, J., and Van Lierde, A. 2005. Recovery of precious metals from electronic scrap by hydrometallurgical processing routes. *Mineral and Metallurgical Processing,* 22: 17–22.

Rankin, W.J. 2011. *Minerals, Metals and Sustainability Meeting Future Material Needs.* Melbourne, Australia: Commonwealth Scientific and Industrial Research Organization (CSIRO).

Ray, R. and Thorpe, R. 2007. A comparison of gasification with pyrolysis for the recycling of plastic containing wastes. *International Journal of Chemical Reactor Engineering,* 5(1): Article A85.

Rocchetti, L., Vegliò, F., Kopacek, B., and Beolchini, F. 2013. Environmental impact assessment of hydrometallurgical processes for metal recovery from WEEE residues using a portable prototype plant. *Environmental Science and Technology,* 47(3): 1581–1588.

Schlummer, M. et al. 2007. Characterisation of polymer fractions from waste electrical and electronic equipment (WEEE) and implications for waste management. *Chemosphere,* 67(9): 1866–1876.

Sheng, P. and Etsell, T. 2007. Recovery of gold from computer circuit board scrap using aqua regia. *Waste Management and Research,* 25: 380–383.

Sousa, M. and Riedewald, F. 2014. Waste printed circuit board pyrolysis with simultaneous sink-float separation of glass and metals by contact with molten salt—A laboratory investigation. *PYRO 2014 Conference,* Birmingham.

Sun, J., Wang, W., Liu, Z., and Ma, C. 2011. Study of the transference rules for bromine in waste printed circuit boards during microwave-induced pyrolysis. *Journal of the Air and Waste Management Association,* 61(5): 535–542.

Swedish EPA. 2011. *Recycling and Disposal of Electronic Waste Health Hazards and Environmental Impacts.* Stockholm, Sweden: Swedish EPA.

Taskmaster®. 2014. *Shredders.* Cited May 31, 2014. Available from http://www.franklinmiller. com/solid-waste-shredders.html.

Taurino, R., Pozzi, P., and Zanasi, T. 2010. Facile characterization of polymer fractions from waste electrical and electronic equipment (WEEE) for mechanical recycling. *Waste Management,* 30(12): 2601–2607.

Tsydenova, O. and Bengtsson, M. 2011. Chemical hazards associated with treatment of waste electrical and electronic equipment. *Waste Management,* 31(1): 45–58.

Umicore SA. 2005. *Exploring Umicore Precious Metals Refining.* Cited October 13, 2014. Available from http://ambientair.rec.org/documents/umicore.pdf.

UNEP. 2005. *E-waste, the hidden side of IT equipment's manufacturing and use,* Nairobi, Kenya: United Nations Environment Programme.

University of United Nations. 2007. *2008 Review of Directive 2002/96 on Waste Electrical and Electronic Equipment (WEEE) Annex to the Final Report,* Bonn, Germany: United Nations University.

USEPA. 2011. *Electronics Waste Management In the United States Through 2009.* Washington, DC: United States Environmental Protection Agency.

Van der Avert, P. and Weckhuysen, B. 2002. Low-temperature destruction of chlorinated hydrocarbons over lanthanide oxide based catalysts. *Angewandte Chemie International Edition,* 41(24): 4730–4732.

Vasile, C. et al. 2007. Feedstock recycling from plastics and thermosets fractions of used computers. II. Pyrolysis oil upgrading. *Fuel,* 86: 477–485.

Veit, H. et al. 2006. Utilization of magnetic and electrostatic separation in the recycling of printed circuit boards scraps by mechanical processing and electrometallurgy. *Journal of Hazardous Materials,* 137(3): 1704–1709.

Weil, E. and Levchik, S. 2004. A review of current flame retardant systems for epoxy resins. *Journal of Fire Sciences,* 22(1): 25–40.

Widmer, R. et al. 2005. Global perspectives on e-waste. *Environmental and Social Impacts of Electronic Waste Recycling,* 25(5): 436–458.

Wu, W., Xu, J., and Ohnishi, R. 2005. Complete hydrodechlorination of chlorobenzene and its derivatives over supported nickel catalysts under liquid phase conditions. *Applied Catalysis B: Environmental,* 60: 129–137.

Xenics NV. 2014. Cited September 20, 2014. Available from http://www.xenics.com/en/ infrared_imaging_applications/infrared_imaging_for_industrial_applications/application_-_ waste_storting.aspx.

Yamane, L., de Moraes, V., Espinosa, D., and Tenório, J. 2011. Recycling of WEEE: Characterization of spent printed circuit boards from mobile phones and computers. *Waste Management,* 31(12): 2553–2558.

Yamawaki, T. 2003. The gasification recycling technology of plastics WEEE containing brominated flame retardants. *Fire and Materials,* 27(6): 315–319.

Yang, X. et al. 2013. Pyrolysis and dehalogenation of plastics from waste electrical and electronic eqipment (WEEE): A review. *Waste Mangement,* 33(2): 462–473.

Zhang, S. and Forssberg, E. 1997. Mechanical separation-oriented characterization of electronic scrap. *Resources, Conservation and Recycling,* 21(4): 247–269.

Zhou, Y. and Qiou, K. 2010. A new technology for recycling materials from waste printed circuit boards. *Journal of Hazardous Materials,* 175(1–3): 823–828.

Zhou, Y., Wu, W., and Qiu, K. 2010. Recovery of materials from waste printed circuit boards by vacuum pyrolysis and vacuum centrifugal separation. *Waste Management,* 30(11): 2299–2304.

11 Recycling of Thermoset Composites

Mikael Skrifvars and Dan Åkesson

CONTENTS

11.1 INTRODUCTION

Thermoset composites are versatile materials that are characterized by high mechanical strength and low weight. Consequently, thermoset materials have found their way into many applications in industries such as marine, automotive, and aerospace. Thermoset materials are however known to be difficult to recycle. The problem can be described as follows. While thermoplastic materials and metals can easily be remelted and processed into new materials, thermosets are cross-linked materials that cannot be melted. Second, composite materials are complexly composed materials consisting of plastics, fibers, and often fillers and core materials. Metals can also be integrated into composites. Furthermore, there is a wide range of different composite materials consisting of very different components. These composites need to be collected, identified, and sorted. This needs to be done in a cost-efficient way. Consequently, there are both technical and economic aspects to consider.

11.2 THE NEED TO RECYCLE

Composites are presently produced in relatively large quantities. It has been estimated that more than one million tons of composites are being produced in Europe every year [1]. There may be both economic and environmental incentives for recycling composites. While many scrap composites have traditionally been disposed in landfill sites, the taxes for landfilling are increasing. It is a reasonable assumption that landfill taxes will continue to increase. This will drive the development of new solutions. Furthermore, thermoset composites are produced from nonrenewable resources and it is obvious that methods to recycle composites are needed. The need to recycle is also driven by legislation. As an example, the EU Directive 2008/98/EC on waste (Waste Framework Directive) states that waste materials containing more than 10% organic material must not be disposed of in landfills. Another directive that also regulates recycling of composite waste is the EU Directive 2000/53/EC (ELV Directive). Here it is stated that in 2015, the material recovery of end-of-life vehicles must be 85% of the average car weight. As composites are important materials in the automotive sector, this means that recycling methods for composites must be introduced soon.

As composites are rather complex materials, and used in many different applications, there are several factors that must be considered when selecting the best method for end-of-life treatment, as shown in Table 11.1. It is evident that depending on the composite waste characteristics, different methods will be used, resulting in different technical strategies for recycling.

TABLE 11.1

Factors to Consider when Selecting the Method for Recycling of End-of-Life Composites

Factors	Characteristics
Waste Type	
Production waste	Short life length
	Mixing with other waste materials
End-of-life products	Long life length
	Material degradation
Waste Volumes	
Geographical distribution	Transportation necessary
Type	Large variation in waste volume, depending on waste type
Composition	
Glass or carbon fibers	Secondary use value
	Possibility to incinerate
Thermoset or thermoplastic	Reprocessing possibilities
Sandwich core material	Limited possibilities to incinerate if PVC present
Application Type	
Size	Must be cut before transportation
Degradation during use	Can result in inferior secondary-use properties
Consumer/industrial use	Consumer waste can be more difficult to collect due to small volumes and combination with other materials, while industrial waste is typically created when dismantling or disposing larger components at a specific time

For composite production, rather large volumes of production waste are gener-ated. This is due to the nature of the production methods used, which often generate trimmings and cut parts that must be scrapped. Some evaluations have estimated the amount of composite production waste as high as 10%–20% of the used materials in the composite production. Resin buckets, brushes, vacuum infusion tubes, and bag films are also generated. Thus, if the sorting is taken care of at the production site, it is rather easy to separate the composite waste that is pure enough to allow for recy-cling. For the individual company, proper recycling of the composite waste will not only save waste handling costs but also be beneficial for the environment.

11.3 COMPOSITE RECYCLING METHODS

The methods for recycling of composites have been divided into the following four groups:

1. *Primary recycling.* It involves conversion of waste into a material with same properties and value.
2. *Secondary recycling.* It involves conversion of waste into a material with inferior properties and lower value.
3. *Tertiary recycling.* It involves conversion into chemicals that can be used in the chemical industry.
4. *Quaternary recycling.* It involves conversion of waste into a fuel that is used in energy production.

Primary, secondary, and tertiary recycling methods result in material recovery, while quaternary recycling methods result in energy recovery.

Another way of classification is based on the following types of recycling process:

1. *Mechanical recycling.* It involves mechanical processing of waste into a grinded material, leading to material recovery.
2. *Thermal recycling.* It involves thermal treatment of waste, which degrades the chemical structure and results in energy or chemicals, leading to energy recovery in most cases.
3. *Chemical recycling.* It involves treatment by solvents or chemicals, which separate the resin and the reinforcement for reuse, leading to material recovery.

These methods result in various fractions, which can be used further in secondary processes, as shown in Figure 11.1.

11.3.1 MECHANICAL RECYCLING OF COMPOSITES

Mechanical recycling is a straightforward way to recycle materials that are homo-geneous and that can easily be processed in a mechanical process. Many materi-als are also recycled using this method, and in principle, it is also possible to use the same for composites. The method can be classified as primary or secondary

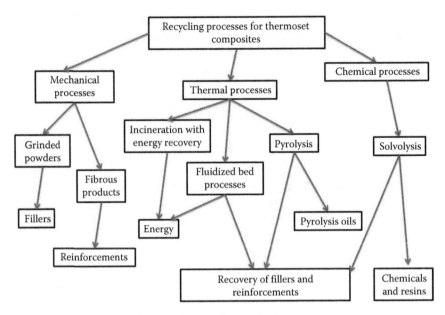

FIGURE 11.1 Overview of composite recycling methods.

recycling, and it involves mechanical fragmentation of the composite material into powder or granulate-type of material. This is usually rather easy as cured thermoset composites are brittle materials, which can easily be broken in a mechanical process. Before fragmentation, large composite structures must be cut into smaller pieces, and after fragmentation, the resulting grinded material must be fractionated, depending on the size and composition. The recycled composite can then be used as a filler or reinforcement in virgin composites, or in other applications such as geotextiles or asphalt. The method has also been implemented in several composite recycling concepts, of which ERCOM Composites Recycling GmbH is the most well-known [2]. ERCOM was founded in 1990 by composite material suppliers and composite users. The aim was to recycle end-of-life composite products made from bulk moulding compounds (BMCs) and sheet moulding compounds (SMCs). These compounds are unsaturated polyester resin-based composites with a high filler and glass fiber content (up to 70 wt/%), and are commonly used in automotive parts and electrical appliance components. ERCOM focused on SMC and BMC components from end-of-life vehicles, which were collected by a mobile shredder unit, which could downsize and grind the composite at the collection site. This reduced the volume and made a cost-effective transportation possible. The shredded composite waste was then transported to a central processing plant in Rastatt, Germany, where it was further grinded and fractionated into a filler type material (see Figure 11.2). Eight different grinded composite fractions were produced, composed both of powders of different particle size and fibers of different length. The intention was to use the recycled composite in new composite products, where it could replace virgin fillers and fibers. In total, 2 million automotive parts were recycled, but in 2004 the recycling ended due to lack of profitability, mainly due to the much lower price for virgin fillers and fibers. The obtained

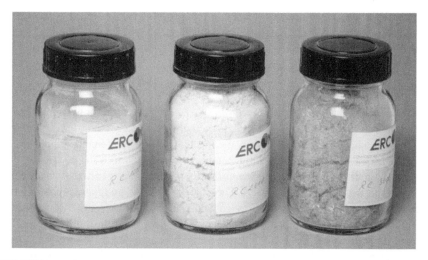

FIGURE 11.2   Fillers made from recycled composites by ERCOM. (Courtesy of M. Skrifvars.)

FIGURE 11.3   Motor boat made from 20 wt% recycled composite. (Courtesy of M. Skrifvars.)

recycled composite was also evaluated in many secondary composite applications, where it was evaluated as a filler or a reinforcement. For example, motor boats were manufactured from recycled composites in a project at SICOMP in Sweden (see Figure 11.3). Up to 20 wt% recycled composite could be incorporated in the boat hull laminate as central layer. Several fully equipped boats were manufactured by a boat manufacturer, and both mechanical properties and performance were similar as for same type of boats made from virgin materials [3]. For both projects, the main reason for lack of success was the low cost of virgin glass fiber reinforcements and fillers, as the recycled composite had a much higher cost. These early trials showed also that the mechanical properties for composites made from 100% recycled composite are lower compared to similar composite made from 100% virgin materials. However, it is possible to partly replace the virgin composite material with recycled composite; typically the load can be up to 20–40 wt%, depending on the composite product.

11.3.2 Energy Recovery of Composites

Combustion is a quaternary recycling method that is most commonly used in waste treatment; for example, household waste is incinerated in many countries for heat and electricity production. For composites, the inert glass fiber reinforcement will reduce the energy content to a high degree, depending on the amount in the composite. The glass also remains as an ash, which must be deposited, and it can also cause some challenges in the combustion process. The matrix, which is an organic material, can of course be combusted and its energy content can be used as fuel. Combustion was evaluated in detail in a large Swedish project (VAMP-18 project), during 1999–2002 [4]. The project involved full-scale combustion of aircraft components, leisure boats, PVC sandwich laminates, and SMC components as well as thermoplastic glass mat composites. These were fragmented by bulldozers at the combustion plant site and then fed into the waste combustor. From these trials, it was possible to evaluate that composite waste can provide 9 MJ/kg to 35 MJ/kg heat, depending on the composition. Second, it could also be seen that the combustion process was not necessarily complete, as large flakes of residual reinforcements could be collected from the ash. These trials showed that glass fiber–containing composite waste could be treated in a conventional waste incineration, but its value as a fuel was rather limited due to the glass fibers present. Many incinerator plants also do not want to receive this type of complex waste due to the need to pretreat the waste, and due to the need to take care of the glass fiber–containing ash.

11.3.3 Energy Recovery with Material Recovery

The incineration can also be performed in a more controlled way so that the residual reinforcement is collected after incineration for material recovery. In this process, the matrix functions as a fuel, while the residual glass fibers can be reused as reinforcement in secondary composites. At Nottingham University, this concept has been developed for glass fiber composites and particularly SMC [5]. The composite waste is first treated in a fluidized bed at 450°C–550°C; this decomposes the polymer matrix to a flue gas, while the glass fibers and fillers can be collected by a cyclone. Metallic components are separated in the fluidized bed. The organic flue gas is burnt in an afterburner, thus producing energy. On average, the collected glass fibers had mechanical properties, which were reduced by 50% compared to the same virgin glass fibers. The recycled glass fibers had very high purity, but had also lost their surface treatment, which is important for glass fibers in composites. Without the size or surface treatment, the compatibility with the matrix is lost, resulting in poor interface and severe reduction in stress transfer from matrix to reinforcement. Feasibility calculations showed that for glass fiber composites, it is necessary to process 10,000 tons of glass fibers per year, if the recycled fiber value is 680% of the virgin glass fiber value. It is therefore rather difficult to achieve profitability when recycling glass fiber composites, but for carbon fiber composites, profitability can be achieved already at much low carbon fiber treatment volumes. This is due to the fact that the price of virgin carbon fibers is high. Thus, the recycled carbon fiber can compete regarding price. Second, the mechanical properties of the recycled carbon fiber are not reduced as much as for the glass fibers. Therefore, the method has been further developed for the recycling of carbon fiber composite waste from end-of-life airplane structures [6].

11.3.4 ENERGY AND MATERIAL RECOVERY IN CEMENT PRODUCTION

A more recent method is to recycle end-of-life composites in cement production as cofuel. Cement production requires large amounts of fuel, as the raw materials used in cement manufacture are heated in the kiln up to 2000°C. Composites can very well be used as a fuel feedstock, as the residual glass fiber can remain in the cement as filler. This method is also endorsed by the European Composites Industry Association, as a sustainable way for end-of-life composite treatment [7]. On average, 1 ton of composite is equal to 400 kg–500 kg coal if heat values are compared. This method is also commercially in use. For example, the companies Holcim and Zajons in Germany have been recycling windmill blades using this method since 2010 [8]. The annual capacity is claimed to be 60,000 tons/year. In Finland, a similar concept is under evaluation, in order to take care of production waste from the Finnish composite industry [9]. The waste is collected and mixed with other energy-rich waste fractions, and then fed into a cement kiln. Here, the main problems include the rather tedious pretreatment and the fact that the production waste has rather low energy value, due to large amounts of pure glass fibers in the end-of-life waste.

11.3.5 PYROLYSIS OF COMPOSITE WASTE

Pyrolysis is a process where an organic material is heated in an inert atmosphere. Pyrolyzing composites will thus result in polymers being degraded to lower molecular weight compounds. As a result of the pyrolysis process, inert fibers such as glass fibers and carbon fibers can be recovered while the polymer is converted into gas and oil. The recovered fibers have the potential to be used for other composite applications, and the gas and the oil could be used as chemical feedstock.

Pyrolysis of scrap composites has been evaluated in number of projects. Several studies of pyrolysis of glass fiber–reinforced composites were done in Spain. Torres et al. studied pyrolysis of SMC [10]. This type of composite typically has a very high content of inorganic material and is consequently not suitable for incineration. Pyrolysis was done in laboratory scale at temperatures ranging from 300°C to 700°C. The obtained pyrolysis oil contained mainly aromatic compound with a high gross calorific values (34–37 MJ/kg). The obtained pyrolysis oil was studied in detail using GC-MS analysis by Torres et al. [11]. The glass fibers obtained after pyrolysis of SMC composites were evaluated as a reinforcement for BMC composite by de Marco et al. [12].

In a recent study by Yun et al. [13], pyrolysis of glass fiber–reinforced plastics was studied under nonisothermal conditions. Samples were heated from 500°C to 900°C with heating rates ranging from 5°C to 20°C/minute using a thermogravimetric analyzer. The main component of the gas was carbon monoxide. Activation energies were determined. Cunliffe et al. [14] studied the pyrolysis of glass fiber–reinforced plastics.

Composites with several different polymer matrixes, including both thermoplastic and thermosetting polymers, were evaluated.

A concept for recycling of wind turbine blades was presented in Denmark by the company ReFiber [15]. Wind turbine blades are soon becoming a main composite waste type, due to upgrading of wind energy plants to longer blade lengths and increased power output. The recycling process involved treatment in a pyrolysis chamber at 600°C, in which the matrix was gasified into a gas, which could be burned at 1100°C to obtain energy. The fillers, reinforcements, and inserts in the blade could be separated from the matrix in the pyrolysis chamber as a rather pure and intact structure. These components could then be used as secondary-type filler or reinforcement in composites, or be processed into wool-like insulation material.

In a recent project, the recycling of glass fiber composites from scrap wind turbine wings was studied by means of microwave pyrolysis. Wind turbine industry is expected to grow during the coming years; however, there is presently no viable method to recycle these materials [16]. Heating by microwaves potentially has some advantages such as even heat distribution, good heat transfer, and good control of the heating process [17]. Microwave pyrolysis has been evaluated for various waste materials that are considered difficult to recycle [17–20]. Scrap wind turbine blades were shredded and microwave pyrolyzed. An example of the pyrolyzed fiber is shown in Figure 11.4.

As can be seen, the fibers were covered with char after the pyrolysis. Tests showed that when pyrolyzing at 360°C, the glass fibers retained about 75% of their mechanical strength [21], as shown in Figure 11.5. The obtained pyrolysis oil had a relatively high calorific value, 36 MJ/kg, and contained mainly aromatic compounds [22]. Both thermoplastic and thermoset composites were prepared from the recovered fibers [21,23].

Several authors have also studied pyrolysis of carbon fiber composites. Nahil and Williams [24] used a fixed bead reactor to pyrolyze composites based on woven fabrics and a phenolic resin. Pyrolysis was carried out at temperatures ranging from

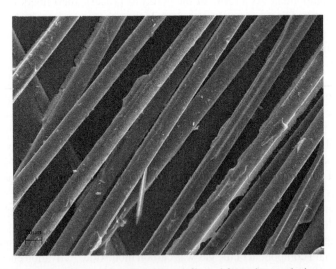

FIGURE 11.4   SEM micrograph of glass fibers collected from the pyrolysis process.

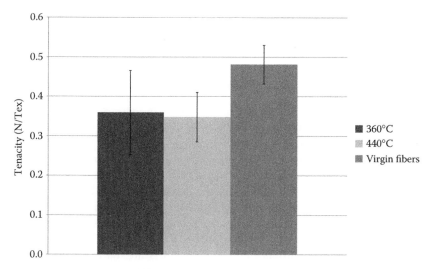

FIGURE 11.5 Tensile strength measured as tenacity (N/tex) of recovered glass fibers compared to virgin glass fibers.

350°C to 700°C. Tests showed that that the obtained pyrolysis oil after pyrolysis at 500°C mainly consisted of aromatic compounds. The gas formed after the pyrolysis was also analyzed. The gas composition varied, depending on the pyrolysis temperature but consisted mainly of CH_4, H_2, CO, CO_2, and other gases (C2–C4). The recovered carbon fiber was also characterized using tensile tests. The quality of the recovered fiber varied, depending on the pyrolysis temperature. Best results were found at 500°C where only a moderate drop of the tensile properties was observed in comparison with the virgin fiber. The tensile strength was reduced from 3.5 GPa to 3.27 GPa and with only a modest drop in the Young's modulus. Other studies have also showed that it is possible to obtain recycled carbon fiber with only very modest reduction of the tensile properties [25,26].

11.3.6 CHEMICAL DEGRADATION OF COMPOSITE WASTE

Chemical degradation of composite waste is in principle a rather difficult way to go, as the thermoset is cured and forms a large molecular network with permanent chemical bonds. Some type of chemical degradation is necessary so that lower molecular weight molecules are formed, which can be dissolved. Ester hydrolysis is one possibility; this degrades the unsaturated polyester chain to shorter segments, which contain the cured styrene units [27]. This has also been demonstrated by using supercritical water as solvent [28]. This removes up to 95% of the resin, and clean fibers with no remaining surface treatment are obtained. The mechanical properties however deteriorate, and as very high temperature and pressure are needed to create the supercritical conditions, this batch process seems to be far away in the future at the moment. For carbon fiber composites, it might become a feasible method, as carbon fibers have much higher value than glass fibers, even when recycled. At the moment, these chemical methods are however not competitive regarding price and performance, and the use of solvents and the severe process conditions must also be considered when evaluating these methods.

11.4   FUTURE PERSPECTIVES FOR COMPOSITE RECYCLING

Over the years, there have been many projects, both in industry and in academics, where recycling of composite materials has been studied. Many studies show that the properties of the recovered materials are rather good. Yet, no project has fully succeeded in finding a feasible solution that could be implemented in large scale. This shows that the recycling of thermoset composites is a complex subject, which also includes economy, logistics, and marketing considerations. The largest volume of waste composites consists of glass fiber composite. The glass fiber is produced from sand and other cheap inorganic components. Thus, recovered glass fibers have very low commercial value. Studies have also shown that recycled glass fibers are of lower technical quality than virgin ones [29]. Carbon fibers are, on the contrary, much more expensive and there is clear economic incentive for recovering them. However, carbon fibers are presently produced in relatively small quantities. Furthermore, these volumes may be spread out between composite producers over large geographical areas. Thus, there are also logistical aspects to be considered. Consequently, large-scale recycling of carbon fibers may not presently be possible.

Perhaps the most straightforward method is mechanical recycling where the composite material is ground into a recyclate. Such a recyclate can be used as filler. However, experience has shown that it is not so easy to succeed with mechanical recycling. The ERCOM project where SMC composites were ground into filler is not active anymore [30]. These materials must compete on the market with existing fillers such as calcium carbonate. Commercially available fillers are typically readily available at relatively low prices and at various grades. Thus, in order to compete successfully, mechanical recycling of composites, the recyclates must be produced at a lower cost than existing fillers. Alternatively, new fillers must be developed with better technical properties.

Combustion of composite end-of-life materials is also a relatively straightforward method. However, the very low energy content in many composite materials renders this method less useful. When glass fiber composites are combusted, the problem remains as to what should be done with the glass waste material. Of the various approaches where incineration is used, the usage of cement kiln seems relatively promising. The resin then serves as an energy source and the glass ends up as part of the cement.

Pyrolysis is a technically complicated method that is still under development. Several studies have been done on pyrolysis of carbon fiber composites. The quality of the recovered carbon fibers is typically very high. The obtained pyrolysis oil from these studies contains mainly aromatic compounds. Thus, this pyrolysis oil has lower value than commercial petroleum-based oils. However, it still seems likely that the gas and the oil that are recovered could find some commercial application. Time will tell if it can be economically viable to recover carbon fibers with pyrolysis. Relatively large volumes are probably necessary.

In summary, it can be concluded that several methods of recycling have been evaluated in both laboratory and industrial scale. The main obstacles to overcome are the high cost for the recycled composite, and therefore a lack of secondary markets for the recycled products. This applies regardless of which method is discussed. It is absolutely necessary to be able to produce products at low cost. The large volumes of thermoset composites consist of glass fiber composites. Since existing fillers on the market are cheap, recyclates must be produced at low cost. Also, markets must obviously be developed for the recycled products. This is an intriguing task.

REFERENCES

1. Krauss, T. and Witten, E., The composites market in Europe: Market developments, challenges, and opportunities. AVK Industrievereinigung Verstärkte Kunststoffe. http://www.carbon-composites.eu/sites/carbon-composites.eu/files/anhaenge/14/10/29/ccev-avk-marktbericht_2014_english.pdf, 2014.
2. Pickering, S.J., Recycling technologies for thermoset composite materials—Current status. *Composites Part A: Applied Science and Manufacturing*, 2006. 37(8): 1206–1215.
3. Pettersson, J. and B. Köllerfors, Pleasure boats manufactured from recycled materials. *Proceedings of the Second North European Engineering and Science conference—Composites and Sandwich Structures*. Edited by Bäcklund, J., J. Zenkert, and B.T. Åström, EMAS, Warrington, 1997: 609–617.
4. Hedlund-Åström, A., Model for end of life treatment of polymer composite material. Doctoral thesis, KTH, School of Industrial Engineering and Management, Stockholm, Sweden, 2005.
5. Pickering, S.J. et al., A fluidised-bed process for the recovery of glass fibres from scrap thermoset composites. *Composites Science and Technology*, 2000. 60(4): 509–523.
6. Turner, T.A., S.J. Pickering, and N.A. Warrior, Development of recycled carbon fibre moulding compounds—Preparation of waste composites. *Composites Part B: Engineering*, 2011. 42(3): 517–525.
7. Mendes, A.A., A.M. Cunha, and C.A. Bernardo, Study of the degradation mechanisms of polyethylene during reprocessing. *Polymer Degradation and Stability*, 2011. 96(6): 1125–1133.
8. Loultcheva, M.K. et al., Recycling of high density polyethylene containers. *Polymer Degradation and Stability*, 1997. 57(1): 77–81.
9. Pinheiro, L.A., M.A. Chinelatto, and S.V. Canevarolo, The role of chain scission and chain branching in high density polyethylene during thermo-mechanical degradation. *Polymer Degradation and Stability*, 2004. 86(3): 445–453.
10. Torres, A. et al., Recycling by pyrolysis of thermoset composites: Characteristics of the liquid and gaseous fuels obtained. *Fuel*, 2000. 79(8): 897–902.
11. Torres, A. et al., GC-MS analysis of the liquid products obtained in the pyrolysis of fibre-glass polyester sheet moulding compound. *Journal of Analytical and Applied Pyrolysis*, 2001. 58–59(0): 189–203.
12. de Marco, I. et al., Recycling of the products obtained in the pyrolysis of fibre-glass polyester SMC. *Journal of Chemical Technology & Biotechnology*, 1997. 69(2): 187–192.
13. Yun, Y.M. et al., Pyrolysis characteristics of GFRP (glass fiber reinforced plastic) under non-isothermal conditions. *Fuel*, 2014. 137: 321–327.
14. Cunliffe, A.M., N. Jones, and P.T. Williams, Recycling of fibre-reinforced polymeric waste by pyrolysis: Thermo-gravimetric and bench-scale investigations. *Journal of Analytical and Applied Pyrolysis*, 2003. 70(2): 315–338.

15. Cruz, S.A. and M. Zanin, Evaluation and identification of degradative processes in post-consumer recycled high-density polyethylene. *Polymer Degradation and Stability*, 2003. 80(1): 31–37.
16. Beauson, J., H. Lilholt, and P. Brøndsted, Recycling solid residues recovered from glass fibre-reinforced composites—A review applied to wind turbine blade materials. *Journal of Reinforced Plastics and Composites*, 2014. 33(16): 1542–1556.
17. Lam, S.S., A.D. Russell, and H.A. Chase, Pyrolysis using microwave heating: A sustainable process for recycling used car engine oil. *Industrial & Engineering Chemistry Research*, 2010. 49(21): 10845–10851.
18. Andersson, M. et al., Microwave assisted pyrolysis of residual fractions of waste electrical and electronics equipment. *Minerals Engineering*, 2012. 29(0): 105–111.
19. Donaj, P. et al., Recycling of automobile shredder residue with a microwave pyrolysis combined with high temperature steam gasification. *Journal of Hazardous Materials*, 2010. 182(1–3): 80–89.
20. Ludlow-Palafox, C. and H.A. Chase, Microwave-induced pyrolysis of plastic wastes. *Industrial & Engineering Chemistry Research*, 2001. 40(22): 4749–4756.
21. Åkesson, D. et al., Microwave pyrolysis as a method of recycling glass fibre from used blades of wind turbines. *Journal of Reinforced Plastics and Composites*, 2012. 31(17): 1136–1142.
22. Åkesson, D. et al., Products obtained from decomposition of glass fiber-reinforced composites using microwave pyrolysis. *Polimery (Warsaw)*, 2013. 58(7/8): 582–586.
23. Åkessan, D. et al., Glass fibres recovered by microwave pyrolysis as a reinforcement for polypropylene. *Polymers & Polymer Composites*, 2013. 21(6): 333–339.
24. Nahil, M.A. and P.T. Williams, Recycling of carbon fibre reinforced polymeric waste for the production of activated carbon fibres. *Journal of Analytical and Applied Pyrolysis*, 2011. 91(1): 67–75.
25. Meyer, L.O., K. Schulte, and E. Grove-Nielsen, CFRP-recycling following a pyrolysis route: Process optimization and potentials. *Journal of Composite Materials*, 2009. 43(9): 1121–1132.
26. Lester, E. et al., Microwave heating as a means for carbon fibre recovery from polymer composites: A technical feasibility study. *Materials Research Bulletin*, 2004. 39(10): 1549–1556.
27. Tesoro, G. and Y. Wu, Chemical products from cured unsaturated polyesters. *Advances in Polymer Technology*, 1993. 12(2): 185–196.
28. Piñero-Hernanz, R. et al., Chemical recycling of carbon fibre reinforced composites in nearcritical and supercritical water. *Composites Part A: Applied Science and Manufacturing*, 2008. 39(3): 454–461.
29. Job, S., Recycling glass fibre reinforced composites—History and progress (Part 1). Reinforce plastics. http://www.reinforcedplastics.com/view/33969/recycling-glass-fibre-reinforced-composites-history-and-progress-part-1/, 2013.
30. Derosa, R., E. Telfeyan, G. Gaustad, and S. Mayes, Strength and microscopic investigation of unsaturated polyester BMC reinforced with SMC-recyclate. *Journal of Thermoplastic Compoite Materials*, 2005. 18(4): 333–349.

12 Recycling of Papers and Fibers

Samuel Schabel, Hans-Joachim Putz, and Winfrid Rauch

CONTENTS

12.1 ADVANTAGES OF PAPER RECYCLING

Paper is a material that was produced from recovered raw materials from its beginning. The invention of papermaking is dated by historians to the time of the first Han dynasty (206 BC to 224 AD) or more precisely to CAI LUN and the year 105 after Christ (Pichol 1999; Sandermann 1997). All historic literature speaks about tatters (used textiles) as a raw material for papermaking as well as hemp, bast fibers, and bark. Some authors even think that paper in those early days was made by the Chinese directly from used textiles (Pichol 1999). For centuries, used textiles were the main raw material for papermaking and in the eighteenth century the raw material supply was the main limitation for increasing the paper production. Some countries prohibited exporting tatters in order to save the raw material for their own needs (Sandermann 1997).

It took as long as to 1840 when, in Germany, Friedrich Gottlob Keller invented a method for making a raw material for papermaking directly from wood. His invention of the grinding process was patented in 1840. In the same century, English chemists Hugh Burgess and Charles Watt in 1851 invented a process for manufacturing chemical pulp by cooking wood chips in soda. After this basic inventions, wood became the main source of papermaking worldwide.

Of course, recovered paper was used from early days of papermaking. This was even regulated by legislation. In 1366 in Venice for example the city councilor legislated that used papers are not allowed to bring to any other locations than the local paper mill of Treviso (Sandermann 1997). But in these days used paper could only be used for making lower quality packaging papers and board since the ink that was used for writing and printing on the paper could not be removed. In the late 1950s, the flotation process known from mineral ore processing was introduced in papermaking as the so-called deinking process. This invention was the start of a worldwide success story of paper recycling with continuously increasing recycling rates, which has not come to an end until today.

In principle paper recycling is a quite simple process. If some standard paper like newsprint is put in water and agitated, the paper web can easily be disintegrated in single fibers and a pulp suspension is formed, which then can be used for making a new sheet of paper. The basic mechanism why paper can be disintegrated that easy and rebuilt from single fibers is the bonding mechanism between the single fibers. Typical papermaking fibers are covered with hydroxyl groups and the oxygen atoms in these hydroxyl groups are able to form hydrogen bonds to the hydrogen atoms of the closest next fibers or to water molecules. If the water is separated from a pulp then hydrogen bonds are built between the fibers and link the fibers to each other. Addition of water easily reverses this process as shown in Figure 12.1.

In addition to securing the raw material supply for paper production, recycling has the big advantage that specific energy consumption for producing recycled paper and also the other environmental impacts are significantly lower compared to virgin fiber production. Both processes for manufacturing papermaking fibers from wood, the groundwood, or mechanical pulping process and the chemical pulping process are energy-intensive. Typical specific energy consumptions for mechanical pulp production are in the range of 2500 kWh/t to 4000 kWh/t and specific energy demand for pulping processes is in the range of 4500 kWh/t to 6000 kWh/t (European Commission 2013).

The specific energy consumption for recycled fiber processing is much lower and, of course, depends also on the quality of the pulp. Simple recycling lines for low-quality packaging grades have a specific energy demand of about 1400 kWh/t and for higher quality graphic paper grades, like copy paper of about 3000 kWh/t (European Commission 2013; Gromke and Detzel 2006; Schabel and Holik 2010).

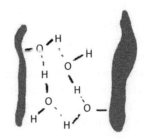

FIGURE 12.1   Schematic principle of fiber–fiber bonding by hydrogen bonds.

The comparison of the environmental impact of producing paper from virgin fiber sources or from recovered paper has been studied in detail for copy paper by the German IFEU Institute (Gromke and Detzel 2006). The environmental advantages of recycling paper are illustrated in Figures 12.2 and 12.3.

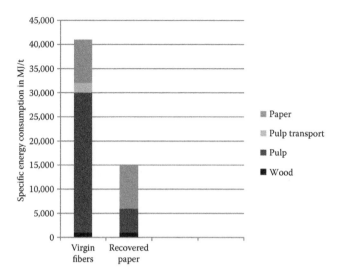

FIGURE 12.2 Comparison of specific energy consumption for the production of copy paper from virgin fiber (chemical pulp imported from South America) and local recovered paper. (Data from Gromke, U. and A. Detzel, *Ökologischer Vergleich von Büropapieren in Abhängigkeit vom Faserrohstoff*, Institut für Energie- und Umweltforschung Heidelberg GmbH, Heidelberg, Germany, 2006.)

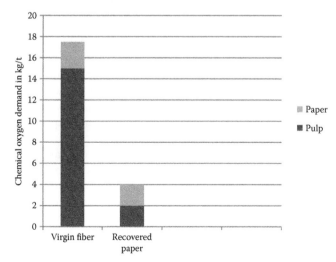

FIGURE 12.3 Comparison of paper production for copy paper from chemical pulp imported from South America and local recovered paper. (Data from Gromke, U. and A. Detzel, *Ökologischer Vergleich von Büropapieren in Abhängigkeit vom Faserrohstoff*, Institut für Energie- und Umweltforschung Heidelberg GmbH, Heidelberg, Germany, 2006.)

TABLE 12.1

Comparison of Specific Resource Consumption for the Production of Different Paper Grades and Savings That Can Be Achieved by Using Recovered Paper as Raw Material

	Kraft Liner	Test Liner/ Recycled Fiber	Fresh Fiber Board	Recycled Fiber Board
Water consumption (m³/t)	15–25	0–9	8–15	0–9
Electrical energy (MWh/t)	1.0–1.5	0.7–0.8	2.3–2.8	0.9–1.0
Steam consumption (GJ/t)	14.0–17.5	6.0–6.5	3.5–13.0	8.0–9.0

Savings	Corrugated paper	Board
Production (t/a)	3,600,000	1,300,000
Water consumption (m³/a)	54,000,000	9,100,000
Electrical energy (MWh/a)	1,800,000	2,080,000
Steam consumption (GJ/a)	34,200,000	0

These results show that recycling paper has a clear advantage not only in the lower specific energy consumption but also in other environmental impacts like chemical oxygen demand, which is relevant for process water consumption and eutrophication.

For packaging paper grades a comparison of water and energy consumption can be found in Table 12.1 (European Commission 2001).

12.2 RECYCLABILITY OF PAPER PRODUCTS

The natural strength of paper is based on hydrogen bonds between the fibers. Because these bondings are easily dissolved in the presence of water with some mechanical treatment, paper is an ideal raw material for recycling. Initially, most papers produced are applicable for recycling, at least if they are not specially treated, for example, wet-strength papers. Fibers separated in water can easily form a new paper sheet, which develops its strength naturally during the dewatering and drying process of the paper by establishing new hydrogen bonds.

Recycling-related problems arise typically not from the base paper, but from the manufactured paper products. Adhesive applications for example are utilized to manufacture magazines and paperbacks, as well as cardboard or corrugated boxes. If adhesive particles are not strong enough under wet conditions, they disintegrate in very small particles during pulping and become insufficiently removable by screening or cleaning processes. Then they may create problems due to their tacky character on paper machines or in converting machines or lead to quality problems of the manufactured paper product. Another example that generates problems can be the ink from the manufactured print products. Flexoprinted newspapers release ink particles that are not sufficiently removable during typical flotation deinking processes, resulting in deinked pulp quality with too low brightness (Putz 2013). From this point of view it has to be postulated that each paper product entering the environment as a consumer product with the intention to recover it for material recycling should be manufactured as a recycling-friendly paper product.

Despite the fact that industrial paper recycling has a very long history, requirements for recycling-friendly paper products are relatively young. Common methods are partially not very well known or not available and sometimes also under discussion due to misunderstandings. One reason for this is that the requirements to be achieved are always related to a single paper product, whereas paper mills usually treat recovered paper mixtures. These mixtures consist of a variety of paper products and it can be argued that the tested paper product is only a minor proportion of the recovered paper mixture. Nevertheless, recyclability tests are always related to the test of a specific paper product and not of a paper and board grade for recycling according to the European Standard EN 643 (BSI 2014).

The oldest method testing recyclability on a laboratory scale is the PTS Method RH 021/95 from 1995 for the evaluation of packaging materials or print products. This method has been revised several times and is available in its actual version from 2012 as PTS Method RH 021/97 (PTS 2012). The method differentiates packaging materials or print products intended for recycling into the following two classes:

- Category I (printed products) or
- Category II (packaging materials)

In the actual version of the method, the deinkability test of print products is in line with INGEDE Method 11, which is explained later, and the assessment of the results according to European Recovered Paper Council (ERPC). The test of packaging material is still in use but because an improved method is already developed on European level and an assessment procedure is in progress, this PTS method will not be discussed in more detail. Further information is given in Putz (2013) and can also be found in the book *Recycled Fiber and Deinking* (Putz 2000).

Actually, for paper products still a distinction has to be made between print and packaging products when testing their recyclability. The background is the difference in their industrial recycling processes. Only graphic (printed white) papers can be used in the commercial recycling processes for manufacturing new white graphic papers or hygiene tissue. When the ink is removed in deinking processes the fibers become white again and can be used to produce bright paper. In contrast, brown or gray fibers from packaging material included in the raw material mixture would stay in its original color and reduce the brightness or create a veining appearance of the new paper produced. Therefore, deinking mills avoid packaging material as part of the used paper and board for recycling. Because for the deinking process on industrial scale chemicals are needed, which are more efficient at higher pulp concentration, the simulating laboratory deinking process has to focus on high-consistency pulping with a suitable chemical recipe and a flotation deinking process for graphic paper products.

Paper mills utilizing recovered packaging material for the production of packaging paper, liner, or medium for corrugated board or folding boxboard can use also graphic paper products in their raw material mixture. Optical properties of the paper and board produced are very often not such important. The industrial recycling process requires no chemicals, removes no ink, and applies low consistency pulping, typically. Therefore, a laboratory recyclability evaluation of packaging products should also apply to low consistency pulping without chemicals.

Based on the later on used numerical assessment criteria, the laboratory recyclability tests are always a test of a single paper or board product (a newspaper of a given issue or a corrugated box for a certain purpose) and never a mixture of paper and board for recycling. According to Table 12.2 a differentiation has to be made between graphic or packaging paper products. Graphic paper products without adhesive applications are tested only on deinkability according to INGEDE Method 11 (INGEDE 2012); for products with adhesive applications the macrosticky potential from the adhesive application is also of interest, which can be tested according to INGEDE Method 12 (INGEDE 2013). For the test on recyclability of packaging products with the ZELLCHEMING Leaflet RECO 1, 2/2014 (ZELLCHEMING 2014) and the EcoPaperLoop Method 1 (Ecopaperloop 2014), two equivalent methods exist. Table 12.2 also gives an overview on the measured parameters that are used in the assessment and the scorecard specifications to be applied according to the publications of the ERPC.

Print products are tested according to two INGEDE methods, which are very well established. They comprise a deinkability test on the one hand and a test on the fragmentation behavior of adhesive applications on the other hand. Both tests have to be

TABLE 12.2
Recyclability Tests for Paper Products

Category of Paper/ Board Product	Graphic Paper Product		Packaging Paper or Board Product
Investigation on	Deinkability	Adhesive application fragmentation	Recyclability
Method used	INGEDE Method 11	INGEDE Method 12	ZELLCHEMING Leaflet RECO 2, 1/2014 EcoPaperLoop Method 1
Investigated and assessed parameters	Luminosity Y Color value a* Dirt speck area >50 and >250 µm Ink elimination Filtrate darkening	Macrosticky area <2.000 µm Size distribution of makrostickies	Coarse reject Flake content Macrosticky area <2.000 µm Optical homogeneity
Assessment according to	ERPC deinkability scorecard	ERPC removability scorecard	ERPC packaging recyclability scorecard (in progress, not yet published)

Note: Color coordinates in the lab system are named as a* and b*.

passed successfully for an overall positive recyclability test result of a print product with adhesive applications. For print products without adhesive applications, only the deinkability assessment is relevant. The actual valid methods are

- INGEDE Method 11: "Assessment of Print Product Recyclability—Deinkability Test" (INGEDE 2012)
- INGEDE Method 12: "Assessment of the Recyclability of Printed Paper Products—Testing of the Fragmentation Behaviour of Adhesive Applications" (INGEDE 2013)

Printed paper products and applied adhesive applications (if present) are treated in laboratory scale, close to industrial conditions. This means in both tests pulping at high consistency by addition of an alkaline deinking chemistry. The procedure of the deinkability test (Figure 12.4) is performed with the printed paper product and includes a simple one-loop flotation process. After flotation deinking, the optical quality of the deinked pulp, the filtrate discoloration, and the ink elimination are evaluated. The test on the fragmentation behavior of adhesive applications (Figure 12.5) is performed with the adhesive application of the print product (or parts of it), complemented by virgin fiber–based copy paper. After pulping the macrosticky area distribution is evaluated, especially below 2000 µm equivalent circle diameter, because larger particles are removable almost completely in industrial screening plants.

For packaging paper products the test method is published as ZELLCHEMING Leaflet RECO 1 (2/2014) (ZELLCHEMING 2014) or as EcoPaperLoop Method (Ecopaperloop 2014; EPL 1). The packaging products are treated in laboratory scale close to industrial conditions with a low consistency pulper and no chemicals (Figure 12.6). For the test, the complete or manifold packaging paper or board

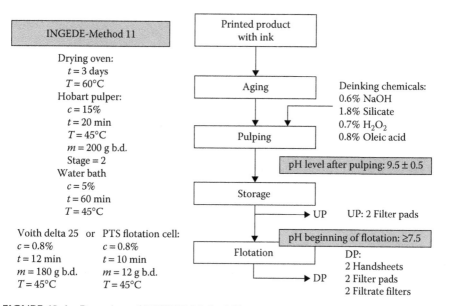

FIGURE 12.4 Procedure of INGEDE Method 11.

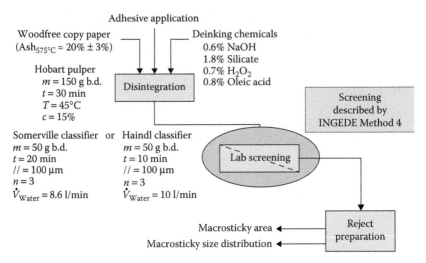

FIGURE 12.5 Procedure of INGEDE Method 12.

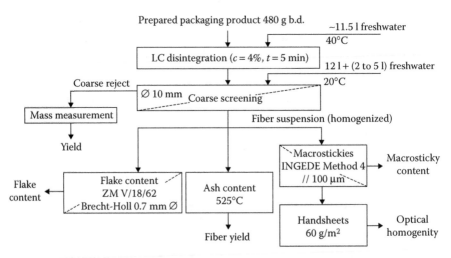

FIGURE 12.6 Procedure of EcoPaperLoop Method 1 (and ZM RECO 1).

product is used. After pulping the coarse reject, flake content, macrosticky area distribution, and the optical homogeneity are evaluated.

All test results of the methods presented are evaluated according to procedures published by the ERPC:

- ERPC document from March 2009: "Assessment of Printed Product Recyclability—Deinkability Score" (ERPC 2009)
- ERPC document from July 2011: "Assessment of the Recyclability of Printed Paper Products—Scorecard for the Removability of Adhesive Applications" (ERPC 2011)

- ERPC document on the assessment of packaging paper and board products is in progress and should be available in 2015 under the given URL (ERPC 2015)

All three assessment procedures have for the various investigated parameters threshold and target values, depending on different categories of printed products or packaging products. All numerical results are transferred into positive score points and added, resulting in one single number—the total score. If a threshold value is missed the print product or the packaging product fails the test and the calculated score of this parameter becomes negative. The rating of the total score points is also identical for all three assessments:

- 71 to 100 score points = Good
- 51 to 70 score points = Fair
- 0 to 50 score points = Tolerable

In the case of at least one single negative score the paper product always failed the test. In the assessment on deinkability a negative test result of a print product means "not suitable for deinking, but the product may be recyclable without deinking." For a negative fragmentation result of an adhesive application in a print product the judgment is "insufficient removable adhesive application." For packaging paper or board products with a negative test result the rating is still under discussion.

Finally it is worth to mention that these test methods and the assessment procedure with the longest history (for printed paper products) now also become part of eco-labels such as the EU flower for printed paper (European Commission 2012), or the German Blue Angel for printing paper predominantly from paper and board for recycling (N.N. 2014), which also covers print products. This upgrading of recyclability tests as one additional parameter for products achieving an eco label will support the manufacturing of more recycling friendly paper products and will help to close the paper recycling loop further.

12.3   COLLECTION SYSTEMS FOR RECOVERED PAPER

The systems used for collection of used paper throughout the world cover a very wide range from nonorganized collections with hand trucks or horse trucks to sophisticated systems using modern technology, which are also attended with comprehensive legislation. Even in Europe in some countries like Bulgaria horse trucks are still used for paper collection (Prinzhorn 2013).

The collection system and the collection culture have, of course, a high impact on the quality of recovered paper. One major aspect is that if paper is collected separately from other materials, contamination can be avoided to a large extent. And a second aspect is the separation of brown and white paper grades. Here it has to be pointed out that brown fibers from packaging products cannot be used as raw material for the production of graphic paper grades. Up till now all technical

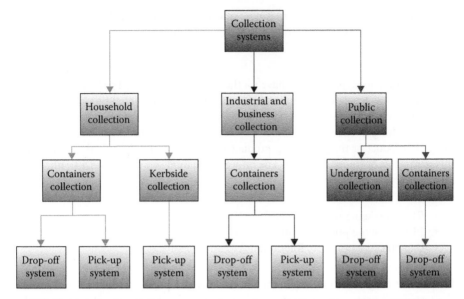

FIGURE 12.7 Basic collection systems for used paper. (Data from PITA, *COST Action E48—The Future of Paper Recycling in Europe: Opportunities and Limitations*, Caric Press, Dorset.)

solutions for bleaching or decolorizing brown fibers are not working effectively on a mixture of brown and white paper grades. Therefore recovered paper for production of white recycling grades should not contain more than about 2.5% of brown fibers.

An overview of the different types of collection systems can be found in Figure 12.7. The most common collection systems are briefly explained as follows (more details can be found in Table 12.3).

In Germany, recovered paper is collected in a special bin ("blue bin"). Waste packaging materials are collected in another bin ("yellow bin"). In addition to these bin systems, there is a bring system with containers, which are distributed throughout the country.

In France used paper and board are collected together with used packaging materials (plastic bottles, metal packaging, liquid packaging, etc.) in one "yellow bin." This is the reason why papers and board in such bins are often contaminated by leftovers of meals or drinks. Very often such organic materials stick on the recovered paper. After collection of these yellow bins there is a special process step, the so called hollow versus flat body sorting, where paper and board are separated from used packaging materials.

In Great Britain in the so-called kerbside collection system, all used packaging materials are collected together with glass and paper and board. This results in poor quality due to breakage of glass. Paper and board from such collections very often have a very low quality and are often not used for the production of recycling paper in Europe. Large quantities are exported to Asia (China). But these low-quality

TABLE 12.3

Benchmark of the Recovered Paper Collection Systems in Different European Countries

	BE	DE	ES	PT	SE	UK	FR
Population	10 574 595	81 757 600	46 030 109	10 637 000	9 360 113	62 041 708	63 961 859
Scheme							
Type of collected flow	Paper/ cardboard	Paper/ cardboard	Paper/ cardboard	Paper/ cardboard	Paper	Kerbside collection + paper/ cardboard	Kerbside collection + paper/ cardboard + paper
Separate collection of graphic papers							
Municipal collection type	Household and business	Household and business	Household and business	Household and business	Household and business	Household	Household and business
Division of collection modalities (kerbside and bring system)	75%–25%	65%–35%	9%–88%	15%–85%	37%–63%	80%–20%	65%–35%
Costs (€/t of graphic paper)	Collection	Collection	Collection	Collection	Collection + sort	Collection	Collection + sort
Operational costs	57	55	85	95	73	59	190
Industrial earnings	116	70	90	80	73	60	80
Average waste costs for association	−59	−15	−5	15	0	−1	111
Global management of graphic papers							
Quantity (kg/cap/annum)							
Occurrence in the municipal cycle	89	70	20	18	66	60	47
Separate collection	52	53	13	4	47	42	23
Papers in the carryover consumer waste	37	18	7	14	19	19	25

(Continued)

TABLE 12.3 (Continued)
Benchmark of the Recovered Paper Collection Systems in Different European Countries

	BE	DE	ES	PT	SE	UK	FR
Population	10 574 595	81 757 600	46 030 109	10 637 000	9 360 113	62 041 708	63 961 859
Costs (€/cap/an)							
Separate collection	−3.08	−0.79	−0.07	0.05	0	−0.04	2.52
Papers in the carryover consumer waste	4.78	2.30	0.41	0.74	1.12	1.88	2.03
Total costs	1.70	1.51	0.35	0.80	1.12	1.84	4.54
Producer's contribution «advanced waste polluter liability»	0	0	0	0	0	0	1.00
Total net costs	1.70	1.51	0.35	0.80	1.12	1.84	3.54
Global management of municipal waste							
Quantity of municipal waste	489	587	652	500	482	526	535
Costs (€/hab/an)	82	130	43	57	75	47	86
Additional information							
Treatment of carryover consumer waste							
Incineration (%)	100%	69%	6%	21%	77%	15%	60%
Organic valorization (%)	0%	31%	40%	9%	23%	8%	
Land filling (%)	0%	0%	54%	70%	0%	77%	40%
Incineration costs (€/t)	130	135	59	138	60	103	94
Organic valorization costs (€/t)		122	40	32	60	40	
Land filling costs			67	30		106	64

Source: Monier, V., and R. Schuster, Benchmark européen de l'économie de gestion des déchets papiers, *ECOFOLIO*, 1–14. http://www.industrie.com/emballage/mediatheque/3/5/0/000008058.pdf, 2012.

recovered paper grades today do not meet the increasing quality demands of Chinese paper mills. This is questioning the British system of exporting unsorted low quality paper and board.

In Poland the share of recovered paper and board is around 170,000 tons in 2010. In Germany in the same year 15,448.000 tons have been recovered. This shows the high potential for increasing the recovered paper utilization in Poland.

In Switzerland there is a system of collecting paper bundles directly from households for decades, which results in excellent quality of the recovered graphic paper.

The paper collection situation in France has been analyzed by Monier and Schuster (2012).

Systems for collecting recovered paper can even differ from region to region in one country.

In England, for example, the "pink bag" is used in Essex where used packaging materials and paper and board are collected in a similar way as in France. In Devon a collection system with separate collection of recovered paper comparable to the system in Germany can be found. This is also the case in Thonon in France.

For these comingled collection systems or separate collection systems often the terms "single stream" and "dual stream" collection are used. In a dual-stream system, recycling material is collected in two streams, where one stream typically is paper and board and the other stream is a mixture of packaging materials (plastic, glass, metal; KCI 2009). In a single-stream collection, paper and board are collected together with all other packaging materials.

There is still a discussion on the efficiency of dual-stream collections in comparison to single-stream collections. One study (Byars 2012) shows findings where single-stream collections can only achieve recycling shares of 13%–39%. Dual-stream collections have a much higher recycling ratio in the range of 30%–40%. There are several other authors who underline the poor quality and the low recycling rates of single-stream collection systems (Morawski 2009, 2010; Pledger 2011).

In North America and Australia, collection systems similar to those in Europe can be found. In other continents collection of waste materials belongs mainly to the informal economy.

Of course, there is also collection of so-called preconsumer recovered paper. According to PITA (2010), "this paper arising from industrial and commercial sources is the easiest, cleanest and most economical to collect." It is defined as paper and board material recovered from manufacturing and converting. Such recovered paper, for instance, not sold issues of magazines or converting residues, is collected by direct pickup from the recovered paper dealers.

According to PITA (2010) no direct correlation could be found between the collection system used and the collection rate in the European countries investigated in this survey. Essential for a good quality of recovered paper is a robust collection system. But the kind of system does not seem to be really important. For obtaining high collection rates not only preconsumer recovered paper has to be collected but also office and household recovered papers.

Even more important than the collection system itself is the environmental awareness and education of people. In a survey of the COST Action E48 on general parameters influencing the future competitiveness of paper recycling, the environmental awareness of people and consumer acceptance were ranked, and the authors of this study (PITA 2010) state that "household collection could be substantially improved by increasing the environmental awareness of consumers."

12.4   DRY SORTING TECHNOLOGIES

Technology used for processing recovered paper before its utilization in paper mills ranges from simple compaction for making transport more effective up to modern sorting plants including optical sorting with camera and near-infrared sensors. In Germany until the late 1970 collected recovered paper was only roughly checked for impurities, for instance, by unloading a truck on the floor of a storage, checking the content, if necessary separating some obvious impurities, and then pressing it to bales or directly reloading on a truck. Since recovered paper was mainly used for packaging grades no other technology was needed and sorting plants looked as simple as shown in that Figure 12.8.

In 1970 in Germany and also in other countries flotation technology was introduced into paper mills for the "deinking" of graphic recovered paper. This new technology created a need of the paper industry for graphic recovered paper grades since only those are deinkable and recovered board or corrugated materials cannot be used for this deinking process. Graphic recovered paper can, of course, be easily collected from industrial paper processing (residues from printing houses, collecting over-issues of magazines) or from households if the consumers are sorting the paper themselves. This system is still in practice in some regions in Switzerland today, where bundles of graphic papers are directly collected from the houses. The quality of such recovered paper is extremely good, but also cost is high and this system requires support by the consumers and is not accepted everywhere. This problem and the higher demand of recovered paper from paper mills lead to the development of the next step of sorting plants. As shown in Figure 12.9 before the pressing station in the sorting plant a conveyor belt can be found, which is loaded by a fork lift and continuously transports the loose recovered paper to a manual sorting station.

FIGURE 12.8   Dry recovered paper sorting plant: pressing station.

FIGURE 12.9   Dry recovered paper sorting plant: continuous manual sorting and pressing.

At such sorting stations the employees separate impurities and corrugated or board materials from the graphic paper. Depending on the speed of the conveyor belt and the layer thickness on the belt the quality of the sorted paper can vary (and, of course, also the production of the line). Figure 12.10 shows such a manual sorting installation.

Over the years due to a change of customers' shopping behavior and the trend toward more products being ordered by customers and shipped directly to their homes, the share of packaging material boards and folding boxboards in the recovered paper collected from households increased. This increasing share caused problems for the manual sorting and resulted in the development of devices for separating typical graphic papers (newspapers, magazines, etc.) from packaging papers. By devices like drum screens, flat screens, or even ballistic separators, larger packaging materials as boxes could be mechanically separated and released for manual sorting. Figure 12.11 shows a typical layout of a sorting plant including screen separation of larger board or box pieces.

Figure 12.12 shows a typical ballistic separator. In this machine, larger boxes and packaging materials are separated from smaller newspapers, magazines, and so on, which can pass the openings.

Since about 2002, optical sensors and image-processing methods have been introduced in recovered paper sorting for further automatization of production lines. Figure 12.13 shows typical processing steps for dry sorting installation with latest technology.

According to Wagner and Schabel (2006) today different material properties and processes are used to separate mixtures of recovered paper. Table 12.4 provides an overview.

The technology of optically analyzing particles on a conveyor belt and then separating them has been used successfully for years for sorting, for example, of plastics as well as for collected waste glass, before it was introduced to the sorting of paper. Figure 12.14 shows a typical concept for such an optical system included in a recovered paper sorting installation.

FIGURE 12.10 Photo from manual sorting of recovered paper.

FIGURE 12.11 Sorting plant including preparation, mechanical prescreening, manual sorting, and bailing.

FIGURE 12.12 Photo of a ballistic separator used for separating larger board and packaging materials from graphic papers.

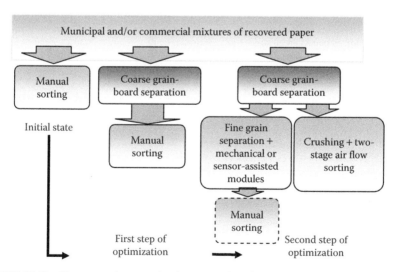

FIGURE 12.13 Process and automatization steps of modern recovered paper sorting.

TABLE 12.4

Material Properties and Processes Used to Separate Mixtures of Recovered Paper

	Screen Classification	Gap Technique (Vertical Adjustment)	Paperspike	Air Separator	Color Sensor	CMYK Sensor	NIR Sensor	Air Stream Sorter
Size	X	X		X				X
Stiffness	X		X					
Color					X	X		
Composition							X	
Weight		X		X				X

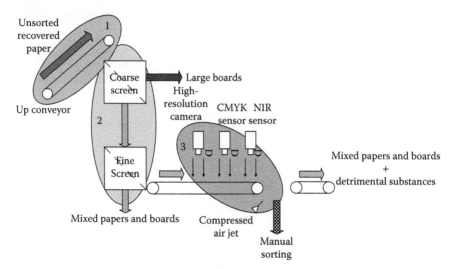

FIGURE 12.14 Automation of recovered paper sorting by coarse and fine screening and recognition with automatic removing by means of compressed air jet.

In Figure 12.14 three different optical systems are used. Today these systems typically are as follows:

- High-resolution color cameras
- Color sensor (CMYK)
- Near-infrared (NIR) sensor

The images recorded with a high-resolution camera are evaluated using the methods of image processing and pattern recognition. The most important property here is the color of the detected object. With a color camera, it is possible to detect brown and gray boxboard as well as dyed throughout papers. According to the manufacturers, distinguishing between gray newspapers and gray boxboard is a problem. Mistakes can be made also in the color. For example, a brown area in a magazine may mistakenly be identified as brown boxboard. It is not possible to distinguish printed boxboard products from color magazines and safely eject them.

The CMYK sensor is able to recognize whether an object was printed with three or four colors. CMYK stands for the colors of cyan, magenta, yellow, and black that are commonly used in color printing (so-called three-color printing manages without black). In view of the fact that printing on boxboard does not usually require an extremely high-quality printed image, this is done mostly by the three-color printing process. Whereas the high-resolution color camera can possibly mistake brown boxboard for brown-printed magazines, safe detection with a CMYK sensor is considerably more likely, as brown board is seldom printed by the costly four-color process. Similarly, the recognition of colored papers is so perfectly possible because these are never printed by the four-color process, whereas identical colors in magazines are printed by the four-color process. The combination of CMYK sensor and high-resolution camera offers a higher degree of recognition safety due to redundancy. Both systems can recognize colors and mutually support each another in the evaluation. Therefore the combination of these sensors offers a relatively safe recognition of boxboard and dyed throughout papers. It also allows the ejection of a certain type of printed boxboard.

NIR sensors detect the adsorption in the infrared wavelength range. They are capable of recognizing the entire spectrum of materials familiar from sorting valuable materials from domestic waste collections. Such materials are especially plastics and beverage cartons. Also sheer layers, such as are commonly used on color packaging (e.g., frozen food packaging), can be recognized, provided that the sensor is calibrated accordingly. Therefore a NIR sensor is used in recovered paper sorting to recognize "foreign" materials and their compounds with boxboard.

Today there are ongoing discussions if the installation of costly optical sensors is economical or not. As it can be seen so far, optical sensors in addition to the mechanic sorting have their advantages with varying input material composition and quality while manual sorting is very flexible. Most sorting plants today, even if they are equipped with mechanical and optical sorting, still have manual sorting at the end for securing the output quality. Figure 12.15 shows a typical layout of modern single line sorting plant.

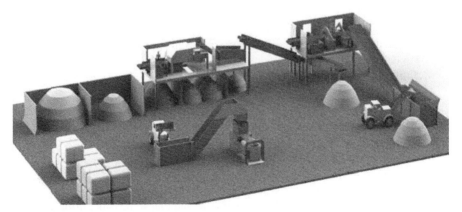

FIGURE 12.15 Layout of a modern single-line dry sorting installation with preparation, mechanical presorting, optical sorting, and manual postsorting and finally bailing.

12.5 CLASSIFICATION OF PAPER FOR RECYCLING IN EUROPE (EN 643)

Until the end of last century, European countries had their own national lists of recovered paper grades. Due to the fact that recovered paper had become more and more an international traded commodity it was more than consequent to implement in 1999 for the first time a common European list of recovered paper grades (EN 643), which was more or less based on the previous German grade list.

The EN 643 was revised and extended in 2014 (BSI 2014). One focus in the revision of EN 643 was to be in accordance with the fundamental changes in waste legislation. The Waste Framework Directive (Directive 2008/98/EC; European Commission 2008) introduces a procedure for defining end-of-waste (EoW) criteria, which a given waste stream needs to fulfill in order to finish to be waste. The EoW criteria require compliance with EN 643, the provision of information on material that has finished to be waste, and the implementation of a quality management system (BSI 2014).

In this context it is important to realize that with the revision of EN 643 the terminology has been changed from "recovered paper and board" to "paper and board for recycling." The new term was created to highlight the intended use of this secondary raw material for paper recycling only (BSI 2014). The term "paper and board for recycling" is defined as natural fiber-based paper and board suitable for recycling and consisting of paper and board in any shape and products made predominantly from paper and board, which may include other constituents that cannot be removed by dry sorting, such as coatings, laminates, spiral bindings, and so on.

Beside an increase of the listed number of total grades from 67 (old EN 643) to 95 (new EN 643), the actual grade list defines for the first time the maximum allowed amount of nonpaper components and of unwanted material for each grade in percent. The general structure of the grade list has been maintained as well as the systematic numerical code with three number combinations for the clear identification of a grade. The existing five different groups of grades provide the first figure of the codification of a grade according to the following list:

1. Ordinary grades (Group 1: 14 grades)
2. Medium grades (Group 2: 20 grades)
3. High grades (Group 3: 28 grades)
4. Kraft grades (Group 4: 9 grades)
5. Special grades (Group 5: 24 grades)

It is necessary to mention that grades of paper and board for recycling from Group 5 can only be recycled by using specific processes, or can cause some particular constraints to recycling in most cases. This means that typical paper and board mills using paper for recycling are not in the position to handle special grades of Group 5. Nevertheless, the inclusion of these grades in the EN 643 is justified by the existence of a significant European market for these grades and that some paper mills are specialized on the recycling of such paper and board grades.

The following two numbers describe the particular grade within the group category. Finally, the two last numbers determine a subgrade within the grade category. In cases where no subgrade exists, the two final numbers are always "00." The following code describes as an example the codification system:

```
1.06.01:       1 = group of ordinary grade
               06 = grade magazines
               01 = subgrade magazines without glue
```

The source of paper and board for recycling is important for paper production and some mills may ask for a declaration from the supplier about the origin of the material, in relation to national regulations or standard requirements. Paper sorted from refuse collection systems is not suitable for use in the paper industry as a general requirement. Paper and board for recycling originating from multimaterial collection systems has to be specifically marked and it is not allowed to mix it unmarked with other grades.

In general, paper and board for recycling should be supplied with no more than the naturally occurring moisture level. Where the moisture is higher than 10% (of air-dried weight), the additional weight above this limit may be claimed back (BSI 2014).

The revised EN 643 comprises also some terms and definitions about material that is limited or not allowed in paper and board for recycling. Prohibited materials are any materials that represent a hazard for health, safety, and environment, such as medical waste, contaminated products, organic waste including foodstuffs, bitumen, and toxic powders and are not permitted at all in any grade of paper and board for recycling (BSI 2014).

In contrast, maximum tolerance levels in weight percentages have been introduced for each paper grade for nonpaper components (0.25%–3.0%) and total unwanted material (0.5%–3.0%). The percentage of nonpaper components is always part of the total amount of unwanted materials and cannot be added together. Nonpaper components are any foreign matter in paper and board for recycling, which is not constituent part of the product and can be separated in principle by dry sorting, such as metal, plastic, glass, textiles, wood, sand, and building materials or synthetic materials. Unwanted materials, sometimes (especially in North America) also called "out throws," may comprise the following:

- Nonpaper components
- Paper and board detrimental to production
- Paper and board not according to grade definition
- Paper products not suitable for deinking (if applicable) (BSI 2014)

The description of *paper and board not according to grade definition* represents paper and board or products thereof, which are not included in the description of the specific grade of paper and board for recycling. *Paper and board detrimental to production* comprises those papers and boards that have been recovered or treated in such a way that they are, for a basic or standard level of technical process equipment, unsuitable as raw material for the manufacture of new paper and board products or are actually damaging machinery equipment or whose presence makes the whole consignment of paper unusable.

Paper products not suitable for deinking are relevant only for all grades intended for deinking during the recycling process. Primarily these are all paper and board products containing brown, unbleached fibers, which are considered as detrimental for the production of deinked pulp. The suitability of paper products for deinking is a print product characteristic and should be determined according to the ERPC "Assessment of Print Product Recyclability—Deinkability Score" (ERPC 2009). Paper products not suitable for deinking belong to unwanted material. At the current state of knowledge, this refers to the most flexographic, inkjet, liquid toner and some UV-cured print products. Additionally, the average age of newspapers is limited according to EN 643 for some deinking grades to six months from the date of issue (BSI 2014).

If the level of nonpaper components and/or unwanted material exceeds the tolerance level in the grade list the load should not be accepted under the indicated grade designation (BSI 2014). Further guidance can be found in the "Guidelines for paper mills for the control of the content of unusable materials in recovered paper" (CEPI 2008). The buyer and the seller of paper and board for recycling always have to agree on the methods used to determine moisture and/or foreign matter in a load of grades delivered.

12.6 BASIC STOCK PREPARATION PROCESSES FOR RECOVERED PAPER PRODUCTION

After collection and dry sorting, recovered paper is transported to paper mills. There it has to be processed again before it can be sent to a paper machine for the production of recycled paper. This is required because even after dry sorting, recovered paper still contains debris and impurities. Those range from sand, stones, or glass pieces to different types of plastics. The reasons are that such impurities and debris are needed for the use of paper products or for their production, like adhesives and printing inks. All these kinds of materials have to be removed in the final so-called stock preparation processes of a paper mill. This processing cannot be done in one single step. It needs a number of process steps for separating impurities with a wide range of physical properties like dimensions, solubility in water, and so on. Figure 12.16 shows different processes with which recovered paper is treated in the dry sorting plants and in stock preparation and which are applied in what the treatment in paper mills. From the figure it can be seen in which size range particles are separated by which of these processes.

All stock preparation processes are working with paper in form of a suspension. This makes the material to be transportable by pumps. Typical solid content of such suspensions, the so-called consistency, is in the range of 1% solids to about 20%–25% solids. This means that in the stock preparation processes,

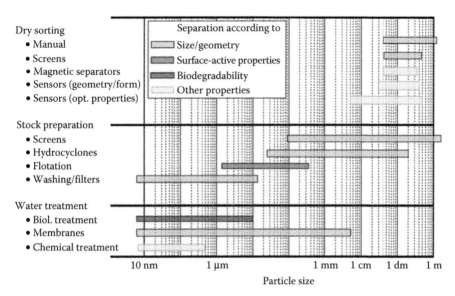

FIGURE 12.16 Operation range of different separation processes for cleaning of recovered paper.

for example, at a consistency of 1% for processing of 1 kg fibers, 99 kg of water has to be pumped.

In order to get an overview over the arrangement of processes and subprocesses in stock preparation systems, Figure 12.17 shows the structure of a recycling line for brown recovered paper grades, which are used for producing test liner.

In this system illustrated in Figure 12.17, after pulping, cleaning, and screening, the pulp is fractionated during two steps in three different size fractions. By this, the long fibers, medium size fibers, and short fibers can be used in different layers if the test liner is produced with a layered structure on the paper machine. In addition to the separation processes, which are used and which will

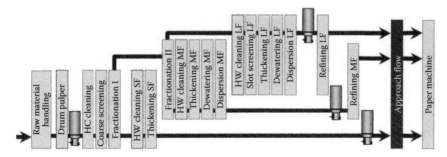

FIGURE 12.17 Schematic demonstration of process arrangements for test liner production. (HC, high consistency; HW, heavy weight; SF, short fiber; MF, medium fiber; LF, long fiber.) (Data from Voith Paper, GmbH & Co.KG, Heidenheim, Germany, 2014.)

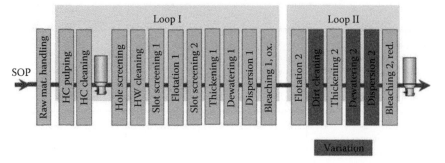

FIGURE 12.18   Schematic demonstration for the stock preparation of graphic paper (so-called deinking). (Data from Voith Paper, GmbH & Co.KG, Heidenheim, Germany, 2014.)

be discussed later, this system also uses refining steps for the long fiber and the medium fiber. This is a surface treatment process that can reactivate the surface of fibers by fibrillating them, which leads to higher specific surface and better binding in the paper.

The stock preparation for graphic recovered paper is a little bit more complex since here also printing inks have to be removed and the clean liner requirements are much higher than for packaging paper. This can be seen in Figure 12.18 where a schematic demonstration for a so-called deinking line is illustrated.

One can see that in such deinking lines, flotation and bleaching are additional process steps compared to stock preparation for brown grades and the whole system is designed in two loops. Each of these loops has its own water circuits and process water is kept within the corresponding loop as much as possible in order to avoid impurities and debris to be carried along in the subsequent loop. In each of the two loops the pulp is treated by flotation, dispersing, and bleaching. These subprocesses have to be applied two times since the quality demands cannot be met with single treatment. Figure 12.19 illustrates the machines and installations of such a two-loop deinking line.

The production capacity of such deinking lines varies a lot. Typical installations have capacities in the range of 500 ton to 2000 ton per day depending on the size of the paper machines these deinking lines are feeding with pulp. During recent years it has been a trend to build paper machines and stock preparation lines bigger and bigger. For the future it is not clear if this trend will continue. There are good arguments for reducing the size of stock preparation lines and paper machines in order to produce paper for a more local demand and avoid transportation of raw materials and products over large distances. Typically, recovered paper mills can be found close to highly populated regions where the used paper of a high number of consumers can be collected and processed with reasonable transport.

In the following sections of this book the most important stock preparation processes are introduced and discussed briefly. For a more detailed description of these processes and more technical data, Schabel and Holik (2010) can be recommended.

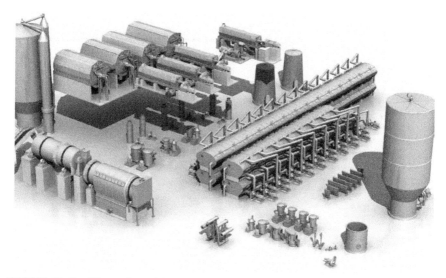

FIGURE 12.19 Illustration of installations and equipment for a two loop deinking line. (Data from Voith Paper, GmbH & Co.KG, Heidenheim, Germany, 2014.)

12.6.1 REPULPING

The first process step has the objective to make a pumpable suspension out of the recovered paper delivered to a paper mill typically in the form of bales or in loose form. This process step is called "pulping" or "repulping" in the case of recovered paper processing. Very often repulping is a batch process where an apparatus like that illustrated in Figure 12.20 is set with paper, water, and sometimes some chemicals like soda or silicate. Then the rotor is started and the introduced mechanical energy creates friction and turbulence inside the vessel, which leads to the disintegration of the paper in single fibers, which are slushed in water and form a pulp suspension.

Typical pulping times range from 15 minutes to 30 minutes and the advantage of batch pulpers is the possibility to adjust process parameters flexible to quality demands or to requirements for variations in incoming paper.

Since continuous processes are always preferred in industry there are also solutions for continuous repulping. This is done in huge drums, so-called drum pulpers, where the paper is sent to a large drum with a diameter in the range of 2.5 m to 3 m at one end together with water and, if necessary, chemicals like soda, alkali, and tensides. The drum rotates and the gentle movement over time leads to the disintegration of paper and the formation of a pulp. Pulping drums are typically followed by screening drums. This can be in one coupled device or in two separate drums installed one after the other. The screening part of the drum has holds that allow the fiber suspension to pass and larger contaminants like plastic bags, beverage cans, and larger stones are separated from the pulp in this very first step. Figure 12.21 shows such a pulping drum.

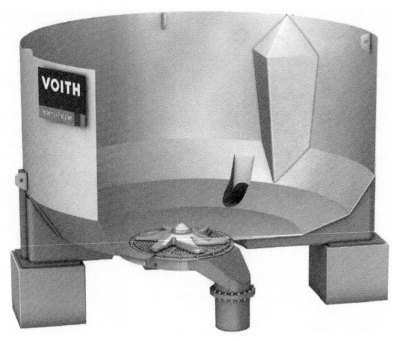

FIGURE 12.20 Illustration of medium consistency pulper for repulping of brown recovered paper grades. (Data from Voith Paper, GmbH & Co.KG, Heidenheim, Germany, 2014.)

FIGURE 12.21 Continuous drum pulper for repulping of graphic paper. (Data from Voith Paper, GmbH & Co.KG, Heidenheim, Germany, 2014.)

12.6.2 Coarse Cleaning

The next step after pulping is called coarse cleaning. In paper-making terminology, "cleaning" means separation of particles with high density from a suspension with hydrocyclones. Hydrocyclones are very old and quite simple devices where the geometry of the device causes the pulp-suspension pump inside it to rotate. Inside a hydrocyclone a vortex flow is formed, which causes centrifugal forces. These centrifugal forces are the reason why particles with higher density than water are moving to the outer wall and then downward where they are collected. Figure 12.22 shows the material separated by a hydrocyclone. For coarse cleaning, hydrocyclones are typically operated at consistencies in the range of 2.5%–4.5% and they can separate high density with a size of several millimeters or bigger like stones, glass, and metal pieces.

The same physical principle is also used for cleaning "at lower consistencies" and with smaller size equipment. Typical heavy weight cleaners or coarse cleaners have diameters of 1.5 m to 3 m and realize centrifugal accelerations of 5 g to 10 g. At lower consistencies and with precleaned pulp, so that plugging of the devices can be avoided, smaller machines with diameters of only a few centimeters up to 20 cm can be used. Then centrifugal forces up to 500 g are possible and high density particles down to a size of about 10 μm to 50 μm can be separated effectively. Figure 12.23 shows the cross section of such a cleaner, which can have a maximum diameter of, for instance, 15 cm.

The separation effect in such device depends strongly on the geometry and on the pressure gradient applied or the throughflow. The physics of the device does not allow to scale the size up or down without varying the separation effect. Because of this physics many devices have to be installed in parallel so that the production of the line can be processed. In typical installations up to 100 of devices are needed for the treatment of the pulp.

A typical arrangement can be seen in Figure 12.24.

FIGURE 12.22   Particles typically separated by heavy weight coarse cleaner.

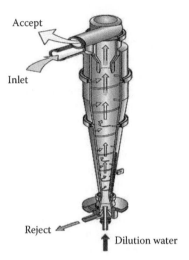

FIGURE 12.23   Cross section of a cleaner. (Data from Voith Paper, GmbH & Co.KG, Heidenheim, Germany, 2014.)

FIGURE 12.24   Typical cleaner arrangement. (Data from Voith Paper, GmbH & Co.KG, Heidenheim, Germany, 2014.)

12.6.3 SCREENING

Screening makes use of geometrical differences between papermaking or pulp fibers and debris particles that are not wanted on the paper machine or in the production process. Screens used in the paper industry can have flat screen plates, which typically have holes with a diameter between 1.0 mm and 4.0 mm or cylindrical with similar holes are slots with a slot size from 150 *μm to 500 μm. The pulp suspension has to be diluted in order to avoid flocculation since flocks could* include debris particles, which then would not be separable by those processes. So typical consistencies for screening processes are in the range of 0.8%–3.5%. The lower consistency is used for highly efficient screening processes where also smaller particles down to 100 μm can be separated from the suspension. Higher consistencies are used when only larger particles have to be separated like the separation of bark particles in a pulp mill. Pressure screens are the type of screens that are most widely used in paper recycling. This means that the whole process is under a small overpressure in order to avoid air cores inside the machines, which can lead to unstable operation. Figure 12.25 shows a picture of a modern pressure screen used for paper recycling.

Figure 12.26 shows a cross section of a screen where the cylindrical basket, a rotor, the feed, and accept and reject pipes can be seen. Since the openings of such screens can be very small, the probability for plugging of those openings by the pulp suspension is very high. In order to avoid this, a moving element, the so-called rotor, in this case with foils with a geometry like airplane wings, is used. When the rotor

FIGURE 12.25   IntegraGuard pressure screen (Voith). (Data from Voith Paper, GmbH & Co.KG, Heidenheim, Germany, 2014.)

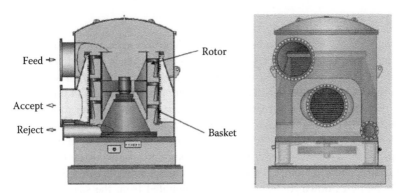

FIGURE 12.26 Cross section and front view of a pressure screen. (Data from Voith Paper, GmbH & Co.KG, Heidenheim, Germany, 2014.)

elements pass a section of the cylindrical screen basket a negative pressure is caused by the hydrodynamics of the rotor foil and the direction of flow is reversed. This reverse flow cleans the screen plate openings for a few milliseconds. Consistency of the pulp, geometry of the screen plate and the rotor, and the circumferential speed of the rotor have to be in the right balance in order to enable efficient screening on one side and good run ability of the machine on the other side.

A key element is the screen plate. There are different manufacturing technologies available for the production of such screen plates. The range of technologies is from machine cutting, welding of bars, laser cutting, and clamping of bars to glueing together structures with adhesives. One major problem for keeping a high precision of the slot dimensions is to avoid heat treatment. Welding or soldering introduces high amounts of heat into the metal, which causes deformation and degreases the precision of the screen basket. Figure 12.27 shows a part and a cross section of a clamped screen plate structure.

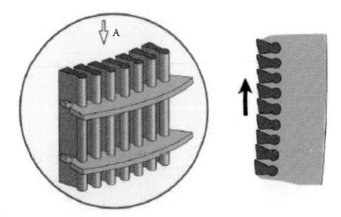

FIGURE 12.27 Screen bar wires embedded in supporting structures and locked by clamping for high-precision screen plate manufacturing. (Data from Voith Paper, GmbH & Co.KG, Heidenheim, Germany, 2014.)

12.6.4 FLOTATION

Flotation is the process most widely used in stock preparation for so-called deinking, which means the separation of printing inks from the pulp suspension and the fibers. Washing processes are also used for this step sometimes but they are not discussed here. The flotation process makes use of the hydrophobic surface of typical printing ink particles. The ink particles are already separated from the fibers in the pulping step by the shear force and friction there. In a flotation device air is now brought into the suspension in form of bubbles in a size range of about 20 μm to 100 μm. These air bubbles also have a hydrophobic surface like ink particles. Now turbulence has to be generated in the flotation device, which creates movement of the particles and causes collisions between ink particles and air bubbles. Because of the hydrophobic surfaces of both types of particles, the printing ink particles mostly stick to an air bubble they have collided with. This agglomerate still has a much lower specific weight than the surrounding water and floats to the top of the device or tank. Of course in the zone of the device where particle air bubble agglomerates are floating, turbulence is unrequested and would potentially disintegrate the agglomerates. In the process many air bubbles and attached ink particles form a dark foam on the top of the suspension surface, which can then be separated by foam collectors or by a simple foam overflow.

Figure 12.28 shows the key elements of such a flotation device. Widely used for aeration of the suspension are injector nozzles, which suck in air, and in a subsequent step diffuser the air is broken down to air bubbles of an appropriate size. The turbulence caused by the step diffuser geometry also causes intense mixing of the suspension and collisions between air bubbles and ink particles. After leaving

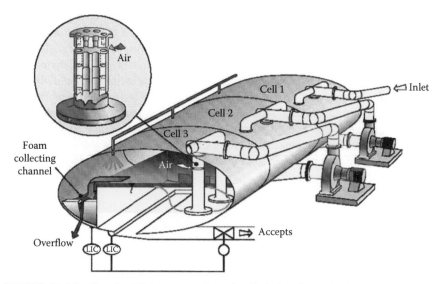

FIGURE 12.28 Sketch of injector nozzle and cell design for a Voith EcoCell flotation device. (Data from Voith Paper, GmbH & Co.KG, Heidenheim, Germany, 2014.)

FIGURE 12.29   Photo of a typical deinking flotation cell. (Data from Voith Paper, GmbH & Co.KG, Heidenheim, Germany, 2014.)

the diffuser nozzle the suspension enters a relatively large tank, which can have an elliptical shape in the design as shown in Figure 12.29 and due to the large cross-sectional area of this device the flow velocity of the suspension goes down and becomes lower than the floating velocity of the air bubbles. This enables a separation of ink and air from the suspension still containing most of the fibers.

Since many of the filler particles used in paper making also have hydrophobic particles this so-called ash is also separated to a significant amount in a flotation device. The rejects of the flotation process, the printing inks, mineral particles from fillers, and some smaller adhesive and other particles cannot be recycled up to today. This "deinking sludge" is dewatered and then mostly incinerated for utilizing the energy of the organic part of this sludge. Typical losses for the deinking process, which is the amount of material separated by the flotation process, are in the range of 20% for deinking processes used for newsprint or other graphic recycling paper and up to 50% for deinking processes used for the production of recycled pulp for tissue paper. In this case not only the ink has to be separated from the pulp but also all the inorganic filler and other particles.

12.6.5   BLEACHING

The final step in some of the stock preparation systems for higher quality graphic recycling paper is a bleaching process. Colors that cannot be separated effectively from the pulp with other mechanic treatments have to be decolorized in order to produce wide recycled pulp from the material. Bleaching is a chemical process where chemical additives are used, which can "destroy" color-creating chemical groups in colors and inks. For oxidative bleaching hydrogen peroxide is often used and for reductive bleaching, sodium dithionite. Some deinking lines even have oxidative and reductive bleaching steps. For starting the bleaching reaction the corresponding chemicals have to be added to the suspension and properly mixed and then the reaction needs time, typically 20 to 30 minutes. This retention time can be achieved by large diameter pipes through which the pulp has to pass. Typical consistencies for the bleaching step are in the range of 20% to 25%, which requires dewatering prior to bleaching by disc filters, screw presses, or extruders. After the bleaching reaction the pulp is diluted again to bring it into a pumpable suspension.

12.7 UTILIZATION OF RECYCLED FIBER PULP IN PAPERMAKING

Chemical pulp was the most important raw material for paper production in the last century. But nowadays it has been replaced by paper and board for recycling. In 2012, 229 million tons of this raw material were used globally (VDP 2014). This volume exceeds the total volume of wood pulp, that is, chemical pulp (151 million tons) and mechanical pulp (30 million tons). These figures show that paper for recycling plays a very important role as raw material for the global paper industry. Processed recycled fiber pulp from paper for recycling acts increasingly as a substitute for virgin fiber pulps.

Figure 12.30 shows the global increase in the use of paper and board for recycling, compared to paper production since 1961 (VDP 2014, 2012, 2002, 1993, 1983;

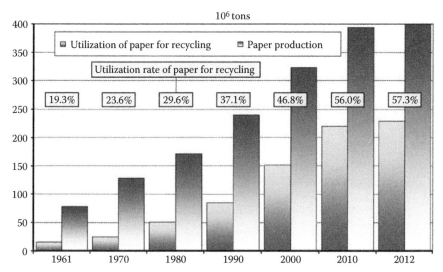

FIGURE 12.30 Global development of the utilization of paper for recycling and paper production (1961–2012). (Data from VDP, Papier 1983—Ein Leistungsbericht, VDP, Bonn, Germany, 1983; VDP, Papier 1993—Ein Leistungsbericht, VDP, Bonn, Germany, 1993; VDP, Papier 2002—Ein Leistungsbericht, VDP, Bonn, Germany, 2002; VDP, Papier 2012—Ein Leistungsbericht, VDP, Bonn, Germany, 2012; VDP, Papier 2014—Ein Leistungsbericht, VDP, Bonn, Germany, 2014; N.N., Review Number 1963, *Pulp and Paper International*, 5, 83–212, 1963; N.N., Review Number 1972, *Pulp and Paper International*, 14, 67–185, 1972.)

N.N. 1972, 1963). Globally the use of paper for recycling increased by approximately 5.4% annually whereas annual paper production growth was only 3.2%.

The paper industry is the exclusive relevant user of paper and board for recycling as a secondary raw material—at least in terms of material recycling. Various processing systems prepare recycled fiber pulp, also named as secondary fibers, for the production of a variety of paper and board grades. These processing systems use different grades of paper and board for recycling, which contain either chemical or mechanical fibers or, mainly, an undefined mixture of both. Paper for recycling grades intended for deinking is typically treated by an ink-removal process to produce a sufficiently bright deinked pulp quality for the production of graphic paper grades or tissue. Brown or mixed grades of paper for recycling are mainly used for the production of new packaging paper and board grades.

Some paper and board grades can be made from secondary fibers exclusively. In Europe, this includes paper grades such as corrugated medium and test liner for the production of corrugated boxes or newsprint. For newsprint as well as for other paper or board grades, blends of recycled and virgin fiber pulps are also used. The proportion of recycled fiber pulp in the raw material furnish can vary from about 5% for fine papers to 100%, depending on the paper grade or geographic region.

Figure 12.31 shows the largest consuming countries of paper and board for recycling. Far ahead on top is China (75 million tons), followed by the United States, Japan, and Germany (26–16 million tons each).

The listed 12 countries used in total 178 million tons, corresponding to 78% of the global consumption of paper and board for recycling.

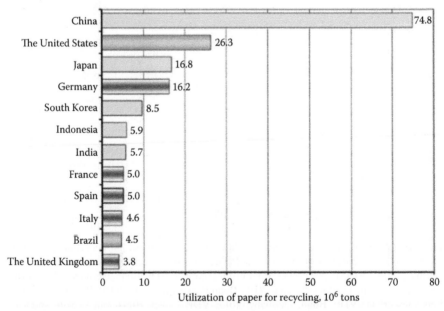

FIGURE 12.31   The 12 largest consumer countries of paper and board for recycling in the world (2012). (Data from VDP, Papier 2014—Ein Leistungsbericht, VDP, Bonn, Germany, 2014.)

Consideration on the use of paper and board for recycling and the recovery of used paper products require definitions. Statistical definitions can be related to the world, a continent, a country, or a certain region (e.g., Europe) or to a certain category of paper products (e.g. graphic papers or newsprint). Differentiation is given by the following statistical parameters:

- Utilization rate of paper for recycling, in percentage, is the amount of paper and board for recycling used as raw material in the paper industry, in tons, divided by paper production, in tons, on an annual basis.
- Collection rate of paper for recycling, in percentage, is the amount of collected paper and board for recycling, in tons, divided by paper consumption, in tons, on an annual basis.

From these definitions it becomes obvious that the utilization rate is related to the usage of paper for recycling in the paper production, whereas the recovery or collection rate is related to the amount of collected paper for recycling related to paper consumption. Both rates can be affected directly either by the paper industry by the use of more or less paper for recycling in paper production or by the waste management industry collecting more or less used paper and board products from consumers, industry, or administrations.

Figure 12.32 shows the utilization rate for the three largest paper production regions in the world where the paper industry uses 83% of the global volume of paper and board for recycling: Asia, Europe, and North America. Paper production in these three regions is annually between 85 and 158 million tons each, resulting in utilization rates of paper for recycling of 72% in Asia, 51% in Europe, and 34% in North America.

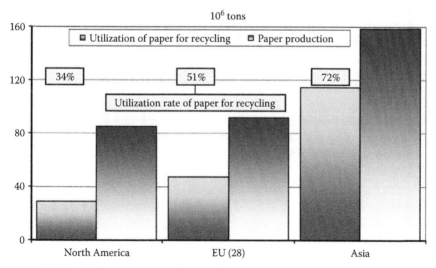

FIGURE 12.32 Utilization of paper for recycling and paper production in Asia, EU countries, and North America (2012). (Data from VDP, Papier 2014—Ein Leistungsbericht, VDP, Bonn, Germany, 2014.)

Utilization rates of paper and board for recycling for different countries should not be compared without further comment. It is important to know the structure of the production program of the different national paper industries in the main product categories of packaging papers and board, graphic papers, and household and hygiene papers as well as specialty papers because the utilization rates for these product segments differ significantly.

In Figure 12.33 for the CEPI countries (Austria, Belgium, Czech Republic, Finland, France, Germany, Hungary, Italy, the Netherlands, Norway, Poland, Portugal, Romania, Slovak Republic, Slovenia, Spain, Sweden, and the United Kingdom) the utilization of paper and board for recycling by the main paper categories is presented and additionally the utilization rates are given (CEPI 2013). The x-axis is a summarizing ordinate representing the total paper production in the CEPI countries of about 91 million tons in 2012, subdivided into the production of the various paper categories. The broader a single paper category, the larger the paper production (e.g., the biggest category, with 30 million tons, is other graphic papers). For each paper product category (e.g. newsprint) the used volume of paper for recycling is shown on the y-axis. In the circles the corresponding utilization rate is given. It becomes obvious that the highest volume of paper for recycling is used in the production of case materials (24 million tons) with a utilization ratio of 94%. About the same utilization rate of 97% is in newsprint, but with a significant lower volume of 7.9 million tons. In the other paper categories between 1.9 and 4.1 million tons of paper and board for recycling are used, resulting in utilization rates between 11% (other graphic papers) and 53% (wrapping and other packaging papers).

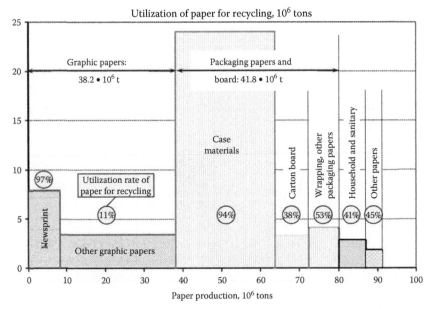

FIGURE 12.33 Utilization of paper for recycling by paper grades in the CEPI countries (2012). (Data from CEPI, *Annual Statistics 2013—European Pulp and Paper Industry*, CEPI, Brussels, Belgium, 2013.)

Traditionally, packaging papers and board have the highest utilization rate of paper and board for recycling. In the CEPI countries this ratio has reached 75% on average of the categories case materials, carton board, wrapping, and other packaging papers. Demanding quality specifications have to be fulfilled by these recycled fiber-based paper and board grades to ensure trouble-free converting, for example, to corrugated board boxes or folding boxes and adequate functional characteristics of the finished products. For food packaging products based on secondary fibers, additional regulations have to be fulfilled for product safety reasons. It is predictable that the utilization ratio will not increase significantly.

The second highest utilization rate has the category other paper and board grades, comprising gypsum liners, special papers for waxing, insulating, and roofing, achieving 45% utilization rate in the CEPI countries. Despite the fact that in this category are also many speciality paper grades such as cigarette paper, filter paper, or banknote paper for which no recycled fibers are used, a slight increase of the utilization rate was observed over the last years.

Sanitary and household papers (41%) are on the third place regarding the utilization rate of paper for recycling. Due to high yield losses in the flotation and washing steps, the proportion of recycled fibers in hygiene paper is not higher than 25% on average. The figure of 25% deinked pulp comes from about 60% yield of 41% utilization rate. In this product category, a decrease of the utilization rate has been observed in the last decade due to changes in the consumer behavior and economical drawbacks such as low process yield and high disposal costs by the use of paper for recycling.

The utilization rate for graphic papers is 30% calculated as average of newsprint and other graphic papers. Due to the wide spectrum of the paper grades produced, a distinction is necessary for this product segment because newsprint has reached a level of 97%, actually. It is expected that it will not considerably increase further. The utilization rate of the other graphic papers is rising slowly but continuously and averages at 11%. Included among this other graphic papers are wood-containing and wood-free papers that are coated or uncoated. Some already use 100% recycled fiber furnish, for example, recycling copy or low weight coated for offset printing papers; others use blends of different raw materials (e.g., super calendered papers), and the final category is based on 100% virgin fiber pulp such as coated art papers.

12.8   FUTURE PERSPECTIVES FOR PAPER RECYCLING

Due to the increasing interest on renewable resources from many industry branches, it can be expected that there will be a need for paper recovery and recycling in the upcoming decades. Consumer habits are changing rapidly, for instance, the habits for accessing latest news. It seems that classical newspapers are requested less and less while actual news is read via online technologies on computers, laptops, or smartphones. But the increasing demand for packaging solutions, where paper and board can offer very interesting material properties, can compensate or even over-compensate the decreasing demand for newspapers and the corresponding paper products.

With the expected growth of the global paper production the share of recycled paper products will increase, since societies have to care more about sustainability and efficient utilization of raw materials. This will lead to more and more closed material loops and a concentration of contaminants in those more or less closed loops. In order to keep the cycles running, the design and production of recyclable or recycling friendly paper and board products will gain importance. This is also true for separation processes, which can help separating impurities and debris from recovered paper. Here the technology push in sorting technology with different sensors can help to solve future challenges. But also an improvement of collection systems will be needed. From today's point of view separate collection of paper and board from households, industry, and commerce seems to be a must to ensure adequate quality of paper for recycling.

REFERENCES

BSI. 2014. BS EN 643:2014 Paper and board—European list of standard grades of paper and board for recycling. London: British Standards Publication.

Byars, Steven A. 2012. Single-stream versus dual stream recycling management: Do the benefits justify the means? Waste360. http://waste360.com/print/recycling-facilities-mrfs/single-stream-versus-dual-stream-recycling-management-do-benefits-justify. (Accessed October 17, 2012).

CEPI. 2008. Guidelines for paper mills for the control of the content of unusable materials in recovered paper. http://www.cepi.org/system/files/public/documents/publications/recycling/2008/UnusableMaterialsContent.pdf (Accessed September 19, 2014).

CEPI. 2013. *Annual Statistics 2013—European Pulp and Paper Industry*. Brussels, Belgium: CEPI.

Ecopaperloop. 2014. Recyclability test for packaging products. http://www.ecopaperloop.com (Accessed September 22, 2014).

ERPC. 2009. Assessment of print product reyclability—Deinkability score: User's manual. http://www.paperforrecycling.eu/uploads/Modules/Publications/ERPC-005-09-115018A.pdf (Accessed September 18, 2014).

ERPC. 2011. Assessment of the recyclability of printed paper products—Scorecard for the removability of adhesive applications. http://www.paperrecovery.org/uploads/Modules/Publications/Removability%20Adhesive%20Applicationsfinal.pdf (Accessed September 21, 2014).

ERPC. 2015. Assessment of the recyclability of packaging products—Recyclability scorecard (in progress). http://www.paperforrecycling.eu/publications/erpc-publications (Accessed September 21, 2014).

European Commission. 2001. Reference document on best available techniques in the pulp and paper industry. Brussels, Belgium: European Commission.

European Commission. 2008. Directive 2008/98/EC of the European Parliament and of the Council of 19 November 2008 on waste and repealing certain Directives, OJ L 312. http://eur-lex.europa.eu/legal-content/EN/TXT/?qid=1408652042838&uri=CELEX:32008L0098 (Accessed September 19, 2014).

European Commission. 2012. Ecologiacal criteria for the award of EU Ecolabel for printed paper. Document C(2012)5364. http://eur-lex.europa.eu/legal-content/EN/ALL/;ELX_SESSIONID=pML9T9VGBnPshLQSDsJCGz5hcqQKSy3rrfb4BML9LqMfkhWRl4F R!1100664340?uri=CELEX:32012D0481 (Accessed September 22, 2014).

European Commission. 2013. Best available techniques (BAT) reference document for the production of pulp, paper and board. Brussels, Belgium: European Commission.

Gromke, Uli, and Andreas Detzel. 2006. *Ökologischer Vergleich von Büropapieren in Abhängigkeit vom Faserrohstoff*. Heidelberg, Germany: Institut für Energie- und Umweltforschung Heidelberg GmbH.

INGEDE. 2012. Assessment of print product recyclability—Deinkability test: INGEDE method 11. http://www.ingede.com/ingindxe/methods/ingede-method-11-2012.pdf (Accessed September 21, 2014).

INGEDE. 2013. Assessment of the recyclability of printed paper products—Testing of the fragmentation behaviour of adhesive applications: INGEDE method 12. http://www.ingede.com/ingindxe/methods/ingede-method-12-2013.pdf (Accessed September 19, 2014).

KCI. 2009. *Material Recycling Facility Technology Review: Prepared for Pinellas County.* Tampa, FL: Kessler Consulting.

Monier, Véronique, and Rita Schuster. 2012. Benchmark européen de l'économie de gestion des déchets papiers. *ECOFOLIO*: 1–14. http://www.industrie.com/emballage/mediatheque/8/5/0/000008058.pdf (Accessed September 18, 2014).

Morawski, Clarissa. 2009. Understanding economic and environmental impacts of single-stream collection systems. http://www.container-recycling.org/index.php/publications/cri-publications (Accessed September 19, 2014).

Morawski, Clarissa. 2010. Single-stream uncovered. *Resource Recycling,* February, 21–25.

N.N. 1963. Review Number 1963. *Pulp and Paper International* 5(9):83–212.

N.N. 1972. Review Number 1972. *Pulp and Paper International* 14(8):67–185.

N.N. 2014. Druck-und Pressepapier überwiegend aus Altpapier. RAL-UZ 72. http://www.blauer-engel.de/produktwelt/buro/druck-und-pressepapiere-berwiegend-aus-altpapier-ausgabe-juli-2014 (Accessed September 22, 2014).

Pichol, Karl. 1999. Zur Invention des Papiers - ein experimentell gestützter Rekonstruktionsversuch der Innovation in China. *Technikgeschichte* 66(2):115–143.

PITA. 2010. *COST Action E48—The Future of Paper Recycling in Europe: Opportunities and Limitations.* Dorset: Caric Press.

Pledger, Lynne. 2011. Concerns with single stream recycling collection. www.cleanwateraction.org/ma (Accessed July 13, 2014).

Prinzhorn, Cord. 2013. A recycled containerboard and packaging producer's perspective. *European PaperWeek*, Brussels, Belgium, November 26–28.

PTS. 2012. *Kennzeichnung der Rezyklierbarkeit von Packmitteln aus Papier, Karton und Pappe sowie von grafischen Druckerzeugnissen. PTS-Methode RH 021/97.* München, Germany: PTS.

Putz, Hans-Joachim. 2000. Recovered paper grades, quality control, and recyclability. In *Recycled Fiber and Deinking*, edited by L. Göttsching and H. Pakarinen, 61–87. Helsinki, Finland: Fapet Oy.

Putz, Hans-Joachim. 2013. Recovered Paper, Recycled Fibers. In *Handbook of Paper and Board*, edited by Herbert Holik, 59–85. Weinheim, Germnay: Wiley-VCH.

Sandermann, Wilhelm. 1997. *Papier: Eine Kulturgeschichte.* 3rd edn. Berlin, Germany: Springer.

Schabel, Samuel, and Herbert Holik. 2010. Unit operations and equipment in recycled fibre processing. In *Recycled and Deinking*, edited by Ulrich Höke and Samuel Schabel, 122–278. Helsinki, Finland: Paperi ja Puu Oy.

VDP. 1983. Papier 1983—Ein Leistungsbericht. Bonn, Germany: VDP.

VDP. 1993. Papier 1993—Ein Leistungsbericht. Bonn, Germany: VDP.

VDP. 2002. Papier 2002—Ein Leistungsbericht. Bonn, Germany: VDP.

VDP. 2012. Papier 2012—Ein Leistungsbericht. Bonn, Germany: VDP.

VDP. 2014. Papier 2014—Ein Leistungsbericht. Bonn, Germany: VDP.

Voith Paper. 2014. *Picture Library.* Heidenheim, Germany: Voith Paper GmbH & Co.KG.

Wagner, Jörg, and Samuel Schabel. 2006. Automatic sorting of recovered paper—Technical solutions and their limitations. *Challenges of Pulp and Papermaking Technology: EUCEPA supported International Symposium*, Bratislava, Slovakia, November 8–10.

ZELLCHEMING. 2014. Prüfung des Rezyklierverhaltens von Verpackungen: ZELLCHEMING-Merkblatt RECO 1. http://www.zellcheming.de/service-center.html (Accessed September 19, 2014).

13 Product Design for Material Recovery*

Taina Flink and Mats Torring

CONTENTS

---

* The contents of this chapter are based on the gathered experience of employees at Stena Recycling AB as well as the sister companies within the Stena Metall Group, internally published material, and the authors' personal experiences in their work with disassembly analyses and consultancy services to support product developers in design for recycling. Many of the guidelines offered in this chapter come from personal observations by the authors and their colleagues at Stena's many recycling, dismantling, and shredding facilities.

13.1   INTRODUCTION

Designing for material recovery is crucial in achieving a more circular economy, where materials from products are used in a technical loop to create new products, without leakages of materials to landfills or incinerators. The term "designing for material recovery" describes the goal of reusing materials better than the more common term "design for recycling" (DfR), since "recycling" is used as a wider term referring to both material and energy recovery. The possibilities to control the product features, including environmental impact, are largest in the early phases of product development. This is why it is pivotal to be aware of how design decisions impact the recyclability of the finished product.

Although it can be useful for designers and product developers to be aware of future technological possibilities, it is by adapting to an existing system that more material will be recovered in reality. The guidelines and design strategies in this chapter are selected, based on the current recycling system and how products can be adapted to better fit industrial recycling methods and processes, in consideration to both economic viability and prevailing technology. For this reason, consideration is not given to recycling possibilities on a research stage or young technologies that might be prevalent in the future.

As the end-of-life stage of a product's life cycle is not the only stage to consider in product development, there are bound to be cases where design decisions have to be made that go against easy material recovery. These decisions might very well be supported in a life cycle analysis, and although this chapter covers only design for material recovery, a holistic approach to sustainable product development is encouraged to achieve the greatest environmental benefit. The guidelines and design input given in this chapter are meant to create awareness of how the design impacts material recovery, and to inspire solutions that can enable recycling in harmony with good product design and sustainable design in other aspects.

13.2   INDIRECT IMPACT ON MATERIAL RECOVERY

In this part of the chapter, some relevant approaches to sustainable design are mentioned, Although their main purpose is not to maximize material recovery, there are side effects that are positive for material recovery and it can therefore be of value to be aware of them.

13.2.1   DESIGN FOR ENVIRONMENT

Designing for the environment means to find solutions that prevent or minimize a product's negative impact on the environment and to adapt products to a sustainable life cycle at the development stage of products. Comprised within this term are all kinds of methods of minimizing the environmental footprint of a product, including DfR and DfD (design for disassembly). The two terms can many times have a similar approach, although in some cases they can be conflicting as well, as DfD often refers to design for manual disassembly, while DfR deals with all means of material separation. Designing a product to be easily disassembled often enables the product to be repaired and serviced, increasing its lifespan. However, an issue with manual disassembly is that manual labor is expensive and it is hardly ever economically viable to completely disassemble products manually in industrialized countries. Legislation makes sure hazardous components are removed manually, and therefore products are partially disassembled manually even in countries where labor is expensive.

Within design for environment (DfE), considerations about the product other than those comprised within DfR and DfD are made, such as

- How and where the raw material is produced
- Energy and material usage, waste, and emissions during production
- Transporting the material and finished products
- Waste, emissions, and consumption of material and energy during usage, service, and repair
- Handling of the finished product

13.2.2 DESIGN FOR REUSE AND REMANUFACTURING

When designing for reuse or remanufacturing, decisions in the development are made in respect not only to this life cycle, but also to the next to follow. This means that a product might be overdimensioned to last longer than the product category generally does, or the materials might be chosen to be of a higher quality than otherwise necessary. Whether the product can be reused as the same product or the parts can be reused and remanufactured to fit another product, considerations should also include design for reparability so the product can be repaired to be reused, and of course DfR since the product will have an end of life, however prolonged the life is. Designing with higher quality and overdimensioning can be fruitful for recycling as well, since the materials might be purer and therefore easier to recycle. It might also mean that the product contains more valuable material, which means that there is a bigger chance that the product will be recycled, especially in countries where recycling is driven solely by economic measures and not so much by legislative measures as in many industrialized countries.

13.2.3 DESIGN FOR MODULARITY

Designing for modularity refers to grouping functions or components in a modular way so that the product can be divided in several sections. Use of modular design simplifies assembly, disassembly, reparability, and serviceability. Modules can be exchanged or serviced separate from the components, which can prolong the longevity of the product since different components, and even different materials, have different durability. Modules can also be removed for reuse or upgrading. Modular design can also reap logistic benefits as well as enable more elaborate business models. Designing for modularity can also ease recycling, if the hazardous and valuable components would be grouped in easily removable modules for instance.

13.2.3.1 Designing for Material Recovery

Simply put, the material recovery rate of a product that consists of several different materials depends on what materials are used, how the materials are treated, and how the materials and components are connected to each other. The recovery rate is also determined by whether it is disassembled manually or mechanically, which in turn is usually dependent on how much hazardous or valuable material the product contains, and in which country it is recycled.

In countries where manual labor is expensive, products containing batteries, larger circuit boards, mercury, or extensive amounts of copper are disassembled manually, at least to the extent that the above-mentioned parts are removed. All other products consisting of several different materials are put through a mechanical treatment process, along with the products that have been partially disassembled manually. This means that designing for manual disassembly is only relevant for products that contain these components, and for the vast majority of products, it is more relevant to design for mechanical processing. But since many products might be reused and end their life cycle in countries where manual disassembly is the prevailing method, the products should work for both scenarios.

13.2.4 MATERIAL CHOICES

Obviously, the choice of materials affects the recyclability rate. But it is not as simple as choosing a material that is theoretically recyclable. There needs to be incentives for recycling the material as well, such as high volumes along with a market for recycled raw materials, legislation, and/or technology for separation and recycling.

Minimizing the number of different materials and reducing the amount of materials used make the losses in mechanical recycling lesser, and make manual disassembly faster.

In Figure 13.1, the compatibility between different materials in mechanical treatment processes has been illustrated. When designing products or components, materials that are not compatible should not be joined together, illustrated by red combinations in the matrix. The red combinations means losses in either material will occur to some extent or their quality will be severely damaged if the two materials have not been separated before smelting. Since the matrix is based on smelting of metal, it only takes into account whether the material on the left is a problem for recycling of the material on the top, and not the other way around. This is why thermoplastics are marked yellow, since the plastics do not necessarily pose problems for the metal melts, but metals in plastic recycling streams would pose big problems.

13.2.4.1 Recycled Material

A key in creating a more circular material loop in products is to create a market for recycled materials, and to enable the material to be relooped (Domini et al.). For metals, most metals are already recycled and the market for recycled materials exists. When

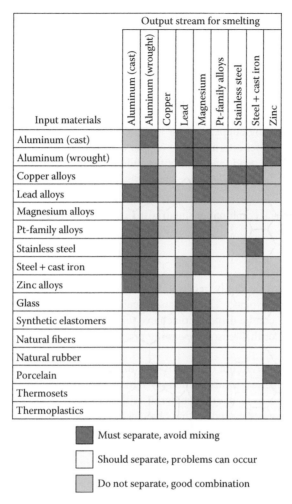

FIGURE 13.1   Compatibility matrix for different materials. (Adapted from Castro, B. 2005. *Design for Resource Efficiency: Preserving the Quality in the [Automotive] Resource Cycles.* PhD thesis, Delft University of Technology, Delft, the Netherlands.)

it comes to plastics, an obvious start for introducing recycled material in a product is to replace components of low performance and low demands on appearance. But it is equally important to make sure at least the most common plastics that have a working recycling process today, such as polystyrene (PS), polyethylene (PE), polypropylene (PP), and acrylonitrile butadiene styrene (ABS) are able to be separated from the other materials in a product.

13.2.4.2   Metal

Metal has the advantage that it is easily recycled and it has an existing recycling system all over the world. It also requires extensive amounts of energy to mine and produce, which makes the environmental gains major when replacing virgin material.

The quality of metals is not reduced by recycling, provided that they are kept pure or mixed with compatible metals. Coatings are rarely a problem provided that they are nontoxic, since they burn off in the smelter.

Some metal mixes are especially unfavorable, while other combinations can even be preferable and improve the material properties. To facilitate recycling of metals, it is important to enable different metals and materials to be separated.

- Avoid combining copper and tin with steel. In applications using steel and electronics, prefer alternatives to copper cables such as fiber optics.
- Iron, antimony, lead, and bismuth pollute copper. These combinations are to be avoided.

Also see the material combinations matrix in Figure 13.1 for preferred and unwanted combinations.

13.2.4.3 Plastic

Thermoplastics are generally recyclable, while thermosets are generally not and should therefore be avoided. Thermoplastics can be remelted, but they require a very pure input, which can be difficult to achieve when separating plastics from a mixed material stream. For this reason, designing products for recovery of the material is a priority for plastics.

Plastics can be sorted out by density in a mixed recycling stream, meaning that any type of changes to a piece's density makes it end up in the wrong fraction, whether it be a piece of another material still stuck to the piece or if the plastic has fillers or materials mixed in that changes its density, such as glass fiber, talc, flame retardants, stabilizers, and so on. A masterbatch is a concentrated mixture of additives enclosed in a carrier resin used to alter the properties of the base plastics. Using more than 5% masterbatch in plastics impedes recycling since the plastic can no longer be separated in a mechanical process (Hultgren 2012). This can also work the other way; if an additive is unwanted in the recycled fraction, an additive that changes the density of the plastic is preferred. This is true for biodegradable polymers that worsen the quality of a recycled plastic, which is why it is important that they stay out of the density range of the commonly recycled plastics. From mixed recycling streams such as electronic waste that is shredded, only PS, PE, PP, and ABS plastics are separated for material recycling, meaning that the density of biodegradable plastics should stay outside the range of $0.89–1.07$ g/cm^3. Coatings on plastics, such as painting, lacquering, plating, or galvanizing, cause the same kind of problems for the recycling of plastics, and if the coatings come off they pollute the material fractions as well. Coatings causing a density change of less than 1% of the materials weight are OK. Always avoid plating since metal is a pollution in plastic recycling. Polymer blends, such as PS-ABS, should also be avoided for the same reason. The blends cannot be separated into PS and ABS by mechanical separation. Furthermore, consider the following when designing products with plastic as a material choice:

- Avoid adding stickers or labels and prefer markings in the material itself. When stickers are necessary, prefer putting them on parts that are energy recovered or on metal parts. When stickers on plastics are necessary, make sure the glue and the label itself is compatible or the same material as the plastic detail.
- Use material markings on plastic parts.
- Do not use chemical substances that are environmentally hazardous, such as substances of very high concern on the Candidate List of REACH or equivalent. REACH is a European Union regulation that applies legal obligations for the use of substances of very high concern (ECHA 2014).
- Avoid designs that expose plastic parts to wear, UV-radiation, dirt, or extreme temperatures to the point that the recyclability is jeopardized.

13.2.4.4 Wood, Cardboard, and Paper

Although cardboard, paper, and wood are renewable materials, they are not compatible with "technical" material such as metal and plastic when used as a minority material. Using plastic coatings or metallic staples is not a problem for the recycling of paper, cardboard, or wood, but the other way around impedes recycling. Cardboard, paper, or wood in products containing plastic or metal will need to be removed manually in order to be recycled, or they will end up in energy recovery or landfill after going through a mechanical separation process. Since the value of these materials as raw materials is not very high, in comparison with metals for instance, it is not always economically viable to remove the materials manually.

13.2.4.5 Glass and Ceramics

Glass and ceramics should be avoided in products with other materials. In a mixed material stream, they are mixed with other brittle materials and landfilled together with them.

Glass as packaging is an exemption from that rule, since they have a separate and existing recycling scheme in many countries and are recycled even though there might be a label or cap from another material. But since glass has a low value on the raw material market and because many applications require very pure glass, glass is neither separated manually nor mechanically from the mixed material flow.

Ceramics are down cycled if they are recycled but mostly landfilled when used in products with mixed materials. Hence, prefer other materials when possible.

13.2.4.6 Composites

Composites will not be recycled when mixed with other materials in a product since they are not separated from the mixed material flow. When composites are used, they should be removed; they should be easily removed manually or should be separated completely in the mechanical treatment process.

13.2.4.7 Textiles and Foams

Textiles and foams are most often not separated manually or mechanically from the mixed material flow, and should therefore be avoided. If they are used, they should be very easy to remove manually or encouraged to be removed by consumers in the countries that have a recycling scheme for textiles or foams.

13.2.4.8   Rare Earth Metals

Rare earth metals (REMs) are strategically important to recycle, although their value is not reflected in the raw material prices. Neodymium, used in magnets, and lanthanum, used in nickel–metal hybrid batteries are examples of REMs that are widely used in electronic products. In the future it will be of importance to recycle REMs, especially due to geopolitical issues, which is why there might be legislation in place to make sure they are recovered. For neodymium magnets, this is likely to mean they have to be removed manually, since magnets are used to separate metals in the mechanical disassembly processes, and the magnets will end up in that fraction. Until better technology is prevalent, it is safer to design products so that components containing REMs are easily removed manually, and so that the product communicates by its design that it contains REMs.

13.2.4.9   Hazardous and Polluting Components

When hazardous and polluting components such as batteries, circuit boards, chords, used filters, or lamps are contained in a product, make sure they are easily accessible and detachable with little manual effort.

13.2.5   JOININGS AND CONNECTIONS

Choosing the joining and connection between different materials and components controls the recyclability to a large extent. They also affect the serviceability, reparability, and reusability.

13.2.5.1   General Guidelines

When designing products or components, materials that are not compatible should preferably not be joined together. When joining material or components, the joining should be of a compatible material as well. For material compatibility, see Figure 13.1 under Section 13.2.4. If noncompatible materials or noncompatible joining materials are necessary, use joining methods that enable separation in both mechanical processes and manual disassembly.

A few simple rules summarize the thinking around designing joinings for recyclability and separability, which are follows:

- The best joining method is one where no joining material is added, such as when pieces are interlocked like pieces of a puzzle.
- Prefer a joining from the same material as the host material, or at least a compatible material.
- Choose a joining that allows a nondestructive disassembly when there is a chance the components could be reused.
- Avoid integration of joining elements by, for instance, casting metal parts in plastics.
- Use joining elements that can be disassembled by standard tools.
- Minimize the number of joining points and place them where they are accessible and easy to find.
- Do not use connections that confine materials permanently in another material to avoid polluting the material streams.

13.2.5.2 Bayonets

Bayonets are a good example of a joining that is easy and fast to disassemble both manually and in the mechanical treatment process. Bayonets work like a wing nut that is twisted to release the connected material.

13.2.5.3 Connecting without Joining Elements

Sliding profiles and different kinds of slots and tabs or hooks are examples of connecting materials by the design of the components instead of by adding a joining element. These kinds of connections are preferable both in manual disassembly and in the mechanical treatment process, since they are fast and easy to disassemble manually and separate completely in mechanical separation processes.

13.2.5.4 Welding and Soldering

Even though welding can be difficult in disassembly since the welds damage the components that are welded together, they do not oppose a problem in material recycling since the welding material is compatible from a recycling perspective with the original material. Therefore, ultrasonic welding can be a great way to join plastic parts together.

- Always place the reusable components so that they do not break when the weld is taken apart in manual disassembly.
- Show how the product should be disassembled by adding easily identifiable breaking lines.
- Avoid soldering different materials together.

13.2.5.5 Glue

Gluing is one of the worst joining methods since it creates nonremovable attachments that only allow for destructive disassembly, except some types of de-bondable glue that can be dissolved chemically, by heat or by electricity. The same rules apply for gluing as for welding, with the following additions:

- The glue can pollute the host material and therefore obstruct recycling. Consider that when choosing glue type or another joining method, for example, welding.
- Many types of glue contain solvents that are harmful to the environment or the human health and are less desirable from a working environment and pollution perspectives.
- In some cases, the surfaces to be glued have to be cleaned with hazardous solvents.

13.2.5.6 Screws

Screws are a good joining method since they enable a nondestructive manual disassembly. But there are some drawbacks, which are as follows:

- Disassembly is time consuming.
- Threads can get damaged and the joints can rust together, obstructing nondestructive disassembly.

- In mechanical separation processes, part of the host material is likely to get stuck to the screw and follow the screw to its corresponding fraction. This means that the host material will have a diminishing recovery rate for every screw used. Therefore, when using screws in products that go through the mechanical treatment process, make sure the screws are compatible with the materials joined together. For example, to design for separation, avoid using a screw of ferrous metal to attach a plastic part or a nonferrous metal part.

Screws can still be a favorable method however, especially if used to join the same material the screw is made of. When using screws as a joining method, consider the following to ease manual disassembly:

- Aim for a short thread length in order to get shorter disassembly times.
- Protect threads and screw heads from corrosion, dirt, and mechanical injuries, especially if they are placed in exposed environments.
- Design the screw joints so that disassembly is done in one direction, preferably with a flat surface on the opposite side.
- Minimize the number of different sizes and types of screws.

An interesting alternative for screws are so-called shape-shifters that change shape when exposed to heat or vibration, such as a screw losing its threads.

13.2.5.7 Rivets

Rivet bands are often difficult to disassemble, both because it is time consuming and because the rivet itself is destroyed, sometimes along with the host materials. This is mainly a problem when the parts or materials could be reused. In mechanical disassembly, rivets contribute to impure fractions, since a piece of the host material is likely to get stuck on the rivet in the separation process, and follow the rivets to its fraction. This is avoided if the rivet is made from the same material as the materials it connects, or if the rivet is designed to stay stuck to the material it is compatible with. A good alternative, at least when joining compatible material, is clinching, where metals are clinched together in a permanent T-like shape, without actually adding a joining element.

13.2.5.8 Snap-Fits

Snap-fits have the advantage that they usually provide a simple and fast assembly and disassembly. They are especially good for housings of electronic products, since housings are usually necessary to remove manually in order to remove batteries, circuit boards, and hazardous components that have to be removed by hand.

- Integrate the snap-fits in the component so that additional material and details are avoided.
- Dimension the snap-fits so that they will not produce a weak point in the design.

- Design gripping points for easier breakup and openings for removal by hand or standard tools.
- Make sure the accessibility to the snap-fits is good.

13.2.5.9 Clips

Clips have the advantage that they can be easily removed by hand and by use of simple tools. The clip itself is destroyed in manual disassembly, but the components it has joined together are usually still intact. The clip can be made from another material than the materials it joins together, provided that it is completely removed during disassembly; otherwise it should be made from the same material it stays attached to during disassembly. In cases where manual disassembly will not be done, the clip should be of the same material as both the materials it joins together.

13.2.5.10 Casting

Casting is not a preferable joining method from a material recovery perspective, since it does not enable a nondestructive manual disassembly and even in a mechanical treatment process the materials will not be fully separable. Especially do not mold in metal parts, circuit boards, or cables in plastic that is material recyclable.

13.2.6 TAKE-BACK SYSTEMS

Take-back systems are a powerful way to affect the recovery rate for a company's products. Not only does it make it easier to reuse or remanufacture components from old products, it also makes it possible to harvest recyclable materials for one's own production and to make sure materials or plastic types that would not otherwise be material recycled due to too small volumes in the larger flow of materials would be material recycled, if the plastic type is theoretically recyclable that is. Another advantage is that the product flow is more homogenous compared to the general waste stream, meaning that the possibilities to set up an automated disassembly process can actually be economically feasible, which is unlikely otherwise.

13.3 CONCLUDING NOTE

The guidelines and design input given in this chapter increase material recovery rates when used in the development of products. Designing products not only for manual disassembly but also for mechanical treatment processes is essential in creating products that can be material recycled in reality. Products that consist of several materials need to use materials that are compatible with each other and the materials need to be easily separated from each other, both by man and by machine. The use of recycled materials in products can contribute to increased recovery rates by creating a bigger market for recycled raw material, which in the long run also can lead to financial incentives to refine recycling processes to retrieve more material and higher quality fractions.

REFERENCES

Castro, B. 2005. *Design for Resource Efficiency: Preserving the Quality in the (Automotive) Resource Cycles.* PhD thesis, Delft University of Technology, Delft, the Netherlands.

Domini, P., Bergendahl, C.G. and Norrblom, H.L. *Handbok för konstruktörer.* Stena Gotthard and Institutet för Verkstadsteknisk Forskning.

The European Chemicals Agency (ECHA). 2014. Regulations and candidate list substances. http://echa.europa.eu/regulations/reach.

Hultgren, N. 2012. *Guidelines and Design Strategies for Improved Product Recyclability.* Master thesis, Chalmers University of Technology, Gothenburg, Sweden.

14 Landfill Mining
On the Potential and Multifaceted Challenges for Implementation

Joakim Krook, Nils Johansson, and Per Frändegård

CONTENTS

14.1 INTRODUCTION

In many regions landfilling has long been seen as a final way to dispose of waste and this is still the most common disposal method globally (Eurostat 2009). Even in countries that have developed waste management and recycling systems, landfilling has remained important, at least until very recently. Regardless of current waste policy and practice, most regions therefore involve a large number of landfills (Jones et al. 2013).

It is a well-known fact that landfills have a host of environmental implications, ranging from local pollution and land management issues to global impacts in terms of landfill gas emissions (Daskalopoulos, Badr, and Probert 1997). This perspective was also strongly reflected in the early landfill mining studies done in the 1990s, which primarily aimed at solving such traditional landfill management issues by excavation and processing of deposited materials (Krook, Svensson, and Eklund 2012). Back then, little emphasis was put on optimizing these projects for recovery of materials and energy resources, although landfill-cover material and, in some rare exceptions, also waste fuel was extracted (Dickinson 1995; Krogmann and Qu 1997). Typically, excavation and screening equipment was used, displaying moderate performance in extracting saleable recyclables apart from landfill-cover material, but capable of fulfilling prime objectives, such as remediation, land reclamation, or creation of new landfill void space (Cossu, Hogland, and Salerni 1996; Hogland 2002; Krogmann and Qu 1997).

In recent years, a revived interest in landfill mining has occurred but this time largely influenced by industrial ecology perspectives, stressing the need for obtaining more resource-effective and cyclical uses of materials and energy resources (Krook and Baas 2013). This emerging research aims at optimizing the recovery of natural resources from landfills by applying more up-to-date material separation and processing technologies and has displayed a significant societal potential for what is still an unconventional strategy (Frändegård et al. 2013; Jones et al. 2013).

Still, most knowledge about the conditions for implementation of landfill mining is largely influenced by an engineering perspective (Krook and Baas 2013). There are, for instance, numerous case studies on the material composition of such sites, technical feasibility issues, and risks of local impacts and accidents during such excavation processes (Krook, Svensson, and Eklund 2012). Although these are all essential aspects for any landfill mining project, there are many other dimensions (e.g., policy and legislation, culture, markets, and organizational issues) that influence the conditions and dissemination of landfill mining (Baas et al. 2010; Jones et al. 2013).

This chapter involves a meta-analysis of landfill-mining literature, which is based on four recent reviews of this research field and their included papers (Baas et al. 2010; Jones et al. 2013; Krook and Baas 2013; Krook, Svensson, and Eklund 2012). The aim is to assess drivers and barriers of generic relevance for the implementation of this emerging strategy. More specifically, state-of-the-art knowledge regarding landfill mining is put in relation to industrial ecology perspectives and transformation theories and analyzed in view of the following specific questions:

- Why should we learn how to mine landfills?
- Why don't we mine the landfills?

Given the engineering approach of most previous studies, emphasis is here on societal and institutional conditions and how such dimensions influence the feasibility and further dissemination of landfill mining. The paper ends with a discussion of possible ways to facilitate realization of landfill mining, giving special attention to the respective roles of knowledge production, knowledge dissemination, and actor collaboration and networking.

14.2 WHY SHOULD WE LEARN HOW TO MINE LANDFILLS?

14.2.1 RESOURCE IMPLICATIONS—THE IMPORTANCE OF CONSIDERING MATERIAL STOCKS

We are now in the twenty-first century, and what history tells us is that the need for natural resources is continuously increasing over time, driven by rising economic standards and growing populations (Ayres, Holmberg, and Anderson 2001). At present, the dominant mode for feeding this market demand is primary production—a sector that is inherently resource-intensive and causes a host of pollution problems (Nriagu 1996). In times of environmental challenges and resource scarcity concerns due to rapidly declining virgin reservoirs, however, the need for societal transformation toward a more effective and cyclical use of natural resources is becoming increasingly apparent and accepted.

One of the key challenges of sustainability is that in the long term, we have to rely more on recycled materials than on primary production, not only due to environmental reasons but also for political and economic implications related to universal changes such as global warming and resource scarcity (Baccini and Brunner 2012). This is especially true for strategically important and finite materials like metals (Tilton and Lagos 2007). From a mass flow perspective, such a challenge is inherently problematic given that as long as there is a rapid growth in consumption (which there is and probably will continue to be for many years to come), recycling of annual waste streams cannot replace a significant share of primary production (Baccini and Brunner 2012; Graedel et al. 2004). Today, for instance, the total copper waste generation is globally a few million tons per year while the annual consumption is approaching 20 million tons (US Geological Survey 2012). So, even if we were capable of material recycling virtually all copper discards—something which at present we are not even close to (UNEP 2012)—most of our raw material supply would still have to be covered by primary production. What this kind of reasoning neglects, however, are agglomerates of previously employed metals and other finite resources that have been excluded from ongoing anthropogenic material cycles (Krook and Baas 2013).

Our continuous need for materials and energy has resulted in a large accumulation of natural resources in buildings, infrastructure, products, and waste deposits. The fact that such technospheric stocks of copper, for example, are now globally comparable in size to the amount remaining in prospected geological ores illustrates the magnitude of this ongoing relocation process of natural capital (Elshkaki et al. 2004; Gordon, Bertram, and Gordon 2006; Halada et al. 2009; Johansson et al. 2013; Lichtensteiger 2002; Müller et al. 2006; Spatari et al. 2005). Perhaps even more notable is that when it comes to base metals such as iron and copper, almost half of the amounts extracted to date are no longer in use. These previously employed resources can be found in different waste deposits and in obsolete products, buildings, and infrastructure networks, which for various reasons have not been collected for waste management but abandoned in their urban or rural location.

Together with other waste deposits, for instance, tailing ponds and slag heaps, landfills constitute the most significant reservoir of obsolete metals in the technosphere (Johansson et al. 2013; Krook and Baas 2013). Again taking copper as an example, the amount of this metal in such waste deposits is globally of a similar magnitude, that is, 390 million tons, as the current in-use stock of copper, that is, 350 million tons (Kapur and Graedel 2006). Landfills thus offer a significant alternative source for metal extraction, which, if exploited, could complement traditional recycling and primary production in feeding the increasing market demand for such coveted materials of modern society. This is especially true given that this pattern of secondary resource accumulation is likely to persist, although the magnitude of all technospheric stocks will significantly increase in the years to come (Kapur 2006; Johansson et al. 2013).

14.2.2 POTENTIAL FOR POLLUTION PREVENTION

In most regions, there are many (old) landfills, a significant share of which lack modern environmental technology and are therefore in need of remediation, sooner or later. Similar to remediation, landfill mining has been argued to constitute a way for addressing such local implications related to waste deposits (Cossu, Hogland, and Salerni 1996; Hogland 2002). This is because once the deposited masses are exhumed, there is a possibility of securing the site by upgrading the landfill infrastructure, for example, through bottom sealing and leachate collection systems, and by taking care of hazardous waste that previously was deposited without precautions. In contrast to traditional remediation, however, excavation, processing, treatment, and recycling of the deposited material also mean that the original pollution source can be permanently eliminated, or at least largely reduced, which in addition facilitates land reclamation (Hull, Krogmann, and Strom 2005; Krogmann and Qu 1997). Another often proclaimed local benefit of landfill mining is the possibility to create new landfill void space and by doing so avoiding the need for occupying "new" land for such activities.

Recent studies imply that for a significant share of the deposited materials it is presumably technically possible to extract and process them into saleable commodities (Frändegård et al. 2013; Van Passel et al. 2013), for example, metals, waste fuel, and construction material (Figure 14.1). If proven feasible, such resource extraction would lead to significant environmental impacts on the global and regional scales. In a Swedish context, it has been estimated that large-scale implementation of landfill mining would in total lead to 30–50 million tons of avoided climate gas emissions compared to leaving the landfills as they are (Frändegård et al. 2013). On the European level, the realization of such a strategy could potentially generate CO_2 equivalent savings of 15–75 million tons per year during a 20- to 30-year period (Jones et al. 2013).

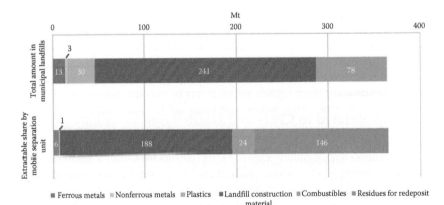

FIGURE 14.1 Estimated total and extractable share, by mobile separation units, of different deposited materials (million tons) in Swedish municipal landfills. (Data from Frändegård, P. et al., *J. Ind. Ecol.*, 17, 742–755, 2013.)

In specific projects, the potential for global and regional environmental impacts is understandably influenced by many factors such as the type of landfill, the applied technology, the available raw material, and energy markets (Frändegård et al. 2013). However, even for a typical Swedish municipal landfill containing limited amounts of highly valuable resources, landfill mining by conventional mobile separation technologies is likely to generate considerable net savings of pollutant emissions of relevance for global warming, acidification, and eutrophication (Krook 2013). For such landfills, several processes such as avoided emissions from long-term methane generation, replaced primary material production, and substituted conventional energy generation seem to contribute rather equally to the environmental potential of landfill mining (Figure 14.2). There are thus reasons to believe that adding a resource recovery component to remediation projects would improve their environmental performance when it comes to such impact categories (Frändegård et al. 2013; Krook 2013). Compared to combined heat and power generation from landfill gas collection, landfill mining also displays a considerably larger potential for global warming mitigation (Van Passel et al. 2013). The most important reason for that are avoided primary production emissions caused by material recycling of the exhumed resources.

Given that municipal landfills typically contain relatively small amounts of highly valuable materials such as metals, that is, a few weight percentage of which the main part are ferrous metals (Frändegård et al. 2013; Krook, Svensson, and Eklund 2012), landfill mining on many industrial deposits such as shredder residue landfills displays an even larger environmental potential (Figure 14.2). For such landfills, the high content of ferrous metals (about 5 wt%) and nonferrous metals (about 8 wt%) means that the net savings of climate, acidification, and eutrophication pollutant emissions become more or less solely related to metal recycling and thus avoid impacts from virgin mining activities (Alm, Christéen, and Collin 2006; Krook 2013).

To sum up, there are quite a few publications displaying a significant potential for landfill mining in regard to global, regional, and local environmental consequences (Krook and Baas 2013). Yet several of the studies indicate a considerable potential for further optimization of the environmental performance through modification and fine-tuning of technology or by applying advanced processing and recovery technologies (Frändegård et al. 2013; Jones et al. 2013; Van Passel et al. 2013). However, landfill mining projects are also related to risks for local impacts, dispersal of hazardous substances, and accidents, something that need to be taken into account for several reasons (Johansson, Krook, and Eklund 2012) and which we will cover later in this article.

14.2.3 Other Potential Socioeconomic Impacts

Given the large agglomerates of materials and energy resources currently situated in European landfills, a successful implementation of landfill mining would improve the material autonomy of the region (Jones et al. 2013). This is not something specific to landfill mining but is a generic benefit of material recycling because such secondary resources are more evenly distributed between (industrial) countries than primary reserves (EEA 2011; Johansson et al. 2013). In contrast to traditional recycling, however, landfill mining could potentially provide the recycling industry with

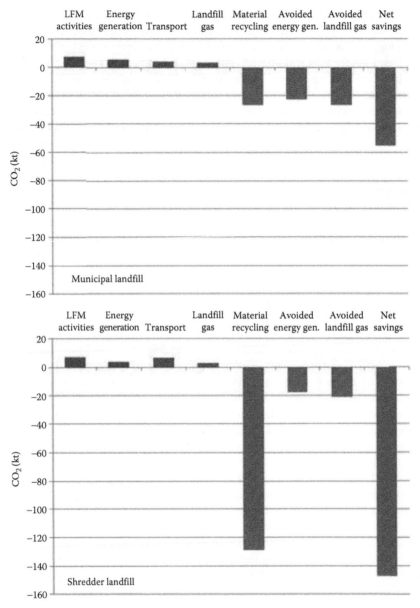

FIGURE 14.2   Added, avoided, and net CO_2 equivalent emissions (in kilotons) divided by type of process for landfill mining (LFM) on a typical 700,000 ton Swedish municipal landfill (top) and on an industrial deposit of the same size, that is, a shredder landfill (bottom). (Calculations based on Frändegård, P. et al., *J. Ind. Ecol.*, 17, 742–755, 2013; Krook, J., Miljökonsekvenser av integrerad sanering och återvinning av kommunala och industriella deponier, Internal report, The Swedish Environmental Protection Agency, Stockholm, Sweden, 2013; Alm, J. et al., Landfill mining at Stena Gotthard's landfill in Halmstad—An environmental and economic evaluation, Internal report, Environmental Technology and Management, Linköping University, Linköping, Sweden, 2006.)

substantial amounts of additional raw materials, making it possible for this sector to grow further and increase its competitiveness as a raw material supplier (Krook and Baas 2013).

The realization of a landfill mining line of business could also constitute a potential way to strengthen local economies by offering new job opportunities related to such projects, but also through spillovers in other upstream and downstream sectors such as cleantech, recycling, and horticulture (Jones et al. 2013). In Europe, there are 150,000–500,000 landfills and these sites are becoming increasingly problematic, especially in densely populated areas where pollution concerns and competition for valuable urban land are common implications (Johansson et al. 2013). For any actor capable of offering feasible concepts, there thus seems to be a tremendously large market, not least in terms of export of landfill mining technology and know-how.

14.3 WHY DO WE NOT MINE LANDFILLS

Recent research on landfill mining implies a significant potential for a new line of business in which natural resources that, for various reasons, have been excluded from ongoing anthropogenic cycles are brought back to work again (Frändegård et al. 2013; Jones et al. 2013; Van Passel et al. 2013). However, the foundation for this potential is still of a theoretical nature and most of the proclaimed economic, environmental, and social benefits seem to materialize on the societal level (Krook and Baas 2013). Accordingly, virtually all full-scale landfill excavation projects so far have been driven by other prime objectives such as remediation and land reclamation—projects that in principle were allowed to be costly for a "greater cause" (Johansson, Krook, and Eklund 2012; Krook, Svensson, and Eklund 2012).

Already back in the 1990s, it was concluded that resource extraction from municipal landfills alone is often not an economic option for private actors such as landfill owners and recycling companies (Dickinson 1995; Fisher and Findlay 1995). The costs, for example, for excavation, material processing, hauling, and site restoration were simply too high in relation to anticipated revenues for extracted materials and energy resources. Despite decades of steadily increasing material prices and improvements in technology, this conclusion from the past still tends to be true, at least when it comes to "low-grade" municipal landfills (Hull, Krogmann, and Strom 2005). In fact, even recent studies assuming high separation and recovery efficiencies and the employment of noncommercial waste-to-energy technologies, namely, gas plasma, come to the conclusion that resource extraction alone cannot justify such projects economically (Van Passel et al. 2013). Such findings imply that although technology indeed remains an important factor for the feasibility of landfill mining, there seem to be other reasons of similar importance for why such initiatives often experience a hard time when exposed to market conditions.

14.3.1 FUNDAMENTALS OF COST-EFFICIENT INDUSTRIAL PRODUCTION

In his article about the supply of raw materials in the Industrial Revolution, Wrigley (1962) describes several essential constituents of cost-efficient industrial production. Although his story is really about the major shift from using distributed organic raw materials to the exploitation of concentrated mineral deposits, it displays many of the inherent fundamental differences between primary and secondary production that we see today. Secondary resources are by definition heterogenic and distributed in society while primary natural resources, at least when it comes to strategically important minerals and metals, occur as more punctiform and uniform reservoirs (Baccini and Brunner 2012). The latter means that once appropriate mining technology is available, production can rapidly be built up to high levels, thereby taking advantage of economy of scale and decreasing the unit costs of production (Ayres 1997; Wrigley 1962). This economic principle was of key importance for the initial evolvement of the modern primary production sector, which then through centuries of practice, specialization, subdivision of production processes, and investments in technology and knowledge has gradually developed into the established mode for resource extraction and cost-efficient production.

The recycling industry, on the other hand, is largely lagging behind, partly because industrial recycling is a relatively new phenomenon, except for the recycling of a few base metals, but primarily since fundamentals of cost-efficient production are more difficult to obtain based on distributed and heterogenic raw materials (Ayres 1997). Simply obtaining access to sufficient amounts of well-defined "raw materials" is often an organizational and logistic challenge in itself for the actors in this sector. Landfills are punctiform accumulations of metals and other valuable resources and are in that sense less distributed than annual discards (Johansson et al. 2013). In comparison to most exploited primary reserves, however, they are in general still small and significantly more distributed (Jones et al. 2013; Swedish Geological Surveys 2009), which in turn influences the possibility of obtaining the same scale effects and thus cost-effectiveness as in the primary production sector.

Resources in landfills are also less accessible than in traditional mines. While the minerals in the Earth's crust are more or less freely accessible in countries such as Sweden (cf. SCS 1998), the material in a landfill is in the ownership of a specific municipality or company and therefore not directly accessible for resource extraction by any external actors. This type of ownership is certainly justifiable from a pollution perspective since responsibility could be assigned to externalities such as leachate of hazardous substances from landfills. From a resource perspective, however, it could pose an obstacle since the metals are not accessible on demand.

The costs of producing commodities from secondary resources are also further increased by their heterogeneity. This very feature influences the level of automation of recycling processes, which still often relies on significant efforts by workers for disassembly, sorting, and control in order to produce saleable commodities, at least to a larger degree than in most primary production units (Ayres 1997; Wrigley 1962). Obviously, the heterogeneity of secondary resources also affects the usefulness and thereby pricing of the produced commodities. Despite recent improvements in waste collection and technology in many industrial countries, it is a well-known fact that material recycling still often involves some sort of cascading of the materials to applications with lower quality demands. And there is still no reason to believe that resources extracted from landfills are any exception here (Johansson et al. 2013). In fact, most of our municipal landfills, which so far have been the main topic of research, consist of a wide spectrum of materials that have been deposited in disarray without any prospects of future recovery initiatives (Hogland 2002; Krook, Svensson, and Eklund 2012). Even if it proves to be technically possible to separate such different materials from each other, these resources have been exposed to long-term degradation processes, often to a larger degree than many of the materials found in annual discards (Fisher and Findlay 1995; Johansson et al. 2013). Some studies also indicate fairly high levels of contamination in deposited materials (e.g., in plastics, paper, soil-type materials, and shredder residues), which in the end might make resource recovery increasingly challenging (Hull, Krogmann, and Strom 2005; Quaghebeur et al. 2013).

14.3.2 Institutional Conditions for Resource Extraction

The differences between primary and secondary production are, however, not just consequences of fundamentals but can also be understood in the context of institutional conditions and political commitment as highlighted by Ayres (1997) and more recently by Johansson, Krook, and Eklund (2014). Ever since the Industrial Revolution, primary production has contributed significantly to our welfare, and still does, and its continued progress is therefore often of highest political priority (Ayres 1997). For instance, huge societal investments in infrastructure destined for transporting extracted primary natural resources to markets have been and are continuously made in many parts of the world—investments of utmost importance for the further growth of this sector (Swedish Government 2013; Wrigley 1962). The institutional conditions for resource extraction have also often been gradually adjusted to meet the needs of resource-intensive and large-scale production, for example, by offering necessities such as cheap and abundant energy and low, or almost no, taxation on virgin natural resources (Ayres 1997). Instead, most taxes in industrial countries are charged on labor, making labor-intensive recycling activities including landfill mining initiatives inherently more costly than primary production.

There is also a growing amount of literature that pinpoints the significant amount of explicit and implicit governmental subsidies to conventional primary production (Davidson 2012; Grudnoff 2012; OECD 2009; Van Beers and de Moor 2001). Although there are some signs of slow but gradual changes, for example, landfill, energy and carbon taxes, one of the major implicit governmental subsidies to this sector is the noninternalization of most environmental externalities related to resource-intensive and polluting primary production (Ayres 1997). Such a practice also means that avoiding, or at least limiting, such externalities through secondary instead of primary production (e.g., reduced carbon footprint and other kinds of pollution, improved regional material autonomy, conservation of strategically important materials such as metals, etc.) do not normally become part of private actors' returns. This so far consolidated principle of governance is of utmost importance for the feasibility of landfill mining, given that most of the anticipated benefits of such initiatives occur on the societal scale (Krook and Baas 2013; Jones et al. 2013). The conclusion of one of the most recent and comprehensive studies on the economics of landfill mining is that the incentives for private actors to engage in such projects largely rely on new or adapted policies taking these externalities, or let us say societal benefits, into account (Van Passel et al. 2013).

At present, there is admittedly also an increasing political commitment to recycling through various strategic policy documents and legislation aiming for more effective collection and recovery of such secondary resources (European Council 1999; SCS 2005). But when it comes to direct (economic) support in terms of grants or capital contributions to cover industrial losses, support export, modernize industry, or through tax exemptions or reductions, the primary production sector is still the main receiver. In Sweden, for instance, the level of such subsidies to the metal mining industry was in 2010 significantly higher than for the metal recycling sector, primarily due to substantial tax exemptions and reductions related to disposal of mining tailings and use of fossil fuels (Johansson, Krook, and Eklund 2014). In principle, this is not something unique to Sweden or metal production, but the tendency of governments, in absolute terms, to mainly support conventional primary production has been reported for several other sectors and countries as well (OECD 2009). Again, given the prominent role and importance of primary production in our societies, this mode of governance is by no means surprising but rather thought of as a necessity for continued economic growth. What it does, however, is tend to hold back the further growth of the recycling industry, not least by keeping costs for labor high, largely neglecting environmental externalities and maintaining prices for virgin raw materials, but then also by indirectly pricing secondary commodities at a relatively low level (Ayres 1997; Kleijn et al. 2001). If we then turn to landfills and more specifically suggest something as unconventional as mining these sites for natural resources, the institutional challenges grow even further (see Johansson, Krook, and Eklund 2012, 2014).

14.3.3   The Landfill Is Stuck in a Dump Regime

Landfills are useless and if they have any value, it is negative, for the surrounding environment and society and for most locals (Daskalopoulos, Badr, and Probert 1997; Nelson, Genereux, and Genereux 1992). This perception of landfills as being an end station for worthless rubbish has for decades, even centuries, been manifested in industrial countries, not least by researchers, legislators, and policy makers. In practice, this means that landfills in fact are far more than what their intrinsic materiality tells us—something that needs to be perfectly understood when considering conditions for landfill mining. More precisely, landfills are embedded in a wider sociotechnical context consisting of knowledge, technology, laws, culture, markets, and policies—multifaceted features that have been developed within and influenced by this perception of landfills as a garbage dump, which we need to be protected from (Johansson, Krook, and Eklund 2012).

The prevailing perception of landfills is strongly mirrored in the policy and legislation surrounding these sites (European Council 1999). In short, this top-down and reinforced regulatory framework aims at isolating landfills from their surroundings, primarily by installing different technical measures. At end of life, final closure is strongly advocated involving capping and postmonitoring for decades. Most of the available knowledge and technology regarding landfills have co-evolved with this perspective (Johansson, Krook, and Eklund 2012). In contrast, for unconventional approaches like landfill mining, our understanding is still limited, involving scattered experimentation by informal constellations while little visible change occurs on the societal level (Krook and Baas 2013).

It is therefore not surprising that reported landfill excavations so far have been initiated by landfill owners (often municipalities) and primarily aimed at regaining the stability of, for some reason, malfunctioning landfills (Krook, Svensson, and Eklund 2012). Even if several of these projects have involved some kind of resource recovery, again mainly landfill cover material, they only represent incremental changes to the dump regime, applying existing knowledge and technology to, for example, remediate or move the landfill to a more suitable site (Johansson, Krook, and Eklund 2012). The choice of such incremental strategies where actors embedded in a sociotechnical system only see the obvious and choose conventional methods is not unusual and within transition theory is referred to as path dependency or lock-in (Geels and Shot 2010).

Suggesting the opposite, to open up these "dangerous" sites, liberate the deposited masses, and bring them back into societal flows again, seems significantly different. Why is that so? After all, it is difficult to separate landfill remediation and landfill mining projects from each other (Cha et al. 1997; Johansson, Krook, and Eklund 2012). The excavation process is in many ways similar and so are the risks for local accidents and impacts related to the very same process (e.g., formation of explosive or poisonous gases and landfill collapses). Furthermore, landfill cover material can be, and often is, recovered in remediation projects while initiatives aiming to optimize material and energy valorization presumably would decrease long-term pollution concerns (Johansson, Krook, and Eklund 2012). In fact, it is unlikely that any landfill mining project within the European regulatory framework could be realized without taking care of hazardous waste and performing adequate site-restoration efforts (Cossu, Hogland, and Salerni 1996; Krogmann and Qu 1997). Instead, the difference between remediation and landfill mining is in the perception of landfills and more precisely in the way the exhumed materials are handled (Johansson, Krook, and Eklund 2012).

Going against any sociotechnical system will cause multidimensional challenges and be met with suspicion and heavily scrutinized by the established regime, its actors, power relations, investments, and knowledge capital (Geels and Shot 2010). In the case of landfill mining, this generates a host of uncertainties, not only regarding legislative implications but also in relation to market and business dimensions—uncertainties that largely prohibit implementation because they make it difficult for actors to foresee the final outcome (Baas et al. 2010; Krook, Svensson, and Eklund 2012).

14.3.4 SOME EXAMPLES OF LEGISLATIVE, MARKET, AND BUSINESS IMPLICATIONS

At present, the types of secondary resources that are recycled just for the sake of their intrinsic raw material value are largely restricted to a few base metals in most regions of the world (EEA 2011). Virtually all other forms of material and energy valorization from waste have been initiated and maintained by top-down legislation, which in principle either makes less-wanted options more expensive or even forbidden (i.e., landfilling) or involves obligations in which producers/consumers are forced to cover costs for collection, transportation, and recycling of their waste

(European Council 1999; SCS 2005). When it comes to resource recovery through landfill mining, however, such legislative stimulus or societal support is not valid, meaning that these projects simply have to bear all costs and benefits on their own (Krook, Svensson, and Eklund 2012), which according to the fundamentals and institutional conditions for secondary production described above is challenging.

In fact, current waste management policy and legislation display several potential barriers for such projects (Krook and Baas 2013; Krook, Svensson, and Eklund 2012). For instance, the objective of the landfill tax is to redirect annual waste flows from landfills to more preferable options from an environmental perspective (SCS 1999), that is, to make alternative options such as waste incineration more cost-efficient for the waste producers. When it comes to landfill mining, however, this very same tax risk creates disincentives for actors to engage in such projects, given that it will be applied to the inevitable redeposition of nonrecoverable materials. For remediation projects, masses in need of such redeposition are deductible from the landfill tax, but the validity of such an exemption during landfill mining projects is largely unclear (SCS 1999). In Sweden, for instance, there is an ongoing governmental review of this tax, involving an assessment of whether landfill mining should be mentioned as an exception in the Waste Tax Act or not.

Current waste policy and legislation also have a number of market implications, influencing the business dimensions of landfill mining. For masses obtained from remediation projects, there are guidelines for potential uses of contaminated soil, making it possible to identify specific applications (SEPA 2009). In landfill mining projects, involving more extensive processing into different separated materials and energy resources, such principles for identifying accessible markets are absent (Baas et al. 2010). In fact, it is questionable if existing recycling companies and incinerators will want or even have the capacity to accept supplementary materials from landfills (Fisher and Findlay 1995; Johansson, Krook, and Eklund 2012; Krook and Baas 2013). This is especially true since such materials will compete with presumably more accessible and high-quality materials obtained from, for example, source separation programs. For instance, why should metal recyclers use their processing units on "low-grade" waste extracted from municipal landfills when there are plenty of other, higher-grade materials available such as incineration slag and different types of industrial discards that could be processed instead?

Another illustrative example of such market implications relates to the handling of the considerable amounts of combustible waste that landfill mining projects are likely to generate, at least when it comes to municipal landfills (Frändegård et al. 2013). Given current landfill bans on such waste and the regulated maximum time for waste storage of three years (European Council 1999; SCS 2001), most landfill mining practitioners will not have much choice other than to deliver it to local incinerators. Here, the landfill tax has opened up a possibility for incinerators to charge significant gate fees for accepting such waste (Baas et al. 2010; Krook, Svensson, and Eklund 2012), which for producers of annual discards are still less costly than landfilling. In contrast, for landfill mining practitioners these will generate significant additional costs.

Obviously, if there is an absolute shortage of waste fuel, the interest in obtaining supplementary combustibles from landfills could theoretically increase (Frändegård et al. 2013). In Sweden, the development of new waste incinerators during the last decade has actually resulted in such overcapacity of incineration (Profu 2010). However, the way in which these plants have dealt with this shortage supports the actor logic of priority-setting based on alternative costs. What they do is simply import combustible waste from other European countries lacking sufficient domestic alternative options to landfilling, and there are plenty of such countries in the region (Krook and Baas 2013). For such imported waste fuel, gate fees are charged, making such a practice the preferable solution while ideas about extracting "free" supplementary fuels from domestic landfills have not yet come to mind. Although recent trends indicate slightly decreasing gate fees for waste incineration, they are still significant—in Sweden ranging from €40 to €80 per ton of combustible waste (Swedish Waste Association 2013).

A suggested approach to overcome such market limitation and competition is to develop on-site processing and energy recovery plants solely constructed for the valorization of landfill mining materials (Jones et al. 2013). However, apart from the fact that such a practice would generate significant additional capital costs to the projects (Van Passel et al. 2013), the implications of alternative costs still seem to remain, namely, why should the owner of such a plant accept landfill mining materials for free when it is far more profitable to run the plant on annually generated combustibles, domestic or imported, for which gate fees can be applied? Potentially, the introduction of novel technologies generating more high-value products from combustible waste (e.g., gas plasma) could decrease the importance of gate fees (Jones et al. 2013). However, such technologies are not yet commercially proven.

Some recent studies argue that our perception of landfills, and thus also waste policy and legislation, has resulted in significant amounts of valuable materials being lost every year by burying them in disorder in landfills or through incineration (Jones et al. 2013). The suggested solution is to change the core function of landfills from being final destinations for waste to temporary storage sites for different separated materials that for various reasons are not yet technically or economically feasible to recycle as material. Such an approach would facilitate future valorization and, if also applicable to landfill mining of old landfills, opens up possibilities for addressing many of the legislative and market implications discussed above. For instance, landfill mining practitioners could then choose to put back separated combustibles and other not yet saleable/fully recoverable resources, thereby avoiding implications of landfill bans, taxes, and treatment gate fees, and wait until the right market and technical conditions for valorization occur. Although this constitutes an extraordinarily interesting concept, in essence it means that more or less the whole landfill policy and legislative framework would need to be pulled up by the roots and reshaped (European Council 1999; SCS 1999, 2001; Wante and Umans 2010). The challenges are immense.

14.4 CONCLUDING DISCUSSION

A large-scale implementation of landfill mining relies on many things other than applicable technology and methods (Krook and Baas 2013). More specifically, a fundamental societal turn in perspective is needed, involving changes in terms of policy, culture, perceptions, and practice. Although landfills are having more and more implications (e.g., competition for land and local pollution), such changes can not only cause instability in the current "dump regime" but must also make unconventional strategies such as landfill mining the most viable option (Johansson, Krook, and Eklund 2012). Otherwise, incremental methods such as remediation and final closure will remain the practice to address such instability caused by malfunctioning landfills (Geels and Shot 2010). In this respect, exogenous changes in relation to our transition toward circular economies are likely to make landfill mining increasingly attractive (Baas et al. 2010; Johansson et al. 2013; Jones et al. 2013; Krook, Svensson, and Eklund 2012). Rapidly growing resource demands and increasing raw material prices, politicization of environmental issues, and resource scarcity concerns are examples of such trends that in time might create strong enough incentives for a broader group of societal actors to fully engage in the realization of landfill mining.

Meanwhile, there is a massive need for knowledge production, regarding how to plan, prospect, organize, and execute landfill mining in order to enhance economic, environmental, and societal performance (Krook and Baas 2013). Admittedly, recent research has displayed a significant societal potential but state of the art is still largely theoretical, implying a need for comprehensive pilot projects developing feasibility and environmental and economic performance further in practice. Learning from the evolution of primary production, such a common knowledge base cannot be developed by scattered, isolated, once-in-a-lifetime projects performed by actors with other core business (e.g., municipalities)—something that in the past has characterized landfill mining research and initiatives (Krook, Svensson, and Eklund 2012). Instead, it needs to involve specialized actors, or constellations of actors, ready to invest in long-term learning processes (Ayres 1997; Wrigley 1962)— because learning how to materialize the concept of landfill mining will indisputably take time.

Fortunately, several transdisciplinary groups from academia, industry, and others have recently emerged in Europe and the level of collaboration among them is gradually evolving. However, except for a few initiatives, for example, in Sweden and Germany, these groups still tend to focus their knowledge production on heterogenic and "low-grade" municipal landfills, while for landfill mining, perhaps more suitable, homogeneous, and metal-rich deposits such as pure shredder and waste incineration slag landfills remain unaddressed (Krook, Svensson, and Eklund 2012). This is risky and might in the end decrease the interest in landfill mining among industry and policy makers, given all the technical, market, and institutional challenges related to realizing such projects cost-efficiently (Hull, Krogmann, and Strom 2005; Krook and Baas 2013; Van Passel et al. 2013). Doing the opposite might therefore prove fruitful and would display similarities with the successful evolution of primary production, in which the most mineral-rich deposits were mined first, and then in tandem with progress in prospecting and mining technology more "low-grade" reservoirs became exploitable (Ayres 1997). In fact, tentative results from ongoing research in Sweden show that for such shredder and slag deposits the economic conditions for landfill mining are significantly improved, although multifaceted challenges still remain in terms of how to best prospect, organize, and execute such projects.

When going through the landfill mining literature from an industrial ecology and transition perspective, there are clear signs that further dissemination would benefit from extended actor collaboration and networking. In contrast to primary production, in which large self-organizing enterprises more or less on their own extract specific elements from rather uniform deposits (Johansson et al. 2013), landfill mining projects will by definition result in multiple outputs of various materials and energy resources (Frändegård et al. 2013; Jones et al. 2013; Van Passel et al. 2013). Optimizing the planning, extraction, and valorization of such complex processes thus seems to rely on the ability to internalize different expert knowledge and resources within the project organization (Krook, Svensson, and Eklund 2012). Such extended collaboration might also constitute a way to secure a market for extracted resources by also including (end-user) material and energy companies. Here, landfill mining could probably benefit by learning more from the field of industrial symbiosis, analyzing conditions for developing mutually beneficial cooperative links between production and service companies, involving exchanges and sharing of material, energy, information, and monetary flows (Baas 2008; Baas et al. 2010).

However, as argued in this chapter, facilitating implementation of landfill mining not only is a matter of knowledge production but also strongly relies on knowledge dissemination and a common understanding about the potential and challenges related to such projects. Recent studies more or less consistently show that adapted policies and institutional conditions are needed as well as political support, where such projects are seen as "green activity" of relevance for tax breaks, investment support, green-energy certificates, and so on (Jones et al. 2013; Krook and Baas 2013; Van Passel et al. 2013). At present, there are in fact some signs of such recognition of landfill mining by nongovernmental organizations, environmental authorities, and other opinion makers (UNEP 2012; SEPA 2012), meaning that policy makers might be open to new ideas. Changing the perception of landfills among policy makers from a "waste dump" to "mines" is indeed not something that will happen overnight and relies on the development of strong and influential cross-sector networks, similar to the Enhanced Landfill Mining consortium in Belgium (Geels and Shot 2010; Johansson, Krook, and Eklund 2012). In order for this societal transformation to take off, there is no alternative for such "advocacy coalitions" but to engage in politics, influence public opinion, and demonstrate that landfill mining can contribute to solving wider societal concerns (Jones et al. 2013; Sabatier 1998). Here, the recently initiated European enhanced landfill mining consortium, EURELCO, consisting of more than 40 members from academia, industry, associations, and authorities is promising, but changing our perception of landfills so radically will be a tremendously challenging endeavor.

REFERENCES

Alm, J., J. Christéen, and G. Collin. 2006. Landfill mining at Stena Gotthard's landfill in Halmstad—An environmental and economic evaluation. Internal report, Environmental Technology and Management, Linköping University, Linköping, Sweden.

Ayres, R.U. 1997. Metal recycling: Economic and environmental implications. *Resources, Conservation and Recycling* 21: 145–173.

Ayres, R.U., U.J. Holmberg, and B. Anderson. 2001. Materials and the global environment: Waste mining in the 21st century. *MRS Bulletin* 26: 477–480.

Baas, L. 2008. Industrial symbiosis in the Rotterdam harbour and industry complex: Reflections on the interconnection of the technosphere with the social system. *Business Strategy and the Environment* 17: 330–340.

Baas, L., J. Krook, M. Eklund, and N. Svensson. 2010. Industrial ecology takes a second look at landfills. *Regional Development Dialogue UNCHR* 31(2): 169–181.

Baccini, P. and P.H. Brunner. 2012. *Metabolism of the Anthroposphere–Analysis, Evaluation, Design*. MIT Press, Cambridge, MA.

Cha, M.C., B.H Yoon, S.Y. Sung, S.P. Yoon, and I.W. Ra. 1997. Mining and remediation works at Ulsan landfill site, Korea. *Proceedings Sardinia, Sixth International Landfill Symposium*. 553–558, Cagliari, Italy.

Cossu, R., W. Hogland, and E. Salerni. 1996. Landfill mining in Europe and the USA. *ISWA Year Book*. 107–114.

Daskalopoulos, E., O. Badr, and S.D. Probert. 1997. Economic and environmental evaluations of waste treatment and disposal technologies for municipal solid waste. *Applied Energy* 58: 209–255.

Davidson, S. 2012. Mining Taxes and Subsidies: Official evidence. Minerals Council of Australia, Canberra, Australia.

Dickinson, W. 1995. Landfill mining comes of age. *Solid Waste Technologies* 9: 42–47.

EEA. 2011. Earnings, jobs and innovation: The role of recycling in a green economy. Report No 8, European Environment Agency, Denmark.

Elshkaki, A., van der Voet, E., van Holderbeke M., and Timmermans, M. 2004. The environmental and economic consequences of the developments of lead stocks in the Dutch economic system. *Resources, Conservation and Recycling* 42: 133–154.

European Council. 1999. Council Directive 1999/31/EC of 26 April 1999 on the Landfill of Waste. Available from: http://eur-lex.europa.eu/legal-content/EN/TXT/?uri=CELEX:31999L0031.

Eurostat. 2009. *Waste Generated and Treated in Europe*. Office for official publications of the European communities, Luxembourg.

Fisher, H. and D. Findlay. 1995. Exploring the economics of mining landfills. *World Wastes* 38: 50–54.

Frändegård, P., J. Krook, N. Svensson, and M. Eklund. 2013. Climate and resource implications of landfill mining. A case study of Sweden. *Journal of Industrial Ecology* 17(5): 742–755.

Geels, F.W. and J.W. Shot. 2010. The dynamics of transitions: A socio-technical perspective. In: J. Grin, J. Rotmans and J. Shot (Eds.). *Transitions to Sustainable Development: New Directions in the Study of Long Term Transformative Change*, Routledge, Oxon.

Gordon, R.B., M. Bertram, and T.E. Graedel. 2006. Metal stocks and sustainability. *PNAS* 103(5): 1209–1214.

Graedel, T.E., D. Van Beers, M. Bertram, K. Fuse, R.B. Gordon, A. Gritsinin, A. Kapur et al. 2004. Multilevel cycle of anthropogenic copper. *Environmental Science & Technology* 38: 1242–1252.

Grudnoff, M. 2012. Pouring Fuel on the Fire: The nature and extent of Federal Government subsidies to the mining industry. Policy Brief No. 38. The Australia Institute, Canberra, Australia.

Halada, K., K. Ijima, M. Shimada, and N. Katagiri. 2009. A possibility of urban mining in Japan. *Journal of Japan Institute of Metals* 73: 151–160.

Hogland, W. 2002. Remediation of an old landfill: Soil analysis, leachate quality and gas production. *Environmental Science & Pollution Research* 1: 49–54.

Hull, R.M., U. Krogmann, and P.F. Strom. 2005. Composition and characteristics of excavated materials from a New Jersey landfill. *Journal of Environmental Engineering* 131: 478–490.

Johansson, N., J. Krook, and M. Eklund. 2012. Transforming dumps into gold mines. Experiences from Swedish case studies. *Environmental Innovation and Social Transitions* 5: 33–48.

Johansson, N., J. Krook, and M. Eklund. 2014. Institutional conditions for Swedish metal production: A comparison of subsidies to metal mining and metal recycling. *Resource Policy* 41: 72–82.

Johansson, N., J. Krook, M. Eklund, and B. Berglund. 2013. An integrated review of concepts and initiatives for mining the technosphere: Towards a new taxonomy. *Journal of Cleaner Production* 55: 35–44.

Jones, P.T., D. Geysen, Y. Tielemans, S. Van Passel, Y. Pontikes, B. Blanpain, M. Quaghebeur, and N. Hoekstra. 2013. Enhanced landfill mining in view of multiple resource recovery: A critical review. *Journal of Cleaner Production* 55: 45–55.

Kapur, A. 2006. The future of the red metal: Discards, energy, water, residues and depletion. *Progress in Industrial Ecology* 3: 209–236.

Kapur, A. and T.E. Graedel. 2006. Copper mines above and below the ground. *Environmental Science & Technology* 40: 3135–3141.

Kleijn, D., F. Berendse, R. Smit, and N. Gilissen. 2001. Agrienvironment schemes do not effectively protect biodiversity in Dutch agricultural landscapes. *Nature* 413: 723–725.

Krogmann, U. and M. Qu. 1997. Landfill mining in the United States. *Proceedings Sardinia, Sixth International Landfill Symposium*: 543–552, Cagliari, Italy.

Krook, J. 2013. Miljökonsekvenser av integrerad sanering och återvinning av kommunala och industriella deponier, Internal report. The Swedish Environmental Protection Agency, Stockholm, Sweden.

Krook, J. and L. Baas. 2013. Getting serious about mining the technosphere: A review of recent landfill mining and urban mining research. *Journal of Cleaner Production* 55: 1–9.

Krook, J., N. Svensson, and M. Eklund. 2012. Landfill mining: A critical review of two decades of research. *Waste Management* 32: 513–520.

Lichtensteiger, T. 2002. *The Petrologic Evaluation*. Springer-Verlag, Berlin, Germany.

Müller, D., T. Wang, B. Duval, and T.E. Graedel. 2006. Exploring the engine of anthropogenic iron cycles. *PNAS* 103: 16111–16116.

Nelson, A., J. Genereux, and M. Genereux. 1992. Price effects of landfills on house values. *Land Economics* 68(4): 359–365.

Nriagu, J.O. 1996. History of global metal pollution. *Science* 272(5259): 223–224.

OECD. 2009. *The Economics of Climate Change Mitigation: Policies and Options for Global Action Beyond 2012*. OECD Publications, Paris, France.

Profu. 2010. Tillgång och efterfrågan på avfallsbehandling 2008–2015. Internal report for the Swedish Waste Association, Gothenburg, Sweden.

Quaghebeur, M., B. Laenen, D. Geysen, P. Nielsen, Y. Pontikes, T. van Gerven, and J. Spooren. 2013. Characterization of landfilled materials: Screening of the enhanced landfill mining potential. *Journal of Cleaner Production* 55: 72–83.

Sabatier, P.A. 1988. An advocacy coalition framework of policy change and the role of policy-oriented learning therein. *Policy Sciences* 21(2/3): 129–168.

SCS. 1998. Miljöbalken 1998:808 [Environmental Code]. Swedish Code of Statues. Swedish Government. Available from: http://www.notisum.se/rnp/sls/lag/19980808.htm.

SCS. 1999. Act on waste tax 1999:673. Swedish Code of Statues. Swedish Government. Available from: http://www.notisum.se/rnp/sls/lag/19990673.htm.

SCS. 2001. Ordinance concerning deposition of waste 2001:512. Swedish Code of Statues. Swedish Government. Available from: http://www.riksdagen.se/sv/Dokument-Lagar/ Lagar/Svenskforfattningssamling/_sfs-2001-512/.

SCS. 2005. Förordning om producentansvar för elektriska och elektroniska produkter. Swedish Code of Statues. Swedish Government. Available from: http://www.notisum.se/rnp/sls/ lag/20050209.htm.

SEPA. 2009. Guidelines for contaminated soil. Report No. 5976, Swedish Environmental Protection Agency, Stockholm, Sweden.

SEPA. 2012. Inspel till svensk mineralstrategi—N2012/1081/FIN. Swedish Environmental Protection Agency NV-02433-12, Stockholm, Sweden.

Spatari, S., M. Bertram, R.B. Gordon, K. Henderson, and T.E. Graedel. 2005. Twentieth-century copper stocks and flows in North America: A dynamic analysis. *Ecological Economics* 54: 37–51.

Swedish Geological Surveys. 2009. Bergverksstatistik 2008. Uppsala, Sweden.

Swedish Government. 2013. Sveriges mineralstrategi. Elanders Mölnlycke, Sweden.

Swedish Waste Association. 2013. Svensk avfallshantering 2013. Malmö, Sweden.

Tilton, J.E. and Lagos, G. 2007. Assessing the long-run availability of copper. *Resources Policy* 32(1): 19–23.

UNEP. 2012. Recycling rates of metals. International Panel for Sustainable Resource Management, Working Group on the Global Metal Flows, Nairobi, Kenya.

US Geological Survey. 2012. *2010 Minerals Yearbook: Copper*. Available from: http://minerals .usgs.gov/minerals/pubs/commodity/copper/.

Van Beers, C. and S. de Moor. 2001. *Public Subsidies and Policy Failures: How Subsidies Distort the Natural Environment. Equity and Trade and How to Reform Them.* Edward Elgar, Cheltenham.

Van Passel, S., M. Dubois, J. Eyckmans, S. de Gheldere, F. Ang, P.T. Jones, and K. Van Acker. 2013. The economics of enhanced landfill mining: Private and societal performance drivers. *Journal of Cleaner Production* 55: 92–102.

Wante, J. and L. Umans. 2010. A European legal framework for enhanced waste management. *Proceedings International Academic Symposium on Enhanced Landfill Mining.* 53–64, Houthalen–Helchteren, Belgium, Germany.

Wrigley, E.A. 1962. The supply of raw materials in the industrial revolution. *The Economic History Review* 15(1): 1–16.

Index

Printed and bound by CPI Group (UK) Ltd, Croydon, CR0 4YY

22/10/2024

01777613-0020